Springer Series in Statistics

More information about this series at http://www.springer.com/series/692

George A.F. Seber

The Linear Model and Hypothesis

A General Unifying Theory

Springer

George A.F. Seber
Department of Statistics
The University of Auckland
Auckland, New Zealand

ISSN 0172-7397 ISSN 2197-568X (electronic)
Springer Series in Statistics
ISBN 978-3-319-34917-6 ISBN 978-3-319-21930-1 (eBook)
DOI 10.1007/978-3-319-21930-1

Springer Cham Heidelberg New York Dordrecht London

Softcover re-print of the Hardcover 1st edition 2015

Printed on acid-free paper

Springer International Publishing AG Switzerland is part of Springer Science+Business Media (www.springer.com)

Preface

In 1966 my monograph The Linear Hypothesis: A General Theory was published as one in a series of statistical monographs by Griffin, London. Part of the book arose out of my PhD thesis, which took a more general approach than usual to linear models. It used the geometrical notion of projections onto vector spaces using idempotent matrices, thus providing an elegant theory that avoided being involved with ranks of matrices. Although not a popular approach at the time, it has since become an integral part of theoretical regression books where least squares estimates, for example, are routinely given a geometrical interpretation.

Over the years I have written extensively on related topics such as linear and nonlinear regression, multivariate analysis, and large sample tests of general hypotheses including, for example, those arising from the multinomial distribution. Given this additional experience and the fact that my original monograph is now out of print, the time has come to rewrite it. This is it! Initially the 1966 monograph was written as an attempt to show how the linear model and hypothesis provide a unifying theme where all hypotheses are either linear or asymptotically so. This means that the linear theory can be applied in a variety of modeling situations and this monograph extends the breadth of these situations. In a monograph of this size, the emphasis is on theoretical concepts, and the reader needs to look elsewhere for practical applications and appropriate software. I appreciate that these days the focus of statistical courses is much more applied. Numerous computationally oriented books have been written, for example, on using the statistical package R that was originally developed in the Statistics Department here at University of Auckland. However I would mention that my books on linear, nonlinear, and multivariate models all have comprehensive chapters on computational details and algorithms, as well as practical examples.

Who is the monograph for? It is pitched at a graduate level in statistics and assumes that the reader is familiar with the basics of regression analysis, analysis of variance, and some experimental designs like the randomized block design, with brief extensions to multivariate linear models. Some previous exposure to nonlinear models and multinomial goodness-of-fit tests will help, and some knowledge of the multivariate normal distribution is assumed. A basic knowledge of the matrix theory is assumed throughout, though proofs of most of the matrix results used are given either in the text or in the Appendix. My aim is to provide the reader with a more global view of modeling and show connections between several major statistical topics.

Chapters 1, 2, 3 and 4 deal with the basic ideas behind the book: Chap. 1 gives some preliminary mathematical results needed in the book; Chap. 2 defines the linear model and hypothesis with examples; Chap. 3 is on estimation; and Chap. 4 is on hypothesis testing, all from a geometrical point of view. Chapter 5 looks at some general properties of the F-test, and in Chap. 6 methods of testing several hypotheses are discussed. Chapters 7, 8 and 9 look at special topics: Chap. 7 is about augmenting hypotheses as in analysis of covariance and missing observations, for example, Chap. 8 looks at nonlinear models and Chap. 9 at multivariate models. Chapters 10, 11 and 12 involve considerable asymptotic theory showing how general hypotheses about sampling from general distributions are asymptotically equivalent to corresponding linear theory. The book finishes with an appendix giving some useful, and in some cases not so common, matrix results with proofs.

Looking back after having been retired for a number of years, I am grateful for the stimulus given to my writing through teaching most of the topics mentioned above at University of Auckland, New Zealand. Teaching certainly clarifies one's understanding of a subject. In conclusion I would like to thank two referees for their helpful comments.

Auckland, New Zealand George A.F. Seber
February 2015

Contents

1	**Preliminaries**	1
1.1	Notation	1
1.2	Linear Vector Spaces	2
1.3	Basis of a Vector Space	3
1.4	Addition and Intersection of Vector Spaces	3
1.5	Idempotent Matrices	4
1.6	Expectation, Covariance, and Variance Operators	6
1.7	Multivariate Normal Distribution	8
1.8	Non-central Distributions	9
1.9	Quadratic Forms	10
1.10	Lagrange Multipliers	17
	References	18
2	**The Linear Hypothesis**	21
2.1	Linear Regression	21
2.2	Analysis of Variance	22
2.3	Analysis of Covariance	24
2.4	General Definition and Extensions	25
3	**Estimation**	27
3.1	Principle of Least Squares	27
3.2	Projection Matrices	28
3.3	Examples	29
3.4	Less Than Full Rank Model	32
3.5	Gauss-Markov Theorem	34
3.6	Estimation of σ^2	35
3.7	Assumptions and Residuals	38
3.8	Multiple Correlation Coefficient	42
3.9	Maximum Likelihood Estimation	44
	References	45

4 **Hypothesis Testing** 47
4.1 The Likelihood Ratio Test 47
4.2 The Hypothesis Sum of Squares 50
4.3 Wald Test 51
4.4 Contrasts 55
4.5 Confidence Regions and Intervals 57
References 60

5 **Inference Properties** 61
5.1 Power of the F-Test 61
5.2 Robustness of the F-Test and Non-normality 64
5.3 Unequal Variances 69
References 71

6 **Testing Several Hypotheses** 73
6.1 The Nested Procedure 73
6.2 Orthogonal Hypotheses 76
6.3 Orthogonal Hypotheses in Regression Models 78
6.4 Orthogonality in Complete Two-Factor Layouts 81
6.5 Orthogonality in Complete p-Factor Layouts 89
6.6 Orthogonality in Randomized Block Designs 93
6.7 Orthogonality in Latin Square Designs 96
6.8 Non-orthogonal Hypotheses 100
References 101

7 **Enlarging the Model** 103
7.1 Introduction 103
7.2 Least Squares Estimation 104
7.3 Hypothesis Testing 106
7.4 Regression Extensions 107
7.5 Analysis of Covariance Extensions 108
7.6 Missing Observations 111
References 116

8 **Nonlinear Regression Models** 117
8.1 Introduction 117
8.2 Estimation 118
8.3 Linear Approximations 120
8.4 Concentrated Likelihood Methods 123
8.5 Large Sample Tests 127
References 127

9 **Multivariate Models** 129
9.1 Notation 129
9.2 Estimation 132
9.3 Hypothesis Testing 135
9.4 Some Examples 141

9.5 Hotelling's Test Statistic 144
9.6 Simultaneous Confidence Intervals 146
References 147

10 Large Sample Theory: Constraint-Equation Hypotheses 149
10.1 Introduction 149
10.2 Notation and Assumptions 150
10.3 Positive-Definite Information Matrix 152
10.3.1 Maximum Likelihood Estimation 153
10.3.2 The Linear Model Approximation 155
10.3.3 The Three Test Statistics 157
10.4 Positive-Semidefinite Information Matrix 161
10.4.1 Hypothesis Testing 163
10.4.2 Lagrange Multipler Test 166
10.5 Orthogonal Hypotheses 171
References 174

11 Large Sample Theory: Freedom-Equation Hypotheses 175
11.1 Introduction 175
11.2 Positive-Definite Information Matrix 176
11.3 Positive-Semidefinite Information Matrix 177
Reference 179

12 Multinomial Distribution 181
12.1 Definitions 181
12.2 Test of $\mathbf{p} = \mathbf{p}_0$ 182
12.3 Score and Wald Statistics 184
12.4 Testing a Freedom Equation Hypothesis 186
12.5 Conclusion 188
References 188

Appendix: Matrix Theory 189
References 201

Index 203

Chapter 1
Preliminaries

1.1 Notation

Linear algebra is used extensively throughout this book and those topics particularly relevant to the development in this monograph are given within the chapters; other results are given in the Appendix. References to the Appendix are labeled with a prefix "A", for example A.3 is theorem 3 in the Appendix. Vectors and matrices are denoted by boldface letters **a** and **A**, respectively, and scalars are denoted by italics. For example, $\mathbf{a} = (a_i)$ is a vector with ith element a_i and $\mathbf{A} = (a_{ij})$ is a matrix with i, jth element a_{ij}. I shall use the same notation with random variables, because using uppercase for random variables and lowercase for their values can cause confusion with vectors and matrices. We endeavor, however, to help the reader by using the lower case letters in the latter half of the alphabet, namely $u, v, \ldots, z$, with the occasional exception (because of common usage) for random variables and the rest of the alphabet for constants. All vectors and matrices contain real elements, that is belong to $\mathbb{R}$, and we denote n-dimensional Euclidean space by $\mathbb{R}^n$.

The $n \times n$ matrix $\text{diag}(a_1, a_2, \ldots, a_n)$ or $\text{diag}(\mathbf{a})$ is a matrix with diagonal elements the elements of $\mathbf{a}' = (a_1, a_2, \ldots, a_n)$ and off-diagonal elements all zero. When the n diagonal elements are all equal to 1 we have the identity matrix $\mathbf{I}_n$. The n-dimensional vector with all its elements unity is denoted by $\mathbf{1}_n$. The trace of a matrix **A**, denoted by trace[**A**], is the sum of its diagonal elements, and the rank of **A** is denoted by rank[**A**]. The determinant of a square matrix **A** is denoted by $\det \mathbf{A}$ or $|\mathbf{A}|$. We shall also use the Kronecker delta, δ_{ij} which is one when $i = j$ and zero otherwise

The length of an n-dimensional vector $\mathbf{x} = (x_i)$ is denoted by $\| \mathbf{x} \|$, so that

$$\| \mathbf{x} \| = \sqrt{(\mathbf{x}'\mathbf{x})} = \sqrt{(x_1^2 + x_2^2 + \cdots + x_n^2)}.$$

G.A.F. Seber, *The Linear Model and Hypothesis*, Springer Series in Statistics,
DOI 10.1007/978-3-319-21930-1_1

We say that two vectors $\mathbf{x}$ and $\mathbf{y}$ in $\mathbb{R}^n$ are orthogonal and write $\mathbf{x} \perp \mathbf{y}$ if $\mathbf{x}'\mathbf{y} = 0$. For an extensive collection of matrix results see Seber (2008).

1.2 Linear Vector Spaces

We shall be interested in particular subsets of $\mathbb{R}^n$ called linear vector spaces that may be defined as follows. A *linear vector space* is a set of vectors $\mathcal{V}$ such that for any two vectors $\mathbf{x}$ and $\mathbf{y}$ belonging to $\mathcal{V}$ and for any real numbers a and b, the vector $a\mathbf{x} + b\mathbf{y}$ also belongs to $\mathcal{V}$. This definition is not the most general one, but it is sufficient for the development given in this book. From now on we shall drop the word "linear" and take it as understood. Since a and b can both be zero, we see that every vector space contains a zero vector. We note that $\mathbb{R}^n$ is also a vector space and we can then say that $\mathcal{V}$ is a *subspace*. To prove two vector spaces are identical we show that one is contained in the other and vice versa, as we see in Theorem 1.1 below.

We now give some examples of useful vector spaces. If $\mathcal{V}$ is a subspace of $\mathbb{R}^n$, then $\mathcal{V}^\perp$, the set of all vectors in $\mathbb{R}^n$ perpendicular to every vector in $\mathcal{V}$ (called the *orthogonal complement* of $\mathcal{V}$), is also a vector subspace. This follows from the fact that if $\mathbf{v} \in \mathcal{V}$, and $\mathbf{x}$ and $\mathbf{y}$ belong to $\mathcal{V}^\perp$, then

$$\mathbf{v}'(a\mathbf{x} + b\mathbf{y}) = a\mathbf{v}'\mathbf{x} + b\mathbf{v}'\mathbf{y} = 0 \quad \text{and} \quad a\mathbf{x} + b\mathbf{y} \in \mathcal{V}^\perp.$$

If $\mathbf{X}$ is an $n \times p$ matrix and $\mathcal{C}[\mathbf{X}]$ is the set of all vectors $\boldsymbol{\theta}$ such that $\boldsymbol{\theta} = \mathbf{X}\boldsymbol{\beta}$ for some $\boldsymbol{\beta}$, that is $\mathcal{C}[\mathbf{X}] = \{\boldsymbol{\theta} : \boldsymbol{\theta} = \mathbf{X}\boldsymbol{\beta}\}$ is the set of all linear combinations of the columns of $\mathbf{X}$, then $\mathcal{C}[\mathbf{X}]$ is a vector space. Also if $\mathcal{N}[\mathbf{X}] = \{\boldsymbol{\phi} : \mathbf{X}\boldsymbol{\phi} = \mathbf{0}\}$, then $\mathcal{N}[\mathbf{X}]$ is also vector space. We find then that associated with every matrix $\mathbf{X}$ there are three vector spaces: (1) the *column space* (also called the range space) $\mathcal{C}[\mathbf{X}]$, (2) the *row space* $\mathcal{C}[\mathbf{X}']$, and (3) the *null space* (sometimes called the kernel) $\mathcal{N}[\mathbf{X}]$ of $\mathbf{X}$; proofs that they are subspaces are left to the reader. Two of these spaces are related by the following theorem used throughout this monograph.

Theorem 1.1 $\mathcal{N}[\mathbf{X}] = \mathcal{C}[\mathbf{X}']^\perp$. *In words, the null space of* $\mathbf{X}$ *is the orthogonal complement of the row space of* $\mathbf{X}$.

Proof If $\boldsymbol{\theta} \in \mathcal{N}[\mathbf{X}]$, then $\mathbf{X}\boldsymbol{\theta} = \mathbf{0}$ and $\boldsymbol{\theta}$ is orthogonal to each row of $\mathbf{X}$. It is therefore orthogonal to any linear combination of the rows of $\mathbf{X}$, so that $\boldsymbol{\theta} \perp \mathcal{C}[\mathbf{X}']$ and $\mathcal{N}[\mathbf{X}] \subset \mathcal{C}[\mathbf{X}']^\perp$. Conversely, if $\boldsymbol{\theta} \perp \mathcal{C}[\mathbf{X}']$ then $\mathbf{X}\boldsymbol{\theta} = \mathbf{0}$, $\boldsymbol{\theta} \in \mathcal{N}[\mathbf{X}]$, and $\mathcal{C}[\mathbf{X}']^\perp \subset \mathcal{N}[\mathbf{X}]$. Hence the result follows.

1.3 Basis of a Vector Space

A set of vectors $\beta_1, \beta_2, \ldots, \beta_p$ is said to span a vector space $\mathcal{V}$ if every vector $\mathbf{v} \in \mathcal{V}$ can be expressed as a linear combination of these vectors, that is, if there exist constants $b_1, b_2, \ldots, b_p$ such that

$$\mathbf{v} = \sum_{i=1}^{p} b_i \beta_i.$$

The vectors $\beta_1, \beta_2, \ldots, \beta_p$ are *linearly independent* if $\sum_{i=1}^{p} b_i \beta_i = \mathbf{0}$ implies that $b_1 = b_2 = \ldots = b_p = 0$. Thus linear independence implies that there is no non-trivial linear relation among the vectors. If the vectors β_i $(i = 1, 2 \ldots, p)$ span $\mathcal{V}$ and are linearly independent, then they are said to form a *basis* of $\mathcal{V}$. Although a basis is not unique, the number of vectors p in it is unique and is called the *dimension* of $\mathcal{V}$ or $\dim \mathcal{V}$. From every basis it is possible to construct an *orthonormal basis* $\alpha_1, \alpha_2, \ldots, \alpha_p$ such that $\alpha_i' \alpha_j = \delta_{ij}$; "ortho" as the vectors are mutually orthogonal and "normal" as they have unit length. The construction can be carried out from a basis using the Gram-Schmidt algorithm (Seber and Lee 2003, 338–339). If $\mathcal{V}$ is a subspace of $\mathbb{R}^n$, it is always possible to enlarge an orthonormal basis of $\mathcal{V}$ to the set $\alpha_1, \alpha_2, \ldots, \alpha_p, \alpha_{p+1}, \alpha_{p+2}, \ldots, \alpha_n$ to form an orthonormal basis for $\mathbb{R}^n$. Thus if $\dim \mathcal{V} = p$, then it is readily seen that $\alpha_{p+1}, \ldots, \alpha_n$ form an orthonormal basis for $\mathcal{V}^\perp$ and $\dim \mathcal{V}^\perp = n - p$.

Since the column space $\mathcal{C}[\mathbf{X}]$ of a matrix $\mathbf{X}$ is the space spanned by its columns, then $\dim \mathcal{C}[\mathbf{X}]$ will be the number of linearly independent columns of $\mathbf{X}$ and therefore the rank of $\mathbf{X}$. The dimension of $\mathcal{N}[\mathbf{X}]$ is known as the *nullity* of $\mathbf{X}$ and is obtained from the rule (A.3)

$$\text{rank} + \text{nullity} = \text{number of columns of } \mathbf{X}.$$

Thus if $\mathbf{X}$ is an $n \times p$ matrix of rank r $(r \leq p, n)$, then we see that $\dim \mathcal{C}[\mathbf{X}] = p$ and $\dim \mathcal{N}[\mathbf{X}] = p - r$.

1.4 Addition and Intersection of Vector Spaces

A vector space $\mathcal{V}$ is said to be the *direct sum* of two vector spaces $\mathcal{V}_1$ and $\mathcal{V}_2$ if every vector $\mathbf{v} \in \mathcal{V}$ can be expressed *uniquely* in the form $\mathbf{v} = \mathbf{v}_1 + \mathbf{v}_2$, where $\mathbf{v}_i \in \mathcal{V}_i$ $(i = 1, 2)$. We represent this symbolically by $\mathcal{V}_1 \bigoplus \mathcal{V}_2$. If we drop the word unique from the definition, we say that $\mathcal{V}$ is the *sum* of $\mathcal{V}_1$ and $\mathcal{V}_2$ and write $\mathcal{V} = \mathcal{V}_1 + \mathcal{V}_2$.

The intersection of two vector spaces $\mathcal{V}_1$ and $\mathcal{V}_2$ is denoted by $\mathcal{V}_1 \cap \mathcal{V}_2$ and is the set of all vectors that belong to both spaces. The reader should check that if $\mathcal{V}_1$ and $\mathcal{V}_2$ are all vector spaces in $\mathbb{R}^n$, then $\mathcal{V}_1 \bigoplus \mathcal{V}_2$, $\mathcal{V}_1 + \mathcal{V}_2$, and $\mathcal{V}_1 \cap \mathcal{V}_2$ are vector spaces. The following theorems will be useful later on.

Theorem 1.2 *If $\mathcal{V}_1$ and $\mathcal{V}_2$ are two vector spaces in $\mathbb{R}^n$, then*

(i) $[\mathcal{V}_1 \cap \mathcal{V}_2]^\perp = \mathcal{V}_1^\perp + \mathcal{V}_2^\perp$.
(ii) If $\mathcal{V}_1^\perp \cap \mathcal{V}_2^\perp = \mathbf{0}$, *then* $[\mathcal{V}_1 \cap \mathcal{V}_2]^\perp = \mathcal{V}_1^\perp \bigoplus \mathcal{V}_2^\perp$.

Proof

(i) We can prove this quite generally by showing that the left hand side is contained in the right hand side, and vice versa; this is left as an exercise. However, the following proof using matrices is instructive as it uses Theorem 1.1 in Sect. 1.2. Let $\mathbf{A}_1$ and $\mathbf{A}_2$ be matrices such that $\mathcal{V}_i = \mathcal{N}[\mathbf{A}_i]$ for $i = 1, 2$. Then

$$\begin{aligned}[\mathcal{V}_1 \cap \mathcal{V}_2]^\perp &= \left\{\mathcal{N}\begin{pmatrix}\mathbf{A}_1\\ \mathbf{A}_2\end{pmatrix}\right\}^\perp \\ &= \mathcal{C}[(\mathbf{A}_1', \mathbf{A}_2')] \quad \text{(cf. Theorem 1.1)} \\ &= \mathcal{C}[\mathbf{A}_1'] + \mathcal{C}[\mathbf{A}_2'] \\ &= \mathcal{V}_1^\perp + \mathcal{V}_2^\perp.\end{aligned}$$

(ii) This follows from the fact that the columns of $\mathbf{A}_1'$ are linearly independent of the columns of $\mathbf{A}_2'$ so that $\mathcal{C}[(\mathbf{A}_1', \mathbf{A}_2')] = \mathcal{C}[\mathbf{A}_1'] \bigoplus \mathcal{C}[\mathbf{A}_2']$.

Theorem 1.3 *If $\mathcal{V}_0$ and $\mathcal{V}_i$ $(i = 1, 2)$ are three vector spaces in $\mathbb{R}^n$ such that $\mathcal{V}_1 \subset \mathcal{V}_0$, then*

$$\mathcal{V}_0 \cap (\mathcal{V}_1 + \mathcal{V}_2) = \mathcal{V}_1 + (\mathcal{V}_0 \cap \mathcal{V}_2).$$

Proof If $\mathbf{v} \in$ LHS (left-hand side), then $\mathbf{v} \in \mathcal{V}_0$ and $\mathbf{v} = \mathbf{v}_1 + \mathbf{v}_2$, where $\mathbf{v}_1 \in \mathcal{V}_1 \subset \mathcal{V}_0$ and $\mathbf{v}_2 \in \mathcal{V}_2$. Hence $\mathbf{v}_2 = \mathbf{v} - \mathbf{v}_1 \in \mathcal{V}_0$ and $\mathbf{v}_2 \in \mathcal{V}_0 \cap \mathcal{V}_2$, so that $\mathbf{v} \in$ RHS and LHS $\subset$ RHS. Conversely, if $\mathbf{v} \in$ RHS, then $\mathbf{v} = \mathbf{v}_1 + \mathbf{v}_2 \in \mathcal{V}_0$, as $\mathbf{v}_1 \in \mathcal{V}_1 \subset \mathcal{V}_0$, and $\mathbf{v}_2 \in \mathcal{V}_0 \cap \mathcal{V}_2 \subset \mathcal{V}_0$. Also $\mathbf{v}_1 + \mathbf{v}_2 \in \mathcal{V}_1 + \mathcal{V}_2$ so that $\mathbf{v} \in$ LHS. Therefore LHS=RHS and the result is proved.

1.5 Idempotent Matrices

We shall see later that symmetric idempotent matrices carry out an important role with regard to projecting vectors orthogonally onto vector spaces, and are therefore called *projection matrices*. The symbol $\mathbf{P}$ will always represent a symmetric idempotent matrix, so that $\mathbf{P}' = \mathbf{P}$ and $\mathbf{PP} = \mathbf{P}^2 = \mathbf{P}$.

Example 1.1 Let $\mathbf{y} = (y_1, y_2, \ldots y_n)'$ and consider $Q_1 = \sum_i (y_i - \overline{y})^2 = \mathbf{y}'\mathbf{A}_1\mathbf{y}$ and $Q_2 = n\overline{y}^2 = \mathbf{y}'\mathbf{A}_2\mathbf{y}$, where $\overline{y}$ is the mean of the y_i. We now show that both $\mathbf{A}_1$ and $\mathbf{A}_2$ are symmetric and idempotent. First $Q_1 = \sum_i y_i^2 - Q_2$ and $Q_2 = n^{-1}(\mathbf{y}'\mathbf{1}_n\mathbf{1}_n'\mathbf{y})$ so that $\mathbf{A}_2 = n^{-1}\mathbf{1}_n\mathbf{1}_n'$ and $\mathbf{A}_2^2 = \mathbf{A}_2$. Also $Q_1 = \mathbf{y}'(\mathbf{I}_n - \mathbf{A}_2)\mathbf{y}$ so that we have

$\mathbf{A}_1^2 = (\mathbf{I}_n - \mathbf{A}_2)^2 = \mathbf{I}_n - 2\mathbf{A}_2 + \mathbf{A}_2 = \mathbf{A}_1$. We note that $\mathbf{y}'\mathbf{y} = Q_1 + Q_2$ or, in terms of matrices, $\mathbf{I}_n = \mathbf{A}_1 + \mathbf{A}_2$. This is a special case of Cochran's theorem discussed later.

Projection matrices have useful properties that are summarized in the following theorem.

Theorem 1.4 *The eigenvalues of a symmetric idempotent matrix* $\mathbf{P}$ *are zero or one, and the number of unit eigenvalues is the rank of* $\mathbf{P}$. *Also,* $\text{rank}\,\mathbf{P} = \text{trace}\,\mathbf{P}$. *Conversely, if* $\mathbf{P}$ *is symmetric and its eigenvalues are zero or one, then* $\mathbf{P}$ *is idempotent.*

Proof Suppose $\mathbf{P}$ is $n \times n$ of rank r. As $\mathbf{P}$ is symmetric there exists an $n \times n$ orthogonal matrix $\mathbf{T}$ (A.7) such that

$$\mathbf{T}'\mathbf{P}\mathbf{T} = \text{diag}(\lambda_1, \lambda_2, \ldots, \lambda_n) = \mathbf{\Lambda} \quad \text{say},$$

where $\lambda_1, \lambda_2, \ldots, \lambda_n$ are the eigenvalues of $\mathbf{P}$. Now

$$\mathbf{\Lambda}^2 = \mathbf{T}'\mathbf{P}\mathbf{T}\mathbf{T}'\mathbf{P}\mathbf{T} = \mathbf{T}'\mathbf{P}\mathbf{P}\mathbf{T} = \mathbf{T}'\mathbf{P}\mathbf{T} = \mathbf{\Lambda},$$

and $\lambda_i^2 = \lambda_i$ for each i. Thus the only possible eigenvalues are zero or one, and the rank of $\mathbf{P}$, which is the number of nonzero eigenvalues, is therefore the number of unit eigenvalues, namely r. As the rank of a matrix is unchanged by premultiplying or post-multiplying by a nonsingular matrix (see A.4(i)), $\text{rank}\,\mathbf{P} = \text{rank}\,\mathbf{\Lambda} = \text{trace}\,\mathbf{\Lambda} = r$. Since by A.1, $\text{trace}[\mathbf{AB}] = \text{trace}[\mathbf{BA}]$,

$$\text{trace}\,\mathbf{P} = \text{trace}[\mathbf{P}\mathbf{T}\mathbf{T}'] = \text{trace}[\mathbf{T}'\mathbf{P}\mathbf{T}] = \text{trace}\,\mathbf{\Lambda},$$

as $\mathbf{T}$ is orthogonal. Conversely, if the eigenvalues are 0 or 1 then $\mathbf{\Lambda}^2 = \mathbf{\Lambda}$, or $\mathbf{T}'\mathbf{P}\mathbf{T}\mathbf{T}'\mathbf{P}\mathbf{T} = \mathbf{T}'\mathbf{P}\mathbf{P}\mathbf{T} = \mathbf{T}'\mathbf{P}\mathbf{T}$, and $\mathbf{P}^2 = \mathbf{P}$. This completes the proof.

Since $\mathbf{a}'\mathbf{P}\mathbf{a} = \mathbf{a}'\mathbf{P}'\mathbf{P}\mathbf{a} = \mathbf{b}'\mathbf{b} \geq 0$, where $\mathbf{b} = \mathbf{P}\mathbf{a}$, we see that $\mathbf{P}$ is nonnegative definite (see the beginning of the Appendix for a definition). This also follows from the fact that the eigenvalues of $\mathbf{P}$ are nonnegative.

Finally we note that $\mathbf{I}_n - \mathbf{P}$ is symmetric and

$$(\mathbf{I}_n - \mathbf{P})(\mathbf{I}_n - \mathbf{P}) = \mathbf{I}_n - 2\mathbf{P} + \mathbf{P}^2 = \mathbf{I}_n - \mathbf{P},$$

which implies that $\mathbf{I}_n - \mathbf{P}$ is also idempotent. Hence, if $\mathbf{c} = (\mathbf{I}_n - \mathbf{P})\mathbf{a}$, then we have $\mathbf{a}'(\mathbf{I}_n - \mathbf{P})\mathbf{a} = \mathbf{c}'\mathbf{c} \geq 0$ and $\mathbf{I}_n - \mathbf{P}$ is nonnegative definite.

Example 1.2 Returning to Example 1.1,

$$\text{rank}[\mathbf{A}_2] = \text{trace}[\mathbf{A}_2] = \text{trace}[n^{-1}\mathbf{1}_n\mathbf{1}_n'] = n^{-1}\,\text{trace}\,\mathbf{1}_n'\mathbf{1}_n = 1$$

and

$$\text{rank}[\mathbf{A}_1] = \text{trace}[\mathbf{A}_1] = n - \text{trace}[\mathbf{A}_2] = n - 1.$$

Example 1.3 It is possible for a matrix to be idempotent without being symmetric. For example, the matrix $\mathbf{X}(\mathbf{X}'\mathbf{V}^{-1}\mathbf{X})^{-1}\mathbf{X}'\mathbf{V}^{-1}$ that arises in generalized least squares regression is idempotent as

$$\mathbf{X}(\mathbf{X}'\mathbf{V}^{-1}\mathbf{X})^{-1}\mathbf{X}'\mathbf{V}^{-1}\mathbf{X}(\mathbf{X}'\mathbf{V}^{-1}\mathbf{X})^{-1}\mathbf{X}'\mathbf{V}^{-1} = \mathbf{X}(\mathbf{X}'\mathbf{V}^{-1}\mathbf{X})^{-1}\mathbf{X}'\mathbf{V}^{-1}.$$

Its properties are similar to those of the symmetric case (see A.13).

1.6 Expectation, Covariance, and Variance Operators

If $\mathbf{Z} = (z_{ij})$ is a matrix (or vector) of random variables, we define the general expectation operator of the random matrix $\mathbf{Z}$ to be $\text{E}[\mathbf{Z}] = (E[z_{ij}])$. Then, by the linear properties of the one-dimensional expectation operator E, we see that $\text{E}[\mathbf{AZB} + \mathbf{C}] = \mathbf{A}\text{E}[\mathbf{Z}]\mathbf{B} + \mathbf{C}$, where $\mathbf{A}$, $\mathbf{B}$, and $\mathbf{C}$ are matrices of appropriate sizes with constant elements. In particular, if $\mathbf{y}$ is a random vector with mean $\boldsymbol{\theta}$, then $\text{E}[\mathbf{Ay}] = \mathbf{A}\boldsymbol{\theta}$.

We can also define the covariance, $\text{Cov}[\mathbf{x}, \mathbf{y}]$, of two random vectors $\mathbf{x} = (x_i)$ and $\mathbf{y} = (y_i)$ as the matrix with (i, j)th element $\text{cov}[x_i, y_j]$. If $\mathbf{x} = \mathbf{y}$, then we write $\text{Cov}[\mathbf{y}, \mathbf{y}] = \text{Var}[\mathbf{y}] = (\text{cov}[y_i, y_j])$. This matrix is known variously as the variance, variance-covariance, or dispersion matrix of $\mathbf{y}$. Its diagonal elements are variances and its off-diagonal elements are covariances.

Theorem 1.5 *Let* $\text{E}[\mathbf{x}] = \boldsymbol{\alpha}$ *and* $\text{E}[\mathbf{y}] = \boldsymbol{\beta}$, *then:*

(i) $\text{Cov}[\mathbf{x}, \mathbf{y}] = \text{E}[(\mathbf{x} - \boldsymbol{\alpha})(\mathbf{y} - \boldsymbol{\beta})']$.
(ii) $\text{Cov}[\mathbf{Ax}, \mathbf{By}] = \mathbf{A}\text{Cov}[\mathbf{x}, \mathbf{y}]\mathbf{B}'$.
(iii) $\text{Var}[\mathbf{By}] = \mathbf{B}\text{Var}[\mathbf{y}]\mathbf{B}'$.
(iv) $\text{Var}[\mathbf{y}]$ *is nonnegative definite, and positive definite if* $\mathbf{a}'\mathbf{y} \neq b$ *for some b and non-zero* $\mathbf{a}$.
(v) If $\mathbf{a}$ *and* $\mathbf{b}$ *are constant vectors of suitable dimensions, then*

$$\text{Cov}[\mathbf{x} - \mathbf{a}, \mathbf{y} - \mathbf{b}] = \text{Cov}[\mathbf{x}, \mathbf{y}].$$

If $\mathbf{c}$ *is a vector of the correct dimension, then*

$$\text{Var}[\mathbf{y} - \mathbf{c}] = \text{Var}[\mathbf{y}].$$

(vi) $\text{Var}[\mathbf{y}] = \text{E}[\mathbf{yy}'] - \text{E}[\mathbf{y}]\text{E}[\mathbf{y}']$.

Proof

(i)

$$\begin{aligned}\mathrm{Cov}[\mathbf{x},\mathbf{y}] &= (\mathrm{cov}[x_i, y_j]) \\ &= (E[(x_i-\alpha_i)(y_j-\beta_j)]) \\ &= \mathrm{E}[(\mathbf{x}-\boldsymbol{\alpha})(\mathbf{y}-\boldsymbol{\beta})'].\end{aligned}$$

(ii) Let $\mathbf{u} = \mathbf{A}\mathbf{x}$ and $\mathbf{v} = \mathbf{B}\mathbf{y}$. Then, by (i),

$$\begin{aligned}\mathrm{Cov}[\mathbf{A}\mathbf{x},\mathbf{B}\mathbf{y}] &= \mathrm{Cov}[\mathbf{u},\mathbf{v}] \\ &= \mathrm{E}[(\mathbf{u}-\mathrm{E}[\mathbf{u}])(\mathbf{v}-\mathrm{E}[\mathbf{v}])'] \\ &= \mathrm{E}[(\mathbf{A}\mathbf{x}-\mathbf{A}\boldsymbol{\alpha})(\mathbf{B}\mathbf{y}-\mathbf{B}\boldsymbol{\beta})'] \\ &= \mathrm{E}[\mathbf{A}(\mathbf{x}-\boldsymbol{\alpha})(\mathbf{y}-\boldsymbol{\beta})'\mathbf{B}'] \\ &= \mathbf{A}\mathrm{E}[(\mathbf{x}-\boldsymbol{\alpha})(\mathbf{y}-\boldsymbol{\beta})']\mathbf{B}' \\ &= \mathbf{A}\mathrm{Cov}[\mathbf{x},\mathbf{y}]\mathbf{B}'.\end{aligned}$$

(iii) From (ii), $\mathrm{Var}[\mathbf{A}\mathbf{y}] = \mathrm{Cov}[\mathbf{A}\mathbf{y},\mathbf{A}\mathbf{y}] = \mathbf{A}\mathrm{Var}[\mathbf{y}]\mathbf{A}'$.

(iv) $\mathbf{a}'\mathrm{Var}[\mathbf{y}]\mathbf{a} = \mathrm{var}[\mathbf{a}'\mathbf{y}] \geq 0$, which is strictly positive for non-zero $\mathbf{a}$ if we don't have $\mathbf{a}'\mathbf{y} = b$ for some b and non-zero $\mathbf{a}$.

(v) From (i),

$$\begin{aligned}\mathrm{Cov}[\mathbf{x}-\mathbf{a},\mathbf{y}-\mathbf{b}] &= \mathrm{E}[\{\mathbf{x}-\mathbf{a}-(\boldsymbol{\alpha}-\mathbf{a})\}\{\mathbf{y}-\mathbf{b}-(\boldsymbol{\beta}-\mathbf{b})\}'] \\ &= \mathrm{Cov}[\mathbf{x},\mathbf{y}].\end{aligned}$$

Then set $\mathbf{x} = \mathbf{y}$ and $\mathbf{a} = \mathbf{b} = \mathbf{c}$.

(vi) Use (i) with $\mathbf{x} = \mathbf{y}$ and expand, namely

$$\begin{aligned}\mathrm{Var}[\mathbf{y}] &= \mathrm{E}[(\mathbf{y}-\boldsymbol{\beta})(\mathbf{y}-\boldsymbol{\beta})'] \\ &= \mathrm{E}[\mathbf{y}\mathbf{y}' - \boldsymbol{\beta}\mathbf{y}' - \mathbf{y}\boldsymbol{\beta}' + \boldsymbol{\beta}\boldsymbol{\beta}] \\ &= \mathrm{E}[\mathbf{y}\mathbf{y}'] - \boldsymbol{\beta}\boldsymbol{\beta}'.\end{aligned}$$

Example 1.4 If $\mathbf{y}$ is an n-dimensional vector with mean $\boldsymbol{\theta}$ and variance-covariance matrix $\boldsymbol{\Sigma} = (\sigma_{ij})$, then

$$\begin{aligned}\mathrm{var}[\overline{y}] &= \mathrm{var}[\mathbf{1}_n'\mathbf{y}/n] \\ &= \mathbf{1}_n'\mathrm{Var}[\mathbf{y}]\mathbf{1}_n/n^2 \\ &= \mathbf{1}_n'\boldsymbol{\Sigma}\mathbf{1}_n/n^2 \\ &= \sum_i\sum_j \sigma_{ij}/n^2.\end{aligned}$$

1.7 Multivariate Normal Distribution

An $n \times 1$ random vector $\mathbf{y} = (y_i)$ is said to have a (non-singular) multivariate normal distribution if its density function is

$$(2\pi)^{-n/2}|\boldsymbol{\Sigma}|^{-1/2} \exp\left\{-\frac{1}{2}(\mathbf{y}-\boldsymbol{\mu})'\boldsymbol{\Sigma}^{-1}(\mathbf{y}-\boldsymbol{\mu})\right\}.$$

We note that $\mathrm{E}[\mathbf{y}] = \boldsymbol{\mu}$ and $\mathrm{Var}[\mathbf{y}] = \boldsymbol{\Sigma}$; we shall write $\mathbf{y} \sim N_n[\boldsymbol{\mu}, \boldsymbol{\Sigma}]$. Since $\boldsymbol{\Sigma}$ is nonsingular, it is positive definite. Some situations arise when $\boldsymbol{\Sigma}$ is singular (e.g., the joint distribution of the residuals in linear regression analysis). In this case the density function does not exist, but then $\mathbf{y}$ can be expressed as $\mathbf{Ax}$, where $\mathbf{x}$ has a non-singular normal distribution of smaller dimension. The main properties of the multivariate normal distribution we shall use are given in the following the theorem.

Theorem 1.6

(i) *If $\mathbf{y} \sim N_n(\boldsymbol{\mu}, \boldsymbol{\Sigma})$, $\mathbf{C}$ is an $m \times n$ matrix of rank m, and $\mathbf{d}$ is an $m \times 1$ vector, then $\mathbf{Cy} + \mathbf{d} \sim N_m(\mathbf{C}\boldsymbol{\mu} + \mathbf{d}, \mathbf{C}\boldsymbol{\Sigma}\mathbf{C}')$: in particular $\mathbf{a}'\mathbf{y}$ is univariate normal.*

(ii) *Let $\mathbf{y} = \mathbf{Tz}$, where $\mathbf{T}$ is an orthogonal matrix, and $\boldsymbol{\Sigma} = \sigma^2\mathbf{I}_n$. Then $\mathbf{z} = \mathbf{T}'\mathbf{y}$, $\mathrm{Var}[\mathbf{z}] = \mathbf{T}'\boldsymbol{\Sigma}\mathbf{T} = \sigma^2\mathbf{I}_n$ and $\mathbf{z} \sim N_n[\mathbf{T}'\boldsymbol{\mu}, \sigma^2\mathbf{I}_n]$, that is the z_i are independently distributed as $N[\mathbf{t}_i'\boldsymbol{\mu}, \sigma^2]$, where $\mathbf{t}_i$ is the ith column of $\mathbf{T}$.*

(iii) *The moment generating function of the multivariate normal vector $\mathbf{y}$ is*

$$\begin{aligned} M(\mathbf{t}) &= E[\exp(\mathbf{t}'\mathbf{y})] \\ &= \exp(\mathbf{t}'\boldsymbol{\mu} + \mathbf{t}'\boldsymbol{\Sigma}\mathbf{t}/2). \end{aligned}$$

This result also holds if $\boldsymbol{\Sigma}$ is singular.

(iv) *A random vector $\mathbf{y}$ with mean $\boldsymbol{\mu}$ and variance-covariance matrix $\boldsymbol{\Sigma}$ has an $N_n(\boldsymbol{\mu}, \boldsymbol{\Sigma})$ distribution if and only if $\mathbf{a}'\mathbf{y}$ has a univariate distribution for every vector $\mathbf{a}$. This can be used to define the multivariate normal distribution for both the non-singular and singular case (when $\boldsymbol{\Sigma}$ is singular).*

(v) *If $\mathbf{y}$ has a singular or non-singular multivariate normal distribution, then the vectors $\mathbf{A}_i\mathbf{y}$ $(i = 1, 2)$ are statistically independent if and only if $\mathrm{Cov}[\mathbf{A}_1\mathbf{y}, \mathbf{A}_2\mathbf{y}] = \mathbf{0}$.*

Proof For detailed summaries of the properties of this distribution see Seber and Lee (2003, chapter 3) and Seber (2008, Section 20.5, 435ff). Property (iv) gives a very useful definition for the multivariate normal as all other properties can be derived from it. It can also be used to provide a similar definition of the Wishart distribution used in multivariate analysis.

In most of this book $\boldsymbol{\Sigma} = \sigma^2\mathbf{I}_n$. The matrix takes this form when the y_i are uncorrelated and have the same variance. In this case we see from the factorization of the density function that the y_i are independently distributed as $N_1[\mu_i, \sigma^2]$. In the future we drop the subscript "1" from the univariate distribution.

Example 1.5 If $y_1, y_2, \ldots, y_n$ is a random sample from $N(\mu, \sigma^2)$ we can prove that $\bar{y}$ is statistically independent of $\sum_i (y_i - \bar{y})^2$ as follows. Now

$$\bar{y} = \mathbf{1}_n'\mathbf{y}/n = \mathbf{A}_1\mathbf{y}$$

and

$$\mathbf{z} = (y_1 - \bar{y}, y_2 - \bar{y}, \ldots, y_n - \bar{y})' = \mathbf{y} - \mathbf{1}_n\mathbf{1}'\mathbf{y}/n = \mathbf{A}_2\mathbf{y},$$

where $\mathbf{A}_2 = \mathbf{I}_n - \mathbf{1}_n\mathbf{1}_n'/n$. Then, by Theorem 1.5(iii),

$$\text{Cov}[\bar{y}, \mathbf{z}] = \text{Cov}[\mathbf{A}_1\mathbf{y}, \mathbf{A}_2\mathbf{y}] = n^{-1}\mathbf{1}_n'\text{Var}[\mathbf{y}]\mathbf{A}_2' = \sigma^2(n^{-1}\mathbf{1}_n')(\mathbf{I}_n - \mathbf{1}_n\mathbf{1}_n'/n) = \mathbf{0}.$$

This implies from Theorem 1.6(v) above that $\bar{y}$ is independent of $\mathbf{z}$, and therefore of $\mathbf{z}'\mathbf{z} = \sum_i (y_i - \bar{y})^2$.

1.8 Non-central Distributions

The random variable x with probability density function

$$f_\nu(x, \delta) = \frac{1}{2^{\nu/2}} e^{-(x+\delta)/2} x^{(\nu/2)-1} \sum_{i=0}^{\infty} \left(\frac{\delta x}{4}\right)^i \frac{1}{i!\Gamma(\frac{\nu}{2} + i)}, \quad x \geq 0,$$

where $\Gamma(a)$ is the Gamma function, is said to have a *non-central Chi-square* distribution with ν degrees of freedom and non-centrality parameter δ; we write $x \sim \chi_\nu^2(\delta)$. The distribution can also be expressed in the form

$$f_\nu(x, \delta) = e^{-\delta/2} \sum_{i=0}^{\infty} \frac{(\delta/2)^i}{i!} f_{\nu+2i}(x, 0),$$

where $f_{\nu+2i}(x, 0)$ is the density function for $\chi_{\nu+2i}^2$, the (central) chi-square distribution with $\nu + 2i$ degrees of freedom.

We note the following properties:

Theorem 1.7

(i) *When $\delta = 0$, the above density reduces to that of χ_ν^2.*
(ii) *$E[x] = \nu + \delta$.*
(iii) *The moment generating function (m.g.f.) of x is*

$$M_x(t, \delta) = (1 - 2t)^{-\nu/2} \exp[\delta t/(1 - 2t)],$$

and it uniquely determines the distribution as it exists as a function of t in an interval containing $t = 0$. *When* $\delta = 0$, *the m.g.f. of the chi-square distribution is* $(1 - 2t)^{-\nu/2}$.

(iv) *The m.g.f. of* λx *is* $\mathrm{E}[\exp(x\lambda t)] = M_x(\lambda t, \delta)$.

(v) *The non-central chi-square can be defined as the distribution of the sum of the squares of n independent univariate normal variables* y_i $(i = 1, 2, \ldots, n)$ *with variances 1 and respective means* μ_i. *Thus if* $\mathbf{y}$ *is distributed as* $N_n(\boldsymbol{\mu}, \sigma^2\mathbf{I}_n)$, *then* $x = \mathbf{y}'\mathbf{y}/\sigma^2 \sim \chi^2_n(\delta)$, *where* $\delta = \boldsymbol{\mu}'\boldsymbol{\mu}/\sigma^2$.

(vi) *The non-central chi-square distribution has the same additive property as the central chi-square, namely, if two random variables are distributed independently as* $\chi^2_{n_1}(\delta_1)$ *and* $\chi^2_{n_2}(\delta_2)$, *then the distribution of their sum is* $\chi^2_{n_1+n_2}(\delta_1 + \delta_2)$.

Proof We shall just give an outline. Using (v), the moment generating function of y_i^2 is readily obtained from which we can find the m.g.f. of $\mathbf{y}'\mathbf{y}$ by multiplying the individual m.g.f.s. together giving us (iii). This m.g.f. can be expanded as a power series of m.g.fs of central chi-square variables and, because of the uniqueness of the underlying density function, we find that the density function is a power series in chi-square density functions, as given above. The result (ii) follows from differentiating the m.g.f., while (iv) is straightforward. The moment generating function of the sum of two independent random variables is the product of their m.g.f.s, which readily leads to (vi).

Since $\delta > 0$, some authors set $\delta = \tau^2$, say. Others use $\delta/2$, which because of (ii) is not so memorable.

If $x \sim \chi^2_m(\delta)$, $y \sim \chi^2_n$, and x and y are statistically independent, then $F = (x/m)/(y/n)$ is said to have a non-central F-distribution with m and n degrees of freedom and non-centrality parameter δ. We write $F \sim F_{m,n}(\delta)$. When $\delta = 0$, we use the usual notation $F_{m,n}$ $(= F_{m,n}(0))$ for the F-distribution. Another statistic that is related to the F-statistic is

$$v = \frac{x}{x + y} = \frac{mF}{mF + n},$$

which has the so-called non-central Beta distribution with a finite domain [0, 1]. For derivations of the above distributions see, for example, Johnson et al. (1994).

1.9 Quadratic Forms

Quadratic forms arise frequently in this book and we begin by finding the mean and variance of a quadratic form.

Theorem 1.8 *Let* $\mathbf{y}$ *be an n-dimensional vector with mean* $\boldsymbol{\mu}$ *and variance-covariance matrix* $\text{Var}[\mathbf{y}] = \boldsymbol{\Sigma}$, *let* $\mathbf{A}$ *be an* $n \times n$ *symmetric matrix and* $\mathbf{c}$ *be an* $n \times 1$ *constant vector. Then:*

(i)

$$E[\mathbf{y}'\mathbf{A}\mathbf{y}] = \text{trace}[\mathbf{A}\boldsymbol{\Sigma}] + \boldsymbol{\mu}'\mathbf{A}\boldsymbol{\mu}.$$

(ii)

$$E[(\mathbf{y}-\mathbf{c})'\mathbf{A}(\mathbf{y}-\mathbf{c})] = \text{trace}[\mathbf{A}\boldsymbol{\Sigma}] + (\boldsymbol{\mu}-\mathbf{c})'\mathbf{A}(\boldsymbol{\mu}-\mathbf{c}).$$

(iii) If $\boldsymbol{\Sigma} = \sigma^2\mathbf{I}_n$ *then,*

$$E[\mathbf{y}'\mathbf{A}\mathbf{y}] = \sigma^2(\textit{sum of coefficients of the } y_i^2) + (\mathbf{y}'\mathbf{A}\mathbf{y})_{\mathbf{y}=\boldsymbol{\mu}}.$$

Proof

(i) This can be derived by simply expanding the quadratic. However, the following proof is instructive.

$$\begin{aligned}
E[\mathbf{y}'\mathbf{A}\mathbf{y}] &= \text{trace}(E[\mathbf{y}'\mathbf{A}\mathbf{y}]) \\
&= E[\text{trace}(\mathbf{A}\mathbf{y}\mathbf{y}')], \quad (\text{since trace}(\mathbf{BC}) = \text{trace}(\mathbf{CB})) \\
&= \text{trace}(\text{E}[\mathbf{A}\mathbf{y}\mathbf{y}']) \\
&= \text{trace}(\mathbf{A}\text{E}[\mathbf{y}\mathbf{y}']) \\
&= \text{trace}[\mathbf{A}(\text{Var}[\mathbf{y}] + \boldsymbol{\mu}\boldsymbol{\mu}')], \quad (\text{by Theorem 1.5(vi)}) \\
&= \text{trace}[\mathbf{A}\boldsymbol{\Sigma}] + \text{trace}[\mathbf{A}\boldsymbol{\mu}\boldsymbol{\mu}'] \\
&= \text{trace}[\mathbf{A}\boldsymbol{\Sigma}] + \boldsymbol{\mu}'\mathbf{A}\boldsymbol{\mu}.
\end{aligned}$$

(ii) Setting $\mathbf{x} = \mathbf{y} - \mathbf{c}$ with mean $\boldsymbol{\mu} - \mathbf{c}$, then $\text{Var}[\mathbf{x}] = \text{Var}[\mathbf{y}]$ (by Theorem 1.5(v)), and the result follows from (i).

(iii) $\text{trace}(\mathbf{A}\boldsymbol{\Sigma}) = \text{trace}(\sigma^2\mathbf{A}) = \sigma^2\,\text{trace}[\mathbf{A}]$, and the result follows from (i).

Example 1.6 Given $y_1, y_2, \ldots, y_n$ a random sample from $N[\mu, \sigma^2]$ we show that

$$Q = \mathbf{y}'\mathbf{A}\mathbf{y} = \frac{1}{2(n-1)}\sum_{i=1}^{n-1}(y_{i+1} - y_i)^2$$

is an unbiased estimate of σ^2. Now

$$\begin{aligned}
Q &= \frac{1}{2(n-1)}[(y_2 - y_1)^2 + (y_3 - y_2)^2 + \ldots + (y_n - y_{n-1})^2] \\
&= \frac{1}{2(n-1)}[y_1^2 + y_n^2 + 2(y_2^2 + \ldots + y_{n-1}^2) - 2y_1y_2 - \ldots - 2y_{n-1}y_n],
\end{aligned}$$

and

$$E[\mathbf{y}'\mathbf{A}\mathbf{y}] = \sigma^2 \,\text{trace}[\mathbf{A}] + (\mathbf{y}'\mathbf{A}\mathbf{y})_{\text{each } y_i=\mu}$$
$$= \sigma^2[2 + 2(n-2)]/[(2(n-1)] = \sigma^2.$$

Example 1.7 Suppose that the random variables $y_1, y_2, \ldots, y_n$ have a common mean μ, common variance σ^2, and the correlation between any pair is ρ. Let $\mathbf{\Sigma} = \text{Var}[\mathbf{y}]$. We now find the expected value of $\mathbf{y}'\mathbf{A}\mathbf{y} = \sum_i (y_i - \overline{y})^2$. Since $\mathbf{A} = (\delta_{ij} - \frac{1}{n})$ and

$$\mathbf{\Sigma} = \sigma^2 \begin{pmatrix} 1 & \rho & \rho & \cdots & \rho \\ \rho & 1 & \rho & \cdots & \rho \\ \cdot & \cdot & \cdot & \cdots & \cdot \\ \rho & \rho & \rho & \cdots & 1 \end{pmatrix},$$

$$E[\mathbf{y}'\mathbf{A}\mathbf{y}] = \text{trace}[\mathbf{A}\mathbf{\Sigma}] + 0$$
$$= \sum_i \sum_j a_{ij}\sigma_{ij}$$
$$= \sigma^2[n - 1 - \rho(n^2 - n)/n]$$
$$= \sigma^2(n-1)(1-\rho).$$

This example show the effect of correlation on the bias of the usual estimate of $s^2 = \sum_i (y_i - \overline{y})^2/(n-1)$ of σ^2. Its expected value is $\sigma^2(1-\rho)$.

Theorem 1.9 *Let $x_1, x_2, \ldots, x_n$ be independent random variables where, for $i = 1, 2, \ldots, n$, x_i has mean θ_i, variance σ^2_{2i}, and third and fourth moments about the mean μ_{3i} and μ_{4i}, respectively (i.e., $\mu_{ri} = E[(x_i - \theta_i)^r]$). If $\mathbf{A}$ is any $n \times n$ symmetric matrix, $\mathbf{d}$ is a column vector of the diagonal elements of $\mathbf{A}$, and $\mathbf{b} = \mathbf{A}\boldsymbol{\theta}$, then:*

(i)

$$\text{var}[\mathbf{x}'\mathbf{A}\mathbf{x}] = \sum_i a_{ii}^2\mu_{4i} + \sum_i \sum_{j,j\neq i} a_{ii}a_{jj}\mu_{2i}\mu_{2j} + 2\sum_i \sum_{j,j\neq i} a_{ij}^2\mu_{2i}\mu_{2j}$$
$$+ 4\sum_i b_i^2\mu_{2i} + 4\sum_i \mu_{3i}b_i a_{ii} - (\sum_i a_{ii}\mu_{2i})^2.$$

(ii) If $\mu_{2i} = \mu_2$, $\mu_{3i} = \mu_3$ and $\mu_{4i} = \mu_4$ for $i = 1, 2, \ldots, n$, then

$$\text{var}[\mathbf{x}'\mathbf{A}\mathbf{x}] = (\mu_4 - 3\mu_2^2)\mathbf{d}'\mathbf{d} + 2\mu_2^2 \,\text{trace}[\mathbf{A}^2] + 4\mu_2\boldsymbol{\theta}'\mathbf{A}^2\boldsymbol{\theta} + 4\mu_3\boldsymbol{\theta}'\mathbf{A}\mathbf{d}.$$

(iii) If the x_i are normally distributed then

$$\text{var}[\mathbf{x}'\mathbf{A}\mathbf{x}] = 2\sum_i\sum_j a_{ij}^2\mu_{2i}\mu_{2j} + 4\sum_i b_i^2\mu_{2i}.$$

(iv) If $\mathbf{x} \sim N_n(\boldsymbol{\theta}, \sigma^2\mathbf{I}_n)$ then

$$\text{var}[\mathbf{x}'\mathbf{A}\mathbf{x}] = 2\sigma^4\,\text{trace}[\mathbf{A}^2] + 4\sigma^2\boldsymbol{\theta}'\mathbf{A}^2\boldsymbol{\theta}.$$

Proof

(i) Now

$$\text{Var}[\mathbf{x}'\mathbf{A}\mathbf{x}] = E[(\mathbf{x}'\mathbf{A}\mathbf{x})^2] - (E[\mathbf{x}'\mathbf{A}\mathbf{x}])^2. \tag{1.1}$$

If $\mathbf{y} = \mathbf{x} - \boldsymbol{\theta}$ so that $\text{E}[\mathbf{y}] = \mathbf{0}$, then

$$\mathbf{x}'\mathbf{A}\mathbf{x} = \mathbf{y}'\mathbf{A}\mathbf{y} + 2\mathbf{b}'\mathbf{y} + \boldsymbol{\theta}'\mathbf{A}\boldsymbol{\theta}.$$

Hence

$$\begin{aligned} E[\mathbf{x}'\mathbf{A}\mathbf{x}] &= E[\mathbf{y}'\mathbf{A}\mathbf{y}] + \boldsymbol{\theta}'\mathbf{A}\boldsymbol{\theta} \\ &= \sum_i\sum_j a_{ij}E[y_iy_j] + \boldsymbol{\theta}'\mathbf{A}\boldsymbol{\theta} \\ &= \sum_i a_{ii}\mu_{2i} + \boldsymbol{\theta}'\mathbf{A}\boldsymbol{\theta}. \end{aligned}$$

Also

$$\begin{aligned} (\mathbf{x}'\mathbf{A}\mathbf{x})^2 &= (\mathbf{y}'\mathbf{A}\mathbf{y})^2 + 4(\mathbf{b}'\mathbf{y})^2 + (\boldsymbol{\theta}'\mathbf{A}\boldsymbol{\theta})^2 \\ &\quad + 2\boldsymbol{\theta}'\mathbf{A}\boldsymbol{\theta}\mathbf{y}'\mathbf{A}\mathbf{y} + 4\boldsymbol{\theta}'\mathbf{A}\boldsymbol{\theta}\mathbf{b}'\mathbf{y} + 4\mathbf{b}'\mathbf{y}\mathbf{y}'\mathbf{A}\mathbf{y}, \end{aligned} \tag{1.2}$$

and $(\mathbf{y}'\mathbf{A}\mathbf{y})^2 = \sum_i\sum_j\sum_k\sum_\ell a_{ij}a_{k\ell}y_iy_jy_ky_\ell$. Since

$$E[y_iy_jy_ky_\ell] = \begin{cases} \mu_{4i}, & i=j=k=\ell \\ \mu_{2i}\mu_{2k}, & i=j, k=\ell \\ \mu_{2i}\mu_{2j}, & i=k, j=\ell \\ \mu_{2i}\mu_{2j}, & i=\ell, j=k \\ 0, & \text{otherwise}, \end{cases}$$

we have

$$E[(\mathbf{y}'\mathbf{A}\mathbf{y})^2] = \sum_i a_{ii}^2\mu_{4i} + \sum_i\sum_{k,k\neq i} a_{ii}a_{kk}\mu_{2i}\mu_{2k} + 2\sum_i\sum_{j,j\neq i} a_{ij}^2\mu_{2i}\mu_{2j}.$$

Also

$$E[(\mathbf{b}'\mathbf{y})^2] = E[(\sum_i b_i y_i)^2] = E[\sum_i \sum_j b_i b_j y_i y_j] = \sum_i b_i^2 \mu_{2i},$$

and

$$E[\mathbf{b}'\mathbf{y}\mathbf{y}'\mathbf{A}\mathbf{y}] = E[\sum_i \sum_j \sum_k b_i y_i a_{jk} y_j y_k] = \sum_i \mu_{3i} b_i a_{ii}.$$

Taking the expected value of (1.2) and substituting into (1.1) leads to our result.

(ii)

$$\begin{aligned}\text{var}[\mathbf{y}'\mathbf{A}\mathbf{y}] &= (\mu_4 - 3\mu_2^2)\sum_i a_{ii}^2 + \mu_2^2 \sum_i \sum_j a_{ii}a_{jj} + 2\mu_2^2 \sum_i \sum_j a_{ij}^2 \\ &\quad + 4\mu_2 \mathbf{b}'\mathbf{b} + 4\mu_3 \mathbf{b}'\mathbf{d} - \mu_2^2 (\sum_i a_{ii})^2 \\ &= (\mu_4 - 3\mu_2^2)\mathbf{d}'\mathbf{d} + 2\mu_2^2 \,\text{trace}[\mathbf{A}^2] + 4\mu_2 \mathbf{b}'\mathbf{b} + 4\mu_3 \mathbf{b}'\mathbf{d},\end{aligned}$$

since $\text{trace}[\mathbf{A}^2] = \sum_i \sum_j a_{ij}a_{ji} = \sum_i \sum_j a_{ij}^2$, which is our result. This result was stated without proof by Atiqullah (1962).

(iii) Since x_i is normally distributed, $\mu_{4i} = 3\mu_{2i}^2$, $\mu_{3i} = 0$, and the result follows.

(iv) Here $\mu_2 = \sigma^2$, and (iv) follows from (iii).

Example 1.8 If $\mathbf{y} \sim N_n[\boldsymbol{\theta}, \boldsymbol{\Sigma}]$, where $\boldsymbol{\Sigma}$ is positive definite, we shall find the variance of $\mathbf{y}'\mathbf{A}\mathbf{y}$, where $\mathbf{A}$ is any symmetric $n \times n$ matrix. Since $\boldsymbol{\Sigma}$ is positive definite, there exist a nonsingular matrix $\mathbf{R}$ such that $\boldsymbol{\Sigma} = \mathbf{R}\mathbf{R}'$ (by A.9(iii)). If $\mathbf{z} = \mathbf{R}^{-1}\mathbf{y}$, then

$$\text{Var}[\mathbf{z}] = \mathbf{R}^{-1}\boldsymbol{\Sigma}\mathbf{R}^{-1\prime} = \mathbf{R}^{-1}\mathbf{R}\mathbf{R}'\mathbf{R}'^{-1} = \mathbf{I}_n$$

and $\mathbf{z} \sim N_n[\mathbf{R}^{-1}\boldsymbol{\theta}, \mathbf{I}_n]$. Using Theorem 1.9(iv),

$$\begin{aligned}\text{var}[\mathbf{y}'\mathbf{A}\mathbf{y}] &= \text{var}[\mathbf{z}'\mathbf{R}'\mathbf{A}\mathbf{R}\mathbf{z}] \\ &= \text{trace}[(\mathbf{R}'\mathbf{A}\mathbf{R})^2] + 4\boldsymbol{\theta}'\mathbf{R}^{-1\prime}(\mathbf{R}'\mathbf{A}\mathbf{R})^2\mathbf{R}^{-1}\boldsymbol{\theta} \\ &= \text{trace}[\mathbf{R}'\mathbf{A}\mathbf{R}\mathbf{R}'\mathbf{A}\mathbf{R}] + 4\boldsymbol{\theta}'\mathbf{R}^{-1\prime}\mathbf{R}'\mathbf{A}\mathbf{R}\mathbf{R}'\mathbf{A}\mathbf{R}\mathbf{R}^{-1}\boldsymbol{\theta} \\ &= \text{trace}[\mathbf{R}'\mathbf{A}\boldsymbol{\Sigma}\mathbf{A}\mathbf{R}] + 4\boldsymbol{\theta}'\mathbf{A}\boldsymbol{\Sigma}\mathbf{A}\boldsymbol{\theta} \\ &= \text{trace}[\mathbf{A}\boldsymbol{\Sigma}\mathbf{A}\boldsymbol{\Sigma}] + 4\boldsymbol{\theta}'\mathbf{A}\boldsymbol{\Sigma}\mathbf{A}\boldsymbol{\theta} \quad \text{(by A.1).}\end{aligned}$$

The following three theorems are used throughout this book.

Theorem 1.10 *Suppose* $\mathbf{y} \sim N_n(\boldsymbol{\mu}, \sigma^2\mathbf{I}_n)$. *If* $\mathbf{P}$ *is symmetric and idempotent of rank* r, *then the quadratic* $\mathbf{y}'\mathbf{P}\mathbf{y}/\sigma^2$ *is distributed as non-central chi-square with* r *degrees of freedom and non-centrality parameter* $\delta = \boldsymbol{\mu}'\mathbf{P}\boldsymbol{\mu}/\sigma^2$.

Proof Suppose **P** is $n \times n$. As **P** is symmetric and idempotent, we may assume without loss of generality that the first r eigenvalues are unity and the rest are zero (Theorem 1.4 in Sect. 1.5). Then there exists an orthogonal matrix **T** such that

$$\mathbf{T}'\mathbf{P}\mathbf{T} = \begin{pmatrix} \mathbf{I}_r & \mathbf{0} \\ \mathbf{0} & \mathbf{0} \end{pmatrix},$$

and

$$\begin{aligned} \mathbf{P} &= \mathbf{T}\begin{pmatrix} \mathbf{I}_r & \mathbf{0} \\ \mathbf{0} & \mathbf{0} \end{pmatrix}\mathbf{T}' \\ &= (\mathbf{t}_i, \mathbf{t}_2, \ldots, \mathbf{t}_r)\begin{pmatrix} \mathbf{t}_1' \\ \vdots \\ \mathbf{t}_r' \end{pmatrix} \\ &= \mathbf{T}_r\mathbf{T}_r', \quad \text{say,} \end{aligned}$$

where $\mathbf{t}_1, \mathbf{t}_2, \ldots, \mathbf{t}_r$ are the first r columns of **T**. Putting $\mathbf{y} = \mathbf{T}\mathbf{z}$ gives us

$$\begin{aligned} \mathbf{y}'\mathbf{P}\mathbf{y} &= \mathbf{z}'\mathbf{T}'\mathbf{P}\mathbf{T}\mathbf{z} \\ &= z_1^2 + z_2^2 + \cdots + z_r^2, \end{aligned}$$

where the z_i are independently distributed as $N[\mathbf{t}_i'\boldsymbol{\mu}, \sigma^2]$ (by Theorem 1.6 (ii) in Sect. 1.6). Hence $\sum_{i=1}^r z_i^2/\sigma^2 \sim \chi_r^2(\delta)$ (by Theorem 1.7(iv)), where

$$\begin{aligned} \delta &= \sum_{i=1}^r (\mathbf{t}_i'\boldsymbol{\mu})^2/\sigma^2 \\ &= \boldsymbol{\mu}'\mathbf{T}_r\mathbf{T}_r'\boldsymbol{\mu}/\sigma^2 \\ &= \boldsymbol{\mu}'\mathbf{P}\boldsymbol{\mu}/\sigma^2. \end{aligned}$$

The converse is also true, as we see in the following theorem.

Theorem 1.11 *Suppose* $\mathbf{y} \sim N_n(\boldsymbol{\mu}, \sigma^2\mathbf{I}_n)$. *If* $\mathbf{y}'\mathbf{A}\mathbf{y}/\sigma^2$, *where* **A** *is symmetric, has a non-central chi-square distribution, then* **A** *is idempotent.*

Proof Let $\mathbf{y}'\mathbf{A}\mathbf{y}$ be any quadratic form and let **S** be the diagonalizing orthogonal matrix. Putting $\mathbf{y} = \mathbf{S}\mathbf{z}$ gives us

$$\mathbf{y}'\mathbf{A}\mathbf{y} = \lambda_1 z_1^2 + \lambda_2 z_2^2 + \cdots + \lambda_n z_n^2,$$

where the λ_i are the eigenvalues of **A**. From the proof of the previous theorem, the z_i are independently distributed as $N[\mathbf{s}_i'\boldsymbol{\mu}, \sigma^2]$ and z_i^2/σ^2 is non-central $\chi_1^2(\delta_i)$, where

$\delta_i = (\mathbf{s}_i'\boldsymbol{\mu})^2/\sigma^2$. The m.g.f. of $\lambda_i z_i^2/\sigma^2$ is, by Theorem 1.7(iii) and (iv),

$$M_i(t, \delta_i) = (1 - 2\lambda_i t)^{-\nu/2} \exp\left(\frac{\delta_i \lambda_i t}{1 - 2\lambda_i t}\right).$$

Hence the m.g.f. of $\mathbf{y}'\mathbf{A}\mathbf{y}/\sigma^2$ is $\prod_i M_i(t, \delta_i)$ which has to be the m.g.f of a non-central chi-square distribution. This can only happen if the λ_i are 0 or 1, so that $\mathbf{A}$ is idempotent. We note that if the first r eigenvalues are 1 and the rest are zero and $\delta = \sum_{i=1}^r \delta_i$, then

$$\prod_i M_i(t, \delta_i) = (1 - 2t)^{-r/2} \exp\left(\frac{\delta t}{1 - 2t}\right),$$

which is the m.g.f. of the non-central chi-square distribution $\chi_k^2(\delta)$.

Theorem 1.12 *Suppose* $\mathbf{y} \sim N_n(\boldsymbol{\mu}, \sigma^2\mathbf{I}_n)$. *Given* $n \times n$ *symmetric idempotent matrices* $\mathbf{A}_i$ $(i = 1, 2)$, *then the quadratics* $\mathbf{y}'\mathbf{A}_i\mathbf{y}/\sigma^2$ $(i = 1, 2)$ *are statistically independent if and only if* $\mathbf{A}_1\mathbf{A}_2 = \mathbf{0}$. *(We note that the assumption of idempotency is not necessary, but the proof is instructive and relevant to its application in this book.)*

Proof It follows from Theorem 1.10 that the quadratics are each distributed as non-central chi-square. Since they are independent, their sum is also non-central chi-square (Theorem 1.7(vi)) so that by Theorem 1.11 $\mathbf{A}_1 + \mathbf{A}_2$ is idempotent. Hence

$$(\mathbf{A}_1 + \mathbf{A}_2) = (\mathbf{A}_1 + \mathbf{A}_2)(\mathbf{A}_1 + \mathbf{A}_2) = \mathbf{A}_1 + \mathbf{A}_1\mathbf{A}_2 + \mathbf{A}_2\mathbf{A}_1 + \mathbf{A}_2,$$

so that $\mathbf{A}_1\mathbf{A}_2 + \mathbf{A}_2\mathbf{A}_1 = \mathbf{0}$. The two equations obtained by multiplying the last equation on the left (right) by $\mathbf{A}_1$ give us

$$\mathbf{A}_1\mathbf{A}_2 + \mathbf{A}_1\mathbf{A}_2\mathbf{A}_1 = \mathbf{0}, \quad \text{and} \quad \mathbf{A}_1\mathbf{A}_2\mathbf{A}_1 + \mathbf{A}_2\mathbf{A}_1 = \mathbf{0},$$

so that $\mathbf{A}_1\mathbf{A}_2 = \mathbf{A}_2\mathbf{A}_1 = \mathbf{0}$. Conversely, given $\mathbf{A}_1\mathbf{A}_2 = \mathbf{0}$, it follows from Theorem 1.5(ii) in Sect. 1.6 that

$$\text{Cov}[\mathbf{A}_1\mathbf{y}, \mathbf{A}_2\mathbf{y}] = \mathbf{A}_1\text{Var}[\mathbf{y}]\mathbf{A}_2' = \sigma^2\mathbf{A}_1\mathbf{A}_2 = \mathbf{0}.$$

Hence by Theorem 1.6(v), $\mathbf{A}_1\mathbf{y}$ and $\mathbf{A}_2\mathbf{y}$ are statistically independent and the quadratics $\mathbf{y}'\mathbf{A}_i\mathbf{y} = (\mathbf{A}_i\mathbf{y})'\mathbf{A}_i\mathbf{y}$ $(i = 1, 2)$ are statistically independent.

1.10 Lagrange Multipliers

For the reader who is not familiar with the use of Lagrange multipliers we introduce a brief section on their use in three situations. First, suppose we wish to find a local maximum or minimum of a function $g(\boldsymbol{\theta})$, where $\boldsymbol{\theta}' = (\theta_1, \theta_2, \ldots, \theta_n)$, is subject to a linear constraint $\mathbf{a}'\boldsymbol{\theta} = 0$, and $\mathbf{a}' = (a_1, a_2, \ldots, a_n)$. We introduce an unknown constant called a Lagrange multiplier λ for the constraint and consider the function

$$f(\boldsymbol{\theta}) = g(\boldsymbol{\theta}) + \lambda(\mathbf{a}'\boldsymbol{\theta}).$$

If we have the notation that

$$\mathbf{D}g(\boldsymbol{\theta}) = \frac{\partial g(\boldsymbol{\theta})}{\partial \boldsymbol{\theta}} = \left(\frac{\partial g(\boldsymbol{\theta})}{\partial \theta_1}, \ldots, \frac{\partial g(\boldsymbol{\theta})}{\partial \theta_n}\right)',$$

then the relative maximum or minimum is then given by differentiating the $f(\boldsymbol{\theta})$ with respect to $\boldsymbol{\theta}$ (cf. A.20), namely

$$\mathbf{D}g(\boldsymbol{\theta}) + \lambda \mathbf{a} = \mathbf{0} \quad \text{and} \quad \mathbf{a}'\boldsymbol{\theta} = 0, \tag{1.3}$$

and solving for $\boldsymbol{\theta}$.

A second situation is when we have k independent linear constraints $\mathbf{a}_i'\boldsymbol{\theta} = 0$ $(i = 1, 2, \ldots, k)$. We then introduce a Lagrange multiplier λ_i for each constraint and optimize the function

$$g(\boldsymbol{\theta}) + \sum_i \lambda_i(\mathbf{a}_i'\boldsymbol{\theta}) = g(\boldsymbol{\theta}) + \boldsymbol{\lambda}'\mathbf{A}\boldsymbol{\theta},$$

where $\mathbf{A}' = (\mathbf{a}_1, \mathbf{a}_2, \ldots, \mathbf{a}_k) = (a_{ji})$. The relative maximum or minimum, $\tilde{\boldsymbol{\theta}}$, is then given by solving

$$\frac{\partial g(\boldsymbol{\theta})}{\partial \theta_j} + \sum_{i=1}^{k} \lambda_i a_{ij} = 0 \quad (j = 1, 2, \ldots, n)$$

and

$$\mathbf{a}_i'\boldsymbol{\theta} = 0 \quad (i = 1, 2, \ldots, k),$$

where a_{ij} is the jth element of $\mathbf{a}_i$. We thus have $(k+n)$ equations in $(k+n)$ unknowns $\boldsymbol{\theta}$ and $\boldsymbol{\lambda}$, and therefore, theoretically, they can be solved. Since $\sum_i \lambda_i a_{ij} = \sum_i a'_{ji}\lambda_i$, the jth element of $\mathbf{A}'\boldsymbol{\lambda}$, the equations can be written in the form

$$\mathbf{D}g(\boldsymbol{\theta}) + \mathbf{A}'\boldsymbol{\lambda} = \mathbf{0} \quad \text{and} \quad \mathbf{A}\boldsymbol{\theta} = \mathbf{0}. \tag{1.4}$$

Finally, if the constraints are nonlinear, say $a_i(\boldsymbol{\theta}) = 0$ for $i = 1, 2, \ldots, k$, and the matrix $\mathbf{A}$ has (i, j)th element $\partial a_i / \partial \theta_j$, then the equations become

$$\mathbf{D}g(\boldsymbol{\theta}) + \mathbf{A}'\boldsymbol{\lambda} = \mathbf{0} \quad \text{and} \quad \mathbf{a}(\boldsymbol{\theta}) = \mathbf{0}, \tag{1.5}$$

where $\mathbf{a}(\boldsymbol{\theta}) = (a_1(\boldsymbol{\theta}), a_2(\boldsymbol{\theta}), \ldots, a_k(\boldsymbol{\theta}))'$.

Sufficient conditions for a local maximum or a local minimum are given by Seber (2008, 516–517). However these conditions are awkward to apply and one usually uses ad hoc methods to determine the nature of the stationary value.

In conclusion, we look at the role of the Lagrange multiplier in applying identifiability conditions. Suppose $\mathbf{g}(\boldsymbol{\theta})$ is any real-valued "well-behaved" function of $\boldsymbol{\theta}$ with domain Δ and range $\mathbb{R}$, and $\mathbf{a}(\boldsymbol{\theta}) = \mathbf{0}$ is now a set of k constraints sufficient for the identifiability of $\boldsymbol{\theta}$. This means that for every $b \in \mathbb{R}$, there exists a unique $\boldsymbol{\theta} \in \Delta$ satisfying the equations $g(\boldsymbol{\theta}) = b$ and $\mathbf{a}(\boldsymbol{\theta}) = \mathbf{0}$. Following Seber (1971, Appendix), let $r = n - k$ and consider the transformation from $\boldsymbol{\theta}$ to

$$\boldsymbol{\phi} = (g(\boldsymbol{\theta}), \theta_2, \ldots, \theta_r, \mathbf{a}'(\boldsymbol{\theta}))' = \mathbf{c}(\boldsymbol{\theta}),$$

say. Now given $b \in \Delta$, then for $\phi_1 = b$, $\phi_i = \theta_i$ $(i = 2, 3, \ldots, r)$, and $\phi_i = 0$ $(i = r + 1, \ldots, n)$, $\boldsymbol{\phi}$ is uniquely determined by the definition of identifiability. This implies that subject to the constraints on $\boldsymbol{\phi}$, the transformation is one-to-one and the matrix of partial derivatives

$$\mathbf{C}_{\theta} = (\partial c_i(\boldsymbol{\theta}) / \partial \theta_j)$$

$$= \begin{bmatrix} \frac{\partial g}{\partial \theta_1} & \cdots & \frac{\partial g}{\partial \theta_n} \\ \mathbf{0} & \mathbf{I}_{r-1} & \mathbf{0} \\ \frac{\partial a_1}{\partial \theta_1} & \cdots & \frac{\partial a_1}{\partial \theta_n} \\ \cdot & \cdots & \cdot \\ \frac{\partial a_k}{\partial \theta_1} & \cdots & \frac{\partial a_k}{\partial \theta_n} \end{bmatrix}$$

is non-singular. Defining $\mathbf{A} = (\partial \mathbf{a}_i / \partial \boldsymbol{\theta}_j)$ as above, then for all $\boldsymbol{\theta}$ such that $\mathbf{a}(\boldsymbol{\theta}) = \mathbf{0}$, the columns of $\mathbf{C}'_{\theta}$ are linearly independent and

$$\mathbf{D}g(\boldsymbol{\theta}) + \mathbf{A}'\boldsymbol{\lambda} = \mathbf{0} \tag{1.6}$$

implies that $\boldsymbol{\lambda} = \mathbf{0}$. This means that in finding the stationary values of g subject to the identifiability constraints $\mathbf{a}(\boldsymbol{\theta}) = \mathbf{0}$, the Lagrange multiplier is zero.

References

Atiqullah, M. (1962). The estimation of residual variance in quadratically balanced least squares problems and the robustness of the *F* test. *Biometrika, 49*, 83–91.

Johnson, N. L., Kotz, S., & Balakrishnan, N. (1994). *Continuous univariate distributions* (Vol. 1, 2nd ed.). New York: Wiley.

Seber, G. A. F. (1971). Estimating age-specific survival rates from bird-band returns when the reporting rate is constant. *Biometrika, 58*(3), 491–497.

Seber, G. A. F. (2008). *A matrix handbook for statisticians*. New York: Wiley.

Seber, G. A. F., & Lee, A. J. (2003). *Linear regression analysis* (2nd ed.). New York: Wiley.

Chapter 2
The Linear Hypothesis

2.1 Linear Regression

Ini this chapter we consider a number of linear hypotheses before giving a general definition. Our first example is found in regression analysis.

Example 2.1 Suppose we have a random variable y with mean θ and we assume that θ is a linear function of p non-random variables $x_0, x_1, \ldots, x_{p-1}$ called regressors or explanatory variables, namely,

$$\theta = \beta_0 x_0 + \beta_1 x_1 + \cdots + \beta_{p-1} x_{p-1},$$

where the β's are unknown constants (parameters). For n values of the x's, we get n observations on y, giving the model G

$$\begin{aligned} y_i &= \theta_i + \varepsilon_i \\ &= x_{i0}\beta_0 + x_{i1}\beta_1 + \cdots + x_{i,p-1}\beta_{p-1} + \varepsilon_i, \quad (i = 1, 2, \ldots, n), \end{aligned}$$

where $E[\varepsilon_i] = 0$; generally $x_{i0} = 1$, which we shall assume unless stated otherwise. This is known as a multiple linear regression model with p parameters, and by putting $x_{ij} = x_i^j$ we see that the polynomial regression model

$$y_i = \beta_0 + \beta_1 x_i + \beta_2 x_i^2 + \cdots + \beta_{p-1} x_i^{p-1} + \varepsilon_i,$$

of degree $p - 1$ for a single variable x is included as a special case. We can also have a mixture of both models. The linearity resides in the parameters.

Two further assumptions about the errors ε_i are generally made: (i) the errors are uncorrelated, or $\text{cov}[\varepsilon_i, \varepsilon_j] = 0$ for all $i \neq j$ and (ii) the errors have the same variance σ^2. If we wish to test the null hypothesis $H : \beta_r = \beta_{r+1} = \cdots = \beta_{p-1} = 0$, then we

G.A.F. Seber, *The Linear Model and Hypothesis*, Springer Series in Statistics,
DOI 10.1007/978-3-319-21930-1_2

need to add a further assumption that the errors are normally distributed. If we define $\mathbf{X} = (x_{ij})$, $\boldsymbol{\beta} = (\beta_0, \beta_1, \ldots, \beta_{p-1})'$, and let $\mathbf{X}_r$ represent the matrix consisting of the first r columns of $\mathbf{X}$, then the model, assumptions, and hypothesis can be written in the form $\mathbf{y} = \boldsymbol{\theta} + \boldsymbol{\varepsilon}$, where $\boldsymbol{\varepsilon} \sim N_n[\mathbf{0}, \sigma^2\mathbf{I}_n]$, $G : \boldsymbol{\theta} = \mathbf{X}\boldsymbol{\beta}$ and $H : \boldsymbol{\theta} = \mathbf{X}_r\boldsymbol{\beta}_r$, where $\boldsymbol{\beta}_r$ is the vector of the first r elements of $\boldsymbol{\beta}$. In this situation $\mathbf{X}$ usually has full rank, that is the rank of $\mathbf{X}$ is p. If we define the two column spaces $\Omega = \mathcal{C}[\mathbf{X}]$ and $\omega = \mathcal{C}[\mathbf{X}_r]$, then it follows from Sect. 1.2 that Ω and ω are vector subspaces of $\mathbb{R}^n$ and $\omega \subset \Omega$. Thus H is the linear hypothesis that $\boldsymbol{\theta}$ belongs to a vector space ω given the assumption G that it belongs to a vector space Ω. We also have that $\mathrm{Var}[\mathbf{y}] = \mathrm{Var}[\mathbf{y} - \boldsymbol{\theta}] = \mathrm{Var}[\boldsymbol{\varepsilon}] = \boldsymbol{\sigma}^2\mathbf{I}_n$ (Theorem 1.5(v)) so that $\mathbf{y} \sim N_n[\mathbf{X}\boldsymbol{\beta}, \sigma^2\mathbf{I}_n]$.

2.2 Analysis of Variance

Example 2.2 We note that some of the x-variables in our regression model can also be so-called *indicator variables*, that is variables taking the values of 0 or 1. For example consider n observations from the straight-line model

$$E[y_i] = \beta_0 + \beta_1 x_i, \quad i = 1, 2, \ldots, n,$$

where $x_i = 0$ for $i = 1, 2, \ldots n_1$ and $x_i = 1$ for $i = n_1 + 1, n_1 + 2, \ldots, n$. If $n - n_1 = n_2$, then $\mathbf{X}\boldsymbol{\beta}$ takes the form

$$\mathbf{X}\boldsymbol{\beta} = \begin{pmatrix} \mathbf{1}_{n_1} & \mathbf{0} \\ \mathbf{1}_{n_2} & \mathbf{1}_{n_2} \end{pmatrix} \begin{pmatrix} \beta_0 \\ \beta_1 \end{pmatrix}.$$

This model splits into two models or samples, namely $E[y_i] = \beta_0$ for $i = 1, 2, \ldots, n_1$ and $E[y_i] = \beta_0 + \beta_1$ for $i = 1, 2, \ldots, n_2$. This would give us a model for comparing the means $\mu_1 (= \beta_0)$ and $\mu_2 (= \beta_0 + \beta_1)$ of two samples of sizes n_1 and n_2 respectively. Testing if $\mu_1 = \mu_2$ is equivalent to testing $\beta_1 = 0$. This type of model where variables enter qualitatively is sometimes referred to as an analysis of variance (ANOVA) model.

Example 2.3 We now consider generalizing the above example to comparing I different samples with J_i observations in the ith sample. Let y_{ij} ($i = 1, 2, \ldots, I$ and $j = 1, 2, \ldots J_i$) be the jth observation from the ith sample, so that we have the model $y_{ij} = \mu_i + \varepsilon_{ij}$. Setting $\mathbf{y} = \boldsymbol{\theta} + \boldsymbol{\varepsilon}$, where

$$\mathbf{y}' = (y_{11}, y_{12}, \ldots y_{1J_1}, y_{21}, y_{22}, \ldots, y_{2J_2}, \ldots, y_{I1}, y_{I2} \ldots, y_{IJ_I}),$$

and $\boldsymbol{\theta}$ is similarly defined, we get $\boldsymbol{\theta} = \mathbf{X}\boldsymbol{\mu}$, where

$$\mathbf{X} = \begin{pmatrix} \mathbf{1}_{J_1} & \mathbf{0} & \cdots & \mathbf{0} \\ \cdots & \mathbf{1}_{J_2} & \cdots & \mathbf{0} \\ \cdot & \cdot & \ddots & \cdot \\ \mathbf{0} & \mathbf{0} & \cdots & \mathbf{1}_{J_I} \end{pmatrix}, \tag{2.1}$$

and $\boldsymbol{\mu} = (\mu_1, \mu_2, \ldots, \mu_I)'$. Suppose we wish to test the hypothesis $H : \mu_1 = \mu_2 = \cdots = \mu_I$ (= μ, say), or $\boldsymbol{\theta} = \mathbf{1}_n\mu$, where $\mathbf{1}_n$ is obtained by adding the columns of $\mathbf{X}$ together. Then, from the previous section, $\Omega = \mathcal{C}[\mathbf{X}]$ and $\omega = \mathcal{C}[\mathbf{1}_n]$.

Alternatively, we can express H in the form

$$\mu_1 - \mu_2 = \mu_2 - \mu_3 = \cdots = \mu_{I-1} - \mu_I = 0,$$

which can be written in matrix form $\mathbf{C}\boldsymbol{\mu} = \mathbf{0}$, where

$$\mathbf{C} = \begin{pmatrix} 1 & -1 & 0 & \cdots & 0 & 0 \\ 0 & 1 & -1 & \cdots & 0 & 0 \\ \cdot & \cdot & \cdot & \cdots & \cdot & \cdot \\ 0 & 0 & 0 & \cdots & 1 & -1 \end{pmatrix}.$$

Since $\boldsymbol{\theta} = \mathbf{X}\boldsymbol{\mu}$ and $\mathbf{X}$ has full rank p, the $p \times p$ matrix $\mathbf{X}'\mathbf{X}$ has rank p and is therefore nonsingular (cf. A.4(ii)). From $\boldsymbol{\theta} = \mathbf{X}\boldsymbol{\mu}$ we can then multiply on the left by $\mathbf{X}'$ and get $\boldsymbol{\mu} = (\mathbf{X}'\mathbf{X})^{-1}\mathbf{X}'\boldsymbol{\theta}$. Hence H takes the form

$$\mathbf{0} = \mathbf{C}\boldsymbol{\mu} = \mathbf{C}(\mathbf{X}'\mathbf{X})^{-1}\mathbf{X}'\boldsymbol{\theta} = \mathbf{B}\boldsymbol{\theta}, \tag{2.2}$$

say, or $\boldsymbol{\theta} \in \omega$, where $\omega = \mathcal{C}[\mathbf{X}] \cap \mathcal{N}[\mathbf{B}]$.

An alternative parametrization can be used for the above example that is more typical of analysis of variance models. Let $\mu = \sum_{i=1}^{I} \mu_i/I$ and define $\alpha_i = \mu_i - \mu$ so that $\mu_i = \mu + \alpha_i$. Then $\sum_{i=1}^{I} \alpha_i = 0$ is an "identifiability condition" (see Sect. 3.4) giving us $I + 1$ parameters or I free parameters still. We now have

$$\mathbf{X}\boldsymbol{\beta} = \begin{pmatrix} \mathbf{1}_{J_1} & \mathbf{1}_{J_1} & \mathbf{0} & \cdots & \mathbf{0} \\ \mathbf{1}_{J_2} & \mathbf{0} & \mathbf{1}_{J_2} & \cdots & \mathbf{0} \\ \cdot & \cdot & \cdot & \cdots & \cdot \\ \mathbf{1}_{J_I} & \mathbf{0} & \mathbf{0} & \cdots & \mathbf{1}_{J_I} \end{pmatrix} \begin{pmatrix} \mu \\ \alpha_1 \\ \alpha_2 \\ \vdots \\ \alpha_I \end{pmatrix}, \tag{2.3}$$

where the first column of $\mathbf{X}$, namely $\mathbf{1}_n$, is the sum of the other columns, and the matrix $\mathbf{X}$ is no longer of full rank.

Example 2.4 We consider one other ANOVA model, the randomized block design where there are J blocks and I treatments randomized in each block. Let y_{ij} with mean θ_{ij} be the observation from the ith treatment in the jth block and, for $i = 1, 2, \ldots, I$, let $\mathbf{y}_i = (y_{i1}, y_{i2}, \ldots, y_{iJ})'$ and $\boldsymbol{\theta}_i = (\theta_{i1}, \theta_{i2}, \ldots \theta_{iJ})'$. Let $\mathbf{y} = (\mathbf{y}_1', \mathbf{y}_2', \ldots \mathbf{y}_I')'$ with $\boldsymbol{\theta}$ and $\boldsymbol{\varepsilon}$ similarly defined. We assume the model

$$y_{ij} = \theta_{ij} + \varepsilon_{ij} = \mu + \alpha_i + \beta_j + \varepsilon_{ij}, \quad (i = 1, 2, \ldots, I : j = 1, 2, \ldots, J),$$

or $\mathbf{y} = \boldsymbol{\theta} + \boldsymbol{\varepsilon}$, where $\boldsymbol{\theta} = \mathbf{X}\boldsymbol{\delta}$, namely

$$\begin{pmatrix} \boldsymbol{\theta}_1 \\ \boldsymbol{\theta}_2 \\ \cdot \\ \boldsymbol{\theta}_I \end{pmatrix} = \begin{pmatrix} \mathbf{1}_J & | & \mathbf{1}_J & \mathbf{0} & \mathbf{0} & \cdots & \mathbf{0} & | & \mathbf{I}_J \\ \mathbf{1}_J & | & \mathbf{0} & \mathbf{1}_J & \mathbf{0} & \cdots & \mathbf{0} & | & \mathbf{I}_J \\ \cdot & | & \cdot & \cdot & \cdot & \cdots & \cdot & | & \cdot \\ \mathbf{1}_J & | & \mathbf{0} & \mathbf{0} & \mathbf{0} & \cdots & \mathbf{1}_J & | & \mathbf{I}_J \end{pmatrix} \begin{pmatrix} \mu \\ \boldsymbol{\alpha} \\ \boldsymbol{\beta} \end{pmatrix},$$

where $\boldsymbol{\alpha} = (\alpha_1, \alpha_2, \ldots, \alpha_I)'$ and $\boldsymbol{\beta} = (\beta_1, \beta_2, \ldots, \beta_J)'$.

We have IJ observations and $1 + I + J$ unknown parameters. Setting $\overline{\theta}_{i\cdot} = \sum_j \theta_{ij}/J$ and $\overline{\theta}_{\cdot\cdot} = \sum_i \sum_j \theta_{ij}/IJ$ etc., we assume from the randomization process that the so-called interactions $\gamma_{ij} = \theta_{ij} - \overline{\theta}_{i\cdot} - \overline{\theta}_{\cdot j} + \overline{\theta}_{\cdot\cdot}$ are all zero, i.e., $\mathbf{C}\boldsymbol{\theta} = \mathbf{0}$ for some matrix $\mathbf{C}$. Since we have $\sum_i \gamma_{ij} = 0$ for $j = 1, 2, \ldots, J$, $\sum_j \gamma_{ij} = 0$ for $i = 1, 2, \ldots, I$, and both sets include $\sum_i \sum_j \gamma_{ij} = 0$, we have $IJ - I - J + 1 = (I-1)(J-1)$ independent constraints so that $\mathbf{C}$ will be $(I-1)(J-1) \times IJ$. The number of parameters that can be estimated is $IJ - (I-1)(J-1) = I + J - 1$, which means we have 2 too many parameters in $\boldsymbol{\delta}$. We need to add two identifiability constraints such as $\sum_i \alpha_i = 0$ and $\sum_j \beta_j = 0$, or $\alpha_I = 0$ and $\beta_J = 0$, for example. By summing columns, we see that the matrix $\mathbf{X}$ above has two linearly dependent columns so that it is $IJ \times (1 + I + J)$ of rank $I + J - 1$. If we set $\alpha_I = 0$ and $\beta_J = 0$ then $\mathbf{X}$ is reduced to $\mathbf{X}_1$, say, with full rank and the same column space as that of $\mathbf{X}$, and $\boldsymbol{\delta}$ is reduced by two elements to $\boldsymbol{\delta}_1$, say. We are usually interested in testing H that there are no differences in the treatments. Then $H : \alpha_1 = \alpha_2 = \cdots = \alpha_{I-1} = 0$ or $\mathbf{C}_1\boldsymbol{\delta}_1 = \mathbf{0}$, say. Using (2.2) with $\boldsymbol{\delta}_1 = (\mathbf{X}_1'\mathbf{X}_1)^{-1}\mathbf{X}_1'\boldsymbol{\theta}$, we now have $\Omega = \mathcal{C}[\mathbf{X}] \cap \mathcal{N}[\mathbf{C}]$ and $\omega = \Omega \cap \mathcal{N}[\mathbf{C}_1(\mathbf{X}_1'\mathbf{X}_1)^{-1}\mathbf{X}_1']$.

2.3 Analysis of Covariance

When we have a mixture of quantitive and qualitative explanatory variables we have a so-called analysis of covariance model. For example

$$y_{ij} = \mu_i + \gamma_i z_{ij} + \varepsilon_{ij} \quad (i = 1, 2, \ldots, I : j = 1, 2, \ldots, J_i)$$

represents observations from I straight-line models. Two hypotheses are of interest, namely H_1 that the lines are parallel (i.e. equal γ_i) and H_2 that the lines have the same intercept on the x-axis (i.e. equal μ_i). If both hypotheses are true, the lines are identical. This model G can usually be regarded as the "sum" of two models with $\Omega = \mathcal{C}[\mathbf{X}] \oplus \mathcal{C}[\mathbf{Z}]$, where $\mathbf{Z} = (z_{ij})$, $\mathbf{X}$ is given by Eq. (2.1) in the previous section, and $\mathcal{C}[\mathbf{X}] \cap \mathcal{C}[\mathbf{Z}] = \mathbf{0}$. Such "augmented" models are discussed in Chap. 7.

2.4 General Definition and Extensions

The above examples illustrate what we mean by a linear hypothesis, and we now give a formal definition. Let $\mathbf{y} = \boldsymbol{\theta} + \boldsymbol{\varepsilon}$, where $\boldsymbol{\theta}$ is known to belong to a vector space Ω, then a linear hypothesis H is a hypothesis which states that $\boldsymbol{\theta} \in \omega$, a linear subspace of Ω. The assumption that $\boldsymbol{\theta} \in \Omega$ we denote by G. For purposes of estimation we add the assumptions $\mathrm{E}[\boldsymbol{\varepsilon}] = \mathbf{0}$ and $\mathrm{Var}[\mathbf{y}] = \mathrm{Var}[\boldsymbol{\varepsilon}] = \sigma^2\mathbf{I}_n$, and for testing H we add the further assumption that $\boldsymbol{\varepsilon}$ has the multivariate normal distribution. We now consider three extensions.

Example 2.5 There is one hypothesis that is basically linear, but does not satisfy the definition. For example, suppose $\boldsymbol{\theta} = \mathbf{X}\boldsymbol{\beta}$, where $\mathbf{X}$ is $n \times p$ of full column rank p, say, and we wish to test $\mathbf{H} : \mathbf{A}\boldsymbol{\beta} = \mathbf{a}$, where $\mathbf{A}$ and $\mathbf{a}$ are known and $\mathbf{a} \neq \mathbf{0}$. Now $(\boldsymbol{\beta} = \mathbf{X}'\mathbf{X})^{-1}\mathbf{X}'\boldsymbol{\theta}$, so that $\omega = \{\boldsymbol{\theta} : \mathbf{A}(\mathbf{X}'\mathbf{X})^{-1}\mathbf{X}'\boldsymbol{\theta} = \mathbf{a}\}$ is not a linear vector space (technically a linear manifold) when $\mathbf{a} \neq \mathbf{0}$. However, if we choose any vector $\mathbf{c}$ such that $\mathbf{A}\mathbf{c} = \mathbf{a}$ (which is possible if the linear equations $\mathbf{A}\boldsymbol{\beta} = \mathbf{a}$ are consistent) and put

$$\mathbf{z} = \mathbf{y} - \mathbf{X}\mathbf{c}, \quad \boldsymbol{\phi} = \boldsymbol{\theta} - \mathbf{X}\mathbf{c} = \mathbf{X}(\boldsymbol{\beta} - \mathbf{c}), \quad \text{and} \quad \boldsymbol{\gamma} = \boldsymbol{\beta} - \mathbf{c},$$

we havc

$$\mathbf{z} = \boldsymbol{\phi} + \boldsymbol{\varepsilon}, \quad G : \boldsymbol{\phi} = \mathbf{X}\boldsymbol{\gamma},$$

and $H : \mathbf{A}\boldsymbol{\gamma} = \mathbf{A}(\boldsymbol{\beta} - \mathbf{c}) = \mathbf{0}$ or $\mathbf{A}(\mathbf{X}'\mathbf{X})^{-1}\mathbf{X}'\boldsymbol{\phi} = \mathbf{A}_1\boldsymbol{\phi} = \mathbf{0}$ is now a linear hypothesis with $\omega = \mathcal{N}[\mathbf{A}_1] \cap \Omega$ and $\Omega = \mathcal{C}[\mathbf{X}]$.

Example 2.6 In some examples the underlying model takes the form $\mathbf{y} = \boldsymbol{\theta} + \boldsymbol{\eta}$, where $\boldsymbol{\eta}$ is $N_n[\mathbf{0}, \sigma^2\mathbf{B}]$ and $\mathbf{B}$ is a known positive-definite matrix. This implies that there exists a nonsingular matrix $\mathbf{V}$ such that $\mathbf{B} = \mathbf{V}\mathbf{V}'$ (by A.9(iii)). Using the transformations $\mathbf{z} = \mathbf{V}^{-1}\mathbf{y}$, $\boldsymbol{\phi} = \mathbf{V}^{-1}\boldsymbol{\theta}$, and $\boldsymbol{\varepsilon} = \mathbf{V}^{-1}\boldsymbol{\eta}$ we can transform the model to $\mathbf{z} = \boldsymbol{\phi} + \boldsymbol{\varepsilon}$, where by Theorem 1.5(iii) in Sect. 1.6,

$$\begin{aligned}\mathrm{Var}[\boldsymbol{\varepsilon}] &= \mathrm{Var}[\mathbf{V}^{-1}\boldsymbol{\eta}] \\ &= \mathbf{V}^{-1}\mathrm{Var}[\boldsymbol{\eta}](\mathbf{V}^{-1})' \\ &= \sigma^2\mathbf{V}^{-1}(\mathbf{V}\mathbf{V}')(\mathbf{V}')^{-1} = \sigma^2\mathbf{I}_n,\end{aligned}$$

as before. To see that linear hypotheses remain linear, let the columns of $\mathbf{W}$ be any basis of Ω. Then

$$\begin{aligned}\Omega &= \{\boldsymbol{\theta} : \boldsymbol{\theta} = \mathbf{W}\boldsymbol{\beta}\}\\ &= \{\boldsymbol{\phi} : \boldsymbol{\phi} = \mathbf{V}^{-1}\mathbf{W}\boldsymbol{\beta}\}\\ &= \mathcal{C}[\mathbf{V}^{-1}\mathbf{W}].\end{aligned}$$

To test $\mathbf{A}\boldsymbol{\beta} = \mathbf{0}$ we note from above that $\boldsymbol{\beta} = (\mathbf{W}'\mathbf{W})^{-1}\mathbf{W}'\boldsymbol{\theta}$ so that we have $H : \mathbf{A}(\mathbf{W}'\mathbf{W})^{-1}\mathbf{W}'\mathbf{V}\boldsymbol{\phi} = \mathbf{0}$ or $\omega = \Omega \cap \mathcal{N}[\mathbf{A}(\mathbf{W}'\mathbf{W})^{-1}\mathbf{W}'\mathbf{V}]$.

Example 2.7 One model of interest is $\mathbf{y} = \boldsymbol{\theta} + \boldsymbol{\varepsilon}$, where $\boldsymbol{\varepsilon} \sim N_n[\mathbf{0}, \mathbf{I}_n]$, $\Omega = \mathbb{R}^n$, and ω is a subspace of $\mathbb{R}^n$. Although this model appears to be impractical, it does arise in the large sample theory used in the last three chapters of this monograph. Large sample models and hypotheses are shown there to be asymptotically equivalent to this simple situation.

Chapter 3
Estimation

3.1 Principle of Least Squares

Suppose we have the model $\mathbf{y} = \boldsymbol{\theta} + \boldsymbol{\varepsilon}$, where $\mathrm{E}[\boldsymbol{\varepsilon}] = \mathbf{0}$, $\mathrm{Var}[\boldsymbol{\varepsilon}] = \sigma^2\mathbf{I}_n$, and $\boldsymbol{\theta} \in \Omega$, a p-dimensional vector space. One reasonable estimate of $\boldsymbol{\theta}$ would be the value $\hat{\boldsymbol{\theta}}$, called the *least squares estimate*, that minimizes the total "error" sum of squares

$$SS = \sum_{i=1}^{n} \varepsilon_i^2 = \| \mathbf{y} - \boldsymbol{\theta} \|^2$$

subject to $\boldsymbol{\theta} \in \Omega$. A clue as to how we might calculate $\hat{\boldsymbol{\theta}}$ is by considering the simple case in which $\mathbf{y}$ is a point P in three dimensions and Ω is a plane through the origin O. We have to find the point Q $(= \hat{\boldsymbol{\theta}})$ in the plane so that PQ^2 is a minimum; this is obviously the case when OQ is the orthogonal projection of OP onto the plane. This idea can now be generalized in the following theorem.

Theorem 3.1 *The least squares estimate $\hat{\boldsymbol{\theta}}$ which minimizes $\| \mathbf{y} - \boldsymbol{\theta} \|^2$ for $\boldsymbol{\theta} \in \Omega$ is the orthogonal projection of $\mathbf{y}$ onto Ω.*

Proof Let $\boldsymbol{\alpha}_1, \boldsymbol{\alpha}_2, \ldots, \boldsymbol{\alpha}_p$ be an orthornormal basis for Ω and let $c_i = \boldsymbol{\alpha}_i'\mathbf{y}$. Then

$$\begin{aligned} \mathbf{y} &= \sum_{i=1}^{p} c_i\boldsymbol{\alpha}_i + (\mathbf{y} - \sum_{i=1}^{p} c_i\boldsymbol{\alpha}_i) \\ &= \mathbf{a} + \mathbf{b}, \text{ say.} \end{aligned}$$

G.A.F. Seber, *The Linear Model and Hypothesis*, Springer Series in Statistics,
DOI 10.1007/978-3-319-21930-1_3

Now premultiplying by $\boldsymbol{\alpha}_j'$ we get

$$\begin{aligned}\boldsymbol{\alpha}_j'\mathbf{b} &= \boldsymbol{\alpha}_j'\mathbf{y} - \sum_{i=1}^{p} c_i\boldsymbol{\alpha}_j'\boldsymbol{\alpha}_i \\ &= c_j - \sum_{i=1}^{p} c_i\delta_{ij} \\ &= 0,\end{aligned}$$

where δ_{ij} is 1 when $i = j$ and 0 otherwise. Thus $\mathbf{a} \in \Omega$, $\mathbf{b} \perp \Omega$, and we have decomposed $\mathbf{y}$ into two orthogonal vectors. This decomposition is unique otherwise there will exist some other decomposition $\mathbf{y} = \mathbf{a}_1 + \mathbf{b}_1$. Then we have $\mathbf{a}_1 - \mathbf{a} = \mathbf{b}_1 - \mathbf{b}$, and since $\mathbf{a}_1 - \mathbf{a} \in \Omega$ and $\mathbf{b}_1 - \mathbf{b} \in \Omega^{\perp}$, both these vectors must be the zero vector; therefore $\mathbf{a}_1 = \mathbf{a}$ and $\mathbf{b}_1 = \mathbf{b}$. The unique vector $\mathbf{a}$ is the orthogonal projection of $\mathbf{y}$ onto Ω, and we now show that $\mathbf{a} = \hat{\boldsymbol{\theta}}$.

Since $\mathbf{a} - \boldsymbol{\theta} \in \Omega$,

$$(\mathbf{y} - \mathbf{a})'(\mathbf{a} - \boldsymbol{\theta}) = \mathbf{b}'(\mathbf{a} - \boldsymbol{\theta}) = 0,$$

and from $\mathbf{y} - \boldsymbol{\theta} = (\mathbf{y} - \mathbf{a}) + (\mathbf{a} - \boldsymbol{\theta})$ we have

$$\begin{aligned}\| \mathbf{y} - \boldsymbol{\theta} \|^2 &= \| \mathbf{y} - \mathbf{a} \|^2 + \| \mathbf{a} - \boldsymbol{\theta} \|^2 + 2(\mathbf{y} - \mathbf{a})'(\mathbf{a} - \boldsymbol{\theta}) \\ &= \| \mathbf{y} - \mathbf{a} \|^2 + \| \mathbf{a} - \boldsymbol{\theta} \|^2 .\end{aligned}$$

Therefore $\| \mathbf{y} - \boldsymbol{\theta} \|^2$ is minimized when $\boldsymbol{\theta} = \mathbf{a}$, and $\hat{\boldsymbol{\theta}} = \mathbf{a}$.

3.2 Projection Matrices

We now show that $\hat{\boldsymbol{\theta}}$ can be found by means of a linear transformation of $\mathbf{y}$.

Theorem 3.2 *If $\hat{\boldsymbol{\theta}}$ is the least squares estimate defined above, then $\hat{\boldsymbol{\theta}} = \mathbf{P}_\Omega\mathbf{y}$, where $\mathbf{P}_\Omega$ is a unique symmetric idempotent matrix of rank p (the dimension of Ω) representing the orthogonal projection of $\mathbb{R}^n$ onto Ω.*

Proof From Theorem 3.1,

$$\begin{aligned}\hat{\boldsymbol{\theta}} &= \mathbf{a} \\ &= \sum_{i=1}^{p} c_i\boldsymbol{\alpha}_i\end{aligned}$$

$$= \sum_{i=1}^{p} \alpha_i(\mathbf{y}'\alpha_i)$$
$$= (\alpha_1, \alpha_2, \ldots, \alpha_p)(\alpha_1, \alpha_2, \ldots, \alpha_p)'\mathbf{y}$$
$$= \mathbf{TT}'\mathbf{y}$$
$$= \mathbf{P}_\Omega\mathbf{y}, \quad \text{say.}$$

Now $\mathbf{P}'_\Omega = \mathbf{P}_\Omega$ and $\mathbf{P}_\Omega\mathbf{P}_\Omega = \mathbf{T}(\mathbf{T}'\mathbf{T})\mathbf{T}' = \mathbf{TT}' = \mathbf{P}_\Omega$. Hence $\mathbf{P}_\Omega$ is symmetric and idempotent. If $\mathbf{P}$ is any other $n \times n$ matrix representing this orthogonal projection then, by the uniqueness of $\mathbf{a}$, $(\mathbf{P} - \mathbf{P}_\Omega)\mathbf{y} = \mathbf{0}$ for all $\mathbf{y}$ so that $\mathbf{P} = \mathbf{P}_\Omega$ and $\mathbf{P}_\Omega$ is unique. Also

$$\text{rank}[\mathbf{P}_\Omega] = \text{rank}[\mathbf{TT}'] = \text{rank}[\mathbf{T}] = p,$$

by (A.4(ii)).

The converse is also true as we see from the following theorem.

Theorem 3.3 *If* $\mathbf{P}$ *is a symmetric* $n \times n$ *idempotent matrix of rank* r*, then it represents an orthogonal projection of* $\mathbb{R}^n$ *onto some* r*-dimensional subspace* $\mathcal{V}$.

Proof From the previous theorem we see that $\mathbf{P}$ can be expressed in the form $(\mathbf{s}_1, \ldots, \mathbf{s}_r)(\mathbf{s}_1, \ldots, \mathbf{s}_r)'$, as with $\mathbf{TT}'$ above. Hence $\mathbf{P}$ represents an orthogonal projection onto the vector space spanned by the orthonormal basis $\mathbf{s}_1, \ldots, \mathbf{s}_r$. If this vector space is $\mathcal{V}$, then $\dim \mathcal{V} = r$ and the proof is complete.

A very useful result that will often be used is the following.

Theorem 3.4 *If* Ω *is any subspace of* $\mathbb{R}^n$ *and* $\mathbf{P}_\Omega$ *represents the orthogonal projection of* $\mathbb{R}^n$ *onto* Ω*, then* $\mathcal{C}[\mathbf{P}_\Omega] = \Omega$.

Proof From Theorem 3.2, $\mathbf{P}_\Omega = \mathbf{TT}'$, where the columns of $\mathbf{T}$ form an orthonormal basis of Ω. If $\boldsymbol{\theta} \in \Omega$, then $\boldsymbol{\theta} = \mathbf{T}\boldsymbol{\alpha}$ for some $\boldsymbol{\alpha}$ and $\mathbf{P}_\Omega\boldsymbol{\theta} = \mathbf{TT}'\mathbf{T}\boldsymbol{\alpha} = \mathbf{T}\boldsymbol{\alpha} = \boldsymbol{\theta}$ and $\boldsymbol{\theta} \in \mathcal{C}[\mathbf{P}_\Omega]$. Conversely, if $\boldsymbol{\theta} \in \mathbf{P}_\Omega$ then $\boldsymbol{\theta} = \mathbf{T}(\mathbf{T}'\boldsymbol{\beta})$ for some $\boldsymbol{\beta}$, and $\boldsymbol{\theta} \in \mathcal{C}[\mathbf{T}] = \Omega$. Thus the two vector spaces are the same.

Since from the previous proof we have $\mathbf{P}_\Omega\boldsymbol{\theta} = \boldsymbol{\theta}$ when $\boldsymbol{\theta} \in \Omega$, we have that $\text{E}[\hat{\boldsymbol{\theta}}] = \mathbf{P}_\Omega\text{E}[\mathbf{y}] = \mathbf{P}_\Omega\boldsymbol{\theta} = \boldsymbol{\theta}$ and $\hat{\boldsymbol{\theta}}$ is an unbiased estimate of $\boldsymbol{\theta}$. In addition $\text{Var}[\hat{\boldsymbol{\theta}}] = \sigma^2\mathbf{P}_\Omega\mathbf{P}'_\Omega = \sigma^2\mathbf{P}_\Omega$.

3.3 Examples

Example 3.1 (Linear Regression) Let $\mathbf{y} = \boldsymbol{\theta} + \boldsymbol{\varepsilon}$, where $\text{E}[\mathbf{y}] = \boldsymbol{\theta} = \mathbf{X}\boldsymbol{\beta}$ and $\mathbf{X}$ is an $n \times p$ matrix of rank p. Here $\Omega = \mathcal{C}[\mathbf{X}] = \{\boldsymbol{\theta} : \boldsymbol{\theta} = \mathbf{X}\boldsymbol{\beta}\}$. Also we assume $\text{Var}[\mathbf{y}] = \sigma^2\mathbf{I}_n$. Now, from the previous section, $\mathbf{P}_\Omega = \mathbf{TT}'$, where the columns of the $n \times p$ matrix $\mathbf{T}$ form an orthonormal basis for Ω, i.e., $\mathbf{T}'\mathbf{T} = \mathbf{I}_p$. Since the columns of $\mathbf{X}$ also form a basis of Ω, $\mathbf{X} = \mathbf{TC}$, where the $p \times p$ matrix $\mathbf{C}$ is a nonsingular matrix. Otherwise if $\mathbf{C}$ is singular, rank[$\mathbf{X}$] $\leq$ rank[$\mathbf{C}$] $< p$ by A.2, which is a contradiction. Hence:

(i) $\mathbf{P}_\Omega = \mathbf{TT}' = \mathbf{XC}^{-1}\mathbf{C}^{-1'}\mathbf{X}' = \mathbf{X}(\mathbf{C}'\mathbf{C})^{-1}\mathbf{X}' = \mathbf{X}(\mathbf{X}'\mathbf{X})^{-1}\mathbf{X}'$. Since $\mathbf{P}_\Omega$ is idempotent, we have $p = \text{rank}[\mathbf{P}_\Omega] = \text{trace}[\mathbf{P}_\Omega]$ by Sect. 1.5.
(ii) $\mathbf{P}_\Omega\mathbf{X} = \mathbf{X}(\mathbf{X}'\mathbf{X})^{-1}\mathbf{X}'\mathbf{X} = \mathbf{X}$.
(iii) By Theorem 1.4 in Sect. 1.5, $\mathbf{I}_n - \mathbf{P}_\Omega$ is symmetric and idempotent and

$$\text{rank}[\mathbf{I}_n - \mathbf{P}_\Omega] = \text{trace}[\mathbf{I}_n - \mathbf{P}_\Omega] = n - \text{trace}[\mathbf{P}_\Omega] = n - p.$$

If $\hat{\boldsymbol{\theta}} = \mathbf{X}\hat{\boldsymbol{\beta}}$, then

$$\hat{\boldsymbol{\beta}} = (\mathbf{X}'\mathbf{X})^{-1}\mathbf{X}'\hat{\boldsymbol{\theta}} = (\mathbf{X}'\mathbf{X})^{-1}\mathbf{X}'\mathbf{P}_\Omega\mathbf{y} = (\mathbf{X}'\mathbf{X})^{-1}\mathbf{X}'\mathbf{y}$$

by (ii), and (see Sect. 1.6)

$$\begin{aligned} \mathrm{E}[\hat{\boldsymbol{\beta}}] &= (\mathbf{X}'\mathbf{X})^{-1}\mathbf{X}'\mathbf{X}\boldsymbol{\beta} = \boldsymbol{\beta} \quad \text{and} \\ \mathrm{Var}[\hat{\boldsymbol{\beta}}] &= (\mathbf{X}'\mathbf{X})^{-1}\mathbf{X}'\mathrm{Var}[\mathbf{y}]\mathbf{X}(\mathbf{X}'\mathbf{X})^{-1} = \sigma^2(\mathbf{X}'\mathbf{X})^{-1}. \end{aligned}$$

These results can also be derived by the more familiar method of minimizing the sum of squares $SS = \| \mathbf{y} - \boldsymbol{\theta} \|^2$ for $\boldsymbol{\theta} \in \Omega$; that is minimizing the sum of squares $SS = (\mathbf{y} - \mathbf{X}\boldsymbol{\beta})'(\mathbf{y} - \mathbf{X}\boldsymbol{\beta}) = \mathbf{y}'\mathbf{y} - 2\boldsymbol{\beta}'\mathbf{X}'\mathbf{y} + \boldsymbol{\beta}'\mathbf{X}'\mathbf{X}\boldsymbol{\beta}$ with respect to $\boldsymbol{\beta}$. If $d/d\boldsymbol{\beta}$ denotes the column vector with ith element $d/d\beta_i$ then we find that (A.20)

$$\frac{d(\boldsymbol{\beta}'\mathbf{X}'\mathbf{y})}{d\boldsymbol{\beta}} = \mathbf{X}'\mathbf{y} \quad \text{and} \quad \frac{d(\boldsymbol{\beta}'\mathbf{X}'\mathbf{X}\boldsymbol{\beta})}{d\boldsymbol{\beta}} = 2\mathbf{X}'\mathbf{X}\boldsymbol{\beta}$$

giving us

$$\frac{d(SS)}{d\boldsymbol{\beta}} = 2\mathbf{X}'\mathbf{X}\boldsymbol{\beta} - 2\mathbf{X}'\mathbf{y} = \mathbf{0}. \tag{3.1}$$

These equations are known as the *least squares* or *normal* equations and have the solution $\hat{\boldsymbol{\beta}} = (\mathbf{X}'\mathbf{X})^{-1}\mathbf{X}'\mathbf{y}$, as before. They also follow directly from the fact that $\mathbf{y} - \hat{\boldsymbol{\theta}} \perp \Omega$, that is

$$\mathbf{0} = \mathbf{X}'(\mathbf{y} - \hat{\boldsymbol{\theta}}) = \mathbf{X}'(\mathbf{y} - \mathbf{X}\hat{\boldsymbol{\beta}}). \tag{3.2}$$

Using the transpose of (3.2) we note that

$$\begin{aligned} &(\mathbf{y} - \mathbf{X}\boldsymbol{\beta})'(\mathbf{y} - \mathbf{X}\boldsymbol{\beta}) \\ &= (\mathbf{y} - \mathbf{X}\hat{\boldsymbol{\beta}} + \mathbf{X}\hat{\boldsymbol{\beta}} - \mathbf{X}\boldsymbol{\beta})'(\mathbf{y} - \mathbf{X}\hat{\boldsymbol{\beta}} + \mathbf{X}\hat{\boldsymbol{\beta}} - \mathbf{X}\boldsymbol{\beta}) \end{aligned}$$

$$= (\mathbf{y} - \mathbf{X}\hat{\beta})'(\mathbf{y} - \mathbf{X}\hat{\beta}) + 2(\mathbf{y} - \mathbf{X}\hat{\beta})'(\mathbf{X}\hat{\beta} - \mathbf{X}\beta) + (\mathbf{X}\hat{\beta} - \mathbf{X}\beta)'(\mathbf{X}\hat{\beta} - \mathbf{X}\beta)$$
$$= (\mathbf{y} - \mathbf{X}\hat{\beta})'(\mathbf{y} - \mathbf{X}\hat{\beta}) + (\hat{\beta} - \beta)'\mathbf{X}'\mathbf{X}(\hat{\beta} - \beta).$$

Since $\mathbf{XX}'$ is positive definite (A.9(vii)), $(\hat{\beta} - \beta)'\mathbf{X}'\mathbf{X}(\hat{\beta} - \beta) > 0$ unless $\beta = \hat{\beta}$, which shows that $\hat{\beta}$ is a unique minimum.

We have, by A.4(i),

$$\text{rank}[(\mathbf{X}'\mathbf{X})^{-1}\mathbf{X}'] = \text{rank}[\mathbf{X}'] = p,$$

so that if $\mathbf{y}$ is multivariate normal we have by Theorem 1.6(i) in Sect. 1.7, $\hat{\beta}$ is $N_p[\beta, \sigma^2(\mathbf{X}'\mathbf{X})^{-1}]$.

Example 3.2 Let $\Omega = \mathcal{N}[\mathbf{A}]$, where the rows of $\mathbf{A}$ are linearly independent. Then $\mathbf{I}_n - \mathbf{P}_\Omega$ represents the orthogonal projection of $\mathbb{R}^n$ onto $\Omega^\perp$ since we have the orthogonal decomposition $\mathbf{y} = \mathbf{P}_\Omega + (\mathbf{I}_n - \mathbf{P}_\Omega)\mathbf{y}$. As $\Omega^\perp = \mathcal{C}[\mathbf{A}']$ (Theorem 1.1 in Sect. 1.2) it follows from Example 3.1 that

$$\mathbf{P}_\Omega = \mathbf{I}_n - \mathbf{P}_{\Omega^\perp} = \mathbf{I}_n - \mathbf{A}'(\mathbf{AA}')^{-1}\mathbf{A}.$$

Example 3.3 Suppose the $y_1, y_1, \ldots, y_n$ are independent observations from $N[\mu, \sigma^2]$. Then $\mathbf{y} = \theta + \varepsilon$, where ε is $N_n[\mathbf{0}, \sigma^2\mathbf{I}_n]$ and $\theta = \mathbf{1}_n\mu$. As $\Omega = \mathcal{C}[\mathbf{1}_n]$, $\mathbf{P}_\Omega = \mathbf{1}_n(\mathbf{1}_n'\mathbf{1}_n)^{-1}\mathbf{1}_n' = n^{-1}\mathbf{1}_n\mathbf{1}_n'$, and therefore

$$\hat{\theta} = \mathbf{P}_\Omega\mathbf{y} = \mathbf{1}_n\bar{y}.$$

Hence, from $\hat{\beta} = (\mathbf{X}'\mathbf{X})^{-1}\mathbf{X}'\theta$,

$$\hat{\mu} = (\mathbf{1}_n'\mathbf{1})^{-1}\mathbf{1}_n'\mathbf{1}_n\bar{y} = \bar{y}.$$

Example 3.4 Suppose that $\mathbf{y} = \mathbf{X}\beta + \eta$ and $\text{Var}[\eta] = \sigma^2\mathbf{B}$, where $\mathbf{X}$ has full rank and $\mathbf{B}$ is a known positive-definite matrix. To find the least squares estimate of β we can use Example 2.6 of Sect. 2.4 and transform the model to the standard form

$$\mathbf{z} = \phi + \varepsilon \quad \text{and} \quad \text{Var}[\varepsilon] = \sigma^2\mathbf{I}_n,$$

using the transformation $\mathbf{z} = \mathbf{V}^{-1}\mathbf{y}$, $\phi = \mathbf{V}^{-1}\theta = \mathbf{V}^{-1}\mathbf{X}\beta$, where $\mathbf{V}$ is given by $\mathbf{B} = \mathbf{VV}'$ (cf. A.9(iii)). We now minimize $(\mathbf{z} - \mathbf{V}^{-1}\mathbf{X}\beta)'(\mathbf{z} - \mathbf{V}^{-1}\mathbf{X}\beta)$ or

$$(\mathbf{y} - \mathbf{X}\beta)'\mathbf{B}^{-1}(\mathbf{y} - \mathbf{X}\beta) = \mathbf{y}'\mathbf{B}\mathbf{y} - 2\beta'\mathbf{X}'\mathbf{B}^{-1}\mathbf{y} + \beta'\mathbf{X}'\mathbf{B}^{-1}\mathbf{X}\beta$$

with respect to β. Differentiating with respect to β and using A.20 we get

$$-2\mathbf{X}'\mathbf{B}^{-1}\mathbf{y} + 2\mathbf{X}'\mathbf{B}^{-1}\mathbf{X}\beta = \mathbf{0},$$

so that the least squares estimate of β is

$$\hat{\beta} = (\mathbf{X}'\mathbf{B}^{-1}\mathbf{X})^{-1}\mathbf{X}'\mathbf{B}^{-1}\mathbf{y},$$

and that of $\boldsymbol{\theta}$ is

$$\hat{\theta} = \mathbf{X}\hat{\beta} = \mathbf{X}(\mathbf{X}'\mathbf{B}^{-1}\mathbf{X})^{-1}\mathbf{X}'\mathbf{B}^{-1}\mathbf{y} = \mathbf{P}\mathbf{y}.$$

We note that $\mathbf{P}^2 = \mathbf{P}$ so that $\mathbf{P}$ represents a projection, but it is an oblique projection rather than an orthogonal one as $\mathbf{P}$ is not symmetric.

In practice it is often simpler to work with the original observations $\mathbf{y}$ and minimize the above modified sum of squares rather than calculate the transformed observations $\mathbf{z}$. The method is referred to as *generalized least squares*. In many applications $\mathbf{B}$ is a diagonal matrix $\mathbf{B} = \text{diag}(w_1, w_2, \ldots, w_n)$, for example when y_i is the mean of a sample of n_i observations so that $w_i = 1/n_i$, and our sum of squares to be minimized takes the form $\sum_{i=1}^{n} w_i^{-1}(y_i - \theta_i)^2$, where the w_i are suitably chosen weights. This method is usually referred to as *weighted least squares*. Sometimes $\mathbf{B}$ is a function of $\boldsymbol{\theta}$ and iterative methods are needed to find the least squares estimates.

3.4 Less Than Full Rank Model

Suppose that $\Omega = \mathcal{C}[\mathbf{X}]$ but now $\mathbf{X}$ is $n \times p$ of rank r $(r < p)$. This means that although $\boldsymbol{\theta}$ is uniquely defined in $\boldsymbol{\theta} = \mathbf{X}\beta$, β is not, as the columns of $\mathbf{X}$ are linearly dependent. In this situation we say that β is *non-identifiable* and the least squares equations (3.1) in Example 3.1 do not have a unique solution for β. To overcome this, we introduce a set of t constraints $\mathbf{H}\beta = \mathbf{0}$ on β satisfying two necessary and sufficient conditions for the identifiability of β: namely (i) for every $\boldsymbol{\theta} \in \Omega$ there exists a β such that $\boldsymbol{\theta} = \mathbf{X}\beta$ and $\mathbf{H}\beta = \mathbf{0}$, and (ii) this β is unique. The first condition is equivalent to $\mathcal{C}[\mathbf{X}'] \cap \mathcal{C}[\mathbf{H}'] = \mathbf{0}$, that is no vector that is a linear combination of the rows of $\mathbf{X}$ is a linear combination of the rows of $\mathbf{H}$ except $\mathbf{0}$ (for a proof see A.11). The second condition is satisfied if the rank of the augmented matrix $\mathbf{G} = (\mathbf{X}', \mathbf{H}')'$ is p, for then the $p \times p$ matrix $\mathbf{G}'\mathbf{G} = \mathbf{X}'\mathbf{X} + \mathbf{H}'\mathbf{H}$ is nonsingular, and adding $\mathbf{X}'\boldsymbol{\theta} = \mathbf{X}'\mathbf{X}\beta$ to $\mathbf{0} = \mathbf{H}'\mathbf{H}\beta$ gives the unique solution $\beta = (\mathbf{G}'\mathbf{G})^{-1}\mathbf{X}'\boldsymbol{\theta}$. Thus, combining these two results, the conditions $\mathbf{H}\beta = \mathbf{0}$ are suitable for identifiability if and only if $\text{rank}[\mathbf{G}] = p$ and $\text{rank}[\mathbf{H}] = p - r$. In general we can assume that there are no redundant identifiability constraints, so that $t = p - r$ and the rows of $\mathbf{H}$ are linearly independent.

From Sect. 1.10 and (1.4), the least squares equations are given by

$$\begin{aligned} 2\mathbf{X}'\mathbf{X}\hat{\beta} - 2\mathbf{X}'\mathbf{y} + \mathbf{H}'\hat{\boldsymbol{\lambda}} &= \mathbf{0} \\ \mathbf{H}\hat{\beta} &= \mathbf{0}, \end{aligned} \tag{3.3}$$

where $\boldsymbol{\lambda}$ is the vector Lagrange multiplier. This leads to the following theorem.

Theorem 3.5 *If the constraints* $\mathbf{H}\boldsymbol{\beta} = \mathbf{0}$ *are suitable for identifiability constraints and* $t = p - r$ *then:*

(i) $\hat{\boldsymbol{\beta}} = (\mathbf{G}'\mathbf{G})^{-1}\mathbf{X}'\mathbf{y}$.
(ii) $\mathbf{P}_\Omega = \mathbf{X}(\mathbf{G}'\mathbf{G})^{-1}\mathbf{X}'$.
(iii) $\hat{\boldsymbol{\lambda}} = \mathbf{0}$.
(iv) $\mathbf{H}(\mathbf{G}'\mathbf{G})^{-1}\mathbf{X}' = \mathbf{0}$.

Proof

(i) Equation (3.2) holds irrespective of the rank of $\mathbf{X}$. Since the constraints $\mathbf{H}\boldsymbol{\beta} = \mathbf{0}$ are suitable for identifiability, there exists a unique $\hat{\boldsymbol{\beta}}$ satisfying $\hat{\boldsymbol{\theta}} = \mathbf{X}\hat{\boldsymbol{\beta}}$, that is satisfying the normal equations (3.1). Hence adding in $\mathbf{H}'\mathbf{H}\hat{\boldsymbol{\beta}} = \mathbf{0}$ leads to $\hat{\boldsymbol{\beta}} = (\mathbf{G}'\mathbf{G})^{-1}\mathbf{X}'\mathbf{y}$.
(ii) $\hat{\boldsymbol{\theta}} = \mathbf{X}\hat{\boldsymbol{\beta}} = \mathbf{X}(\mathbf{G}'\mathbf{G})^{-1}\mathbf{X}'\mathbf{y} = \mathbf{P}_\Omega\mathbf{y}$. As $\mathbf{P}_\Omega$ is unique (Theorem 3.2), we have $\mathbf{P}_\Omega = \mathbf{X}(\mathbf{G}'\mathbf{G})^{-1}\mathbf{X}'$.
(iii) Since (3.1) holds for $\hat{\boldsymbol{\beta}}$, $\mathbf{H}'\hat{\boldsymbol{\lambda}} = \mathbf{0}$ (by (3.3)). Hence $\hat{\boldsymbol{\lambda}} = \mathbf{0}$ as the columns of $\mathbf{H}'$ are linearly independent.
(iv) From $\mathbf{0} = \mathbf{H}\hat{\boldsymbol{\beta}} = \mathbf{H}(\mathbf{G}'\mathbf{G})^{-1}\mathbf{X}'\mathbf{y}$ for all $\mathbf{y}$, we have $\mathbf{H}(\mathbf{G}'\mathbf{G})^{-1}\mathbf{X}' = \mathbf{0}$, and this completes the theorem.

We note that as $\mathbf{H}\boldsymbol{\beta} = \mathbf{0}$,

$$\mathrm{E}[\hat{\boldsymbol{\beta}}] = (\mathbf{G}'\mathbf{G})^{-1}\mathbf{X}'\mathbf{X}\boldsymbol{\beta} = (\mathbf{G}'\mathbf{G})^{-1}\mathbf{G}'\mathbf{G}\boldsymbol{\beta} = \boldsymbol{\beta}$$

and $\hat{\boldsymbol{\beta}}$ is unbiased.

The essence of the above theory is that we wish to find *a* solution to the normal equations (3.1) and we do this by imposing the identifiability constraints $\mathbf{H}\boldsymbol{\beta} = \mathbf{0}$ without changing Ω (as indicated by $\hat{\boldsymbol{\lambda}} = \mathbf{0}$, a special case of (1.6) in Sect. 1.10). However, another method of finding a solution is to use a weak (generalized) inverse of $\mathbf{X}'\mathbf{X}$. A weak inverse of a matrix $\mathbf{L}$ is any matrix $\mathbf{L}^-$ satisfying $\mathbf{L}\mathbf{L}^-\mathbf{L} = \mathbf{L}$ (See Seber 2008, chapter 7). Now using the normal equations,

$$\mathbf{X}'\mathbf{y} = \mathbf{X}'\mathbf{X}\hat{\boldsymbol{\beta}} = \mathbf{X}'\mathbf{X}(\mathbf{X}'\mathbf{X})^-\mathbf{X}'\mathbf{X}\hat{\boldsymbol{\beta}} = \mathbf{X}'\mathbf{X}[(\mathbf{X}'\mathbf{X})^-\mathbf{X}'\mathbf{y}],$$

so that $\hat{\boldsymbol{\beta}} = (\mathbf{X}'\mathbf{X})^-\mathbf{X}'\mathbf{y}$ is a solution of the normal equations. Since

$$\mathbf{P}_\Omega\mathbf{y} = \hat{\boldsymbol{\theta}} = \mathbf{X}\hat{\boldsymbol{\beta}} = \mathbf{X}(\mathbf{X}'\mathbf{X})^-\mathbf{X}'\mathbf{y}$$

for all $\mathbf{y}$, we see that $\mathbf{P}_\Omega = \mathbf{X}(\mathbf{X}'\mathbf{X})^-\mathbf{X}'$, by the uniqueness of $\mathbf{P}_\Omega$ (Theorem 3.2). From Theorem 3.5(ii), $\mathbf{X}'\mathbf{X}(\mathbf{G}'\mathbf{G})^{-1}\mathbf{X}'\mathbf{X} = \mathbf{X}'\mathbf{P}_\Omega\mathbf{X} = \mathbf{X}'\mathbf{X}$, so that $(\mathbf{G}'\mathbf{G})^{-1}$ is a weak inverse of $\mathbf{X}'\mathbf{X}$. It can be shown that another weak inverse of $\mathbf{X}'\mathbf{X}$ is $\mathbf{B}_{11}$ defined by

$$\begin{pmatrix} \mathbf{X}'\mathbf{X} & \mathbf{H}' \\ \mathbf{H} & \mathbf{0} \end{pmatrix}^{-1} = \begin{pmatrix} \mathbf{B}_{11} & \mathbf{B}_{12} \\ \mathbf{B}_{21} & \mathbf{0} \end{pmatrix}.$$

Instead of introducing identifiability constraints, another approach to this problem of identifiability of the β is to find out what functions of β are estimable. A linear function $\mathbf{c}'\beta$ is said to be *estimable* if it has a linear unbiased estimate $\mathbf{a}'\mathbf{y}$. Thus

$$\mathbf{a}'\mathbf{X}\beta = E[\mathbf{a}'\mathbf{y}] = \mathbf{c}'\beta$$

is an identity in β, and $\mathbf{c}' = \mathbf{a}'\mathbf{X}$. Hence $\mathbf{c}'\beta$ is an estimable function of β if and only if $\mathbf{c}$ is a linear combination of the rows of $\mathbf{X}$. Since $\mathbf{c}'\beta = \mathbf{a}'\boldsymbol{\theta}$, the class of estimable functions is simply the class of all linear functions $\mathbf{a}'\boldsymbol{\theta}$ of the mean vector. We note that if $\mathbf{c}$ is linearly independent of the rows of $\mathbf{X}$, then $\mathbf{c}'\beta$ is not estimable. Thus it follows from A.11 that the identifiability constraints $\mathbf{H}\beta = \mathbf{0}$ are simply obtained from a set of non-estimable functions $\mathbf{h}_i'\beta$, where the $\mathbf{h}_i'$ form the rows of $\mathbf{H}$.

We note that if $\mathbf{c}'\beta$ is estimable, and $\hat{\beta}$ is any solution of the normal equations, then $\mathbf{c}'\hat{\beta}$ is unique. To show this we first note that $\mathbf{c} = \mathbf{X}'\mathbf{a}$ for some $\mathbf{a}$, so that $\mathbf{c}'\beta = \mathbf{a}'\mathbf{X}\beta = \mathbf{a}'\boldsymbol{\theta}$. Similarly, $\mathbf{c}'\hat{\beta} = \mathbf{a}'\mathbf{X}\hat{\beta} = \mathbf{a}'\hat{\boldsymbol{\theta}}$, which is unique.

3.5 Gauss-Markov Theorem

Having given a method of estimating $\boldsymbol{\theta}$, namely by a least squares procedure, we now ask if there are better ways of estimating $\boldsymbol{\theta}$. Our question is partly answered by the following theorem (due to Gauss) that proves that the least squares estimate is best in a certain sense.

Theorem 3.6 *If* $\mathrm{E}[\mathbf{y}] = \boldsymbol{\theta}$, $\mathrm{Var}[\mathbf{y}] = \sigma^2\mathbf{I}_n$, $\boldsymbol{\theta} \in \Omega$, *and* $c = \mathbf{a}'\boldsymbol{\theta}$, *then among the class of linear unbiased estimates of* c *there exists a unique estimate* $\hat{c} = \mathbf{a}'\mathbf{P}_\Omega\mathbf{y}$ *which has minimum variance. Thus if* $\mathbf{b}'\mathbf{y}$ *is any other linear unbiased estimate of* c, *then* $\mathrm{var}[\mathbf{b}'\mathbf{y}] > \mathrm{var}[\hat{c}]$.

Proof Since $\mathbf{P}_\Omega\boldsymbol{\theta} = \boldsymbol{\theta}$ and $\mathbf{a} = \mathbf{P}_\Omega\mathbf{a} + (\mathbf{I}_n - \mathbf{P}_\Omega)\mathbf{a}$,

$$\begin{aligned} c &= E[\mathbf{a}'\mathbf{y}] \\ &= E[(\mathbf{P}_\Omega\mathbf{a})'\mathbf{y}] + \mathbf{a}'(\mathbf{1}_n - \mathbf{P}_\Omega)\boldsymbol{\theta} \\ &= E[\hat{c}]. \end{aligned}$$

Thus $\hat{c}$ is a linear unbiased estimate of c.

If $\mathbf{b}'\mathbf{y}$ is any other linear unbiased estimate of c, then by a similar argument $(\mathbf{P}_\Omega\mathbf{b})'\mathbf{y}$ is also an unbiased estimate. Now

$$\begin{aligned} 0 &= E[(\mathbf{P}_\Omega\mathbf{a} - \mathbf{P}_\Omega\mathbf{b})'\mathbf{y}] \\ &= (\mathbf{P}_\Omega\mathbf{a} - \mathbf{P}_\Omega\mathbf{b})'\boldsymbol{\theta} \end{aligned}$$

for every $\boldsymbol{\theta} \in \Omega$, and hence $(\mathbf{P}_\Omega \mathbf{a} - \mathbf{P}_\Omega \mathbf{b}) \in \Omega^\perp$. But this vector belongs to Ω so that $\mathbf{P}_\Omega \mathbf{a} = \mathbf{P}_\Omega \mathbf{b}$ for every $\mathbf{b}$, and this projection of $\mathbf{b}$, namely $\mathbf{P}_\Omega \mathbf{b}$, leads to a unique $\hat{c}$. Also

$$\begin{aligned}
\text{var}[\mathbf{b}'\mathbf{y}] &= \sigma^2 \parallel \mathbf{b} \parallel^2 \\
&= \sigma^2(\parallel \mathbf{P}_\Omega \mathbf{b} \parallel^2 + \parallel (\mathbf{I}_n - \mathbf{P}_\Omega)\mathbf{b} \parallel^2) \\
&\geq \sigma^2 \parallel \mathbf{P}_\Omega \mathbf{b} \parallel^2 \\
&= \sigma^2 \parallel \mathbf{P}_\Omega \mathbf{a} \parallel^2 \\
&= \text{var}[\,\hat{c}\,],
\end{aligned}$$

with equality only if $\mathbf{b} = \mathbf{P}_\Omega \mathbf{b} = \mathbf{P}_\Omega \mathbf{a}$. Thus $\hat{c}$ is the unique unbiased estimate of c with minimum variance for the class of unbiased estimates.

If we are interested in the single elements θ_i, then we choose $\mathbf{a} = \mathbf{e}_i$, where $\mathbf{e}_i$ has 1 in the ith position and zeros elsewhere. We therefore have that the linear unbiased estimate of θ_i with minimum variance is $\mathbf{e}_i'\mathbf{P}_\Omega \mathbf{y} = \mathbf{e}_i'\hat{\boldsymbol{\theta}} = \hat{\theta}_i$, the least squares estimate of θ_i.

3.6 Estimation of σ^2

Let $\mathbf{y} = \boldsymbol{\theta} + \boldsymbol{\varepsilon}$ where $\text{E}[\boldsymbol{\varepsilon}] = \mathbf{0}$ and $\text{Var}[\boldsymbol{\varepsilon}] = \sigma^2 \mathbf{I}_n$. Since

$$E[(\mathbf{y} - \boldsymbol{\theta})'(\mathbf{y} - \boldsymbol{\theta})] = E[\sum_i \varepsilon_i^2] = n\sigma^2,$$

we would expect the residual sum of squares $RSS = \parallel \mathbf{y} - \hat{\boldsymbol{\theta}} \parallel^2$ to provide some estimate of σ^2. Let $\mathbf{R} = \mathbf{I}_n - \mathbf{P}_\Omega$. Since $\mathbf{R}$ is idempotent,

$$\begin{aligned}
RSS &= \parallel \mathbf{y} - \mathbf{P}_\Omega \mathbf{y} \parallel^2 \\
&= \mathbf{y}'\mathbf{R}^2\mathbf{y} \\
&= \mathbf{y}'\mathbf{R}\mathbf{y} \\
&= (\mathbf{y} - \boldsymbol{\theta})'\mathbf{R}(\mathbf{y} - \boldsymbol{\theta}) \\
&= \boldsymbol{\varepsilon}'\mathbf{R}\boldsymbol{\varepsilon},
\end{aligned}$$

as $\mathbf{R}\boldsymbol{\theta} = \mathbf{0}$. From Theorem 1.8(iii) in Sect. 1.9 we have

$$E[RSS] = \sigma^2 \,\text{trace}[\mathbf{R}] = \sigma^2(n - \text{trace}[\mathbf{P}_\Omega]) = \sigma^2(n - p). \tag{3.4}$$

Therefore an unbiased estimate of σ^2 is given by

$$s^2 = \mathbf{y}'\mathbf{R}\mathbf{y}/(n-p). \tag{3.5}$$

If $\mathbf{y} \sim N_n(\boldsymbol{\theta}, \sigma^2\mathbf{I}_n)$, then, from Theorem 1.9(iv) ,

$$\text{var}[\mathbf{y}'\mathbf{R}\mathbf{y}] = 2\sigma^4 \text{trace}[\mathbf{R}^2] = 2\sigma^4 \text{trace}[\mathbf{R}] = 2\sigma^4(n-p),$$

so that $\text{var}[s^2] = 2\sigma^4/(n-p)$.

We now ask what optimal properties RSS might have as an estimate of $\sigma^2(n-p)$. Since $\sigma^2 \geq 0$ it seems reasonable to restrict our class of estimates to those that are unbiased, non-negative, and quadratic (Rao 1952; Atiqullah 1962). Thus if $\mathbf{y}'\mathbf{A}\mathbf{y}$ is such an estimate, then $\mathbf{y}'\mathbf{A}\mathbf{y} \geq 0$ for all $\mathbf{y}$ so that $\mathbf{A}$ is non-negative definite. Also, from Theorems 1.8(i) in Sect. 1.9 and 1.9(ii), we have

$$E[\mathbf{y}'\mathbf{A}\mathbf{y}] = \sigma^2 \text{trace}[\mathbf{A}] + \boldsymbol{\theta}'\mathbf{A}\boldsymbol{\theta} = (n-p)\sigma^2 \tag{3.6}$$

and

$$\text{var}[\mathbf{y}'\mathbf{A}\mathbf{y}] = 2\sigma^4 \text{trace}[\mathbf{A}^2] + 4\sigma^2\boldsymbol{\theta}'\mathbf{A}^2\boldsymbol{\theta}. \tag{3.7}$$

Hence from (3.6), $\text{trace}[\mathbf{A}] = n-p$ and $\boldsymbol{\theta}'\mathbf{A}\boldsymbol{\theta} = 0$ for all $\boldsymbol{\theta} \in \Omega$. Since $\mathbf{A}$ is non-negative definite, $\mathbf{A} = \mathbf{V}\mathbf{V}'$, where $\mathbf{V}$ has linearly independent columns (A.9(iii)). Given $\boldsymbol{\theta} \in \Omega$, $\boldsymbol{\theta} = \mathbf{P}_\Omega\boldsymbol{\alpha}$ for some $\boldsymbol{\alpha}$. Then $\boldsymbol{\theta}'\mathbf{A}\boldsymbol{\theta} = 0$ implies that $0 = \boldsymbol{\alpha}'\mathbf{P}_\Omega\mathbf{V}\mathbf{V}'\mathbf{P}_\Omega\boldsymbol{\alpha} = \| \mathbf{V}'\mathbf{P}_\Omega\boldsymbol{\alpha} \|^2$ for all $\boldsymbol{\alpha}$ so that $\mathbf{V}'\mathbf{P}_\Omega = \mathbf{0}$ and $\mathbf{A}\mathbf{P}_\Omega = \mathbf{V}\mathbf{V}'\mathbf{P}_\Omega = \mathbf{0}$.

If $\mathbf{A} = \mathbf{R} + \mathbf{D}$, then $\mathbf{D}$ is symmetric and $\text{trace}[\mathbf{A}] = \text{trace}[\mathbf{R}] + \text{trace}[\mathbf{D}]$ so that $\text{trace}[\mathbf{D}] = 0$. Also $\mathbf{0} = \mathbf{A}\mathbf{P}_\Omega = \mathbf{R}\mathbf{P}_\Omega + \mathbf{D}\mathbf{P}_\Omega$ so that $\mathbf{D}\mathbf{P}_\Omega = \mathbf{0}$ and $\mathbf{D}\mathbf{R} = \mathbf{D}$. Now $\mathbf{D} = \mathbf{D}' = \mathbf{R}'\mathbf{D}' = \mathbf{R}\mathbf{D}$ and

$$\begin{aligned}
\mathbf{A}^2 &= (\mathbf{R}+\mathbf{D})(\mathbf{R}+\mathbf{D}) \\
&= \mathbf{R}^2 + \mathbf{R}\mathbf{D} + \mathbf{D}\mathbf{R} + \mathbf{D}^2 \\
&= \mathbf{R} + 2\mathbf{D} + \mathbf{D}^2.
\end{aligned}$$

Since $\text{trace}[\mathbf{D}] = 0$, we take the trace of the above equation and get

$$\text{trace}[\mathbf{A}^2] = (n-p) + \text{trace}[\mathbf{D}^2] = (n-p) + \sum_i \sum_j d_{ij}^2.$$

This has a minimum when $\mathbf{D} = \mathbf{0}$ and $\mathbf{A} = \mathbf{R}$ so that s^2 has minimum variance.

Atiqullah (1962) introduced the concept of a quadratically balanced design matrix $\mathbf{X}$ of full rank as one for which $\mathbf{P}_\Omega = \mathbf{X}(\mathbf{X}'\mathbf{X})^{-1}\mathbf{X}'$ has equal diagonal elements. He showed that s^2 has minimum variance of all non-negative unbiased quadratic estimates irrespective of the distribution of $\mathbf{y}$ if $\mathbf{X}$ is quadratically balanced. Quadratic balance is mentioned again in Sect. 5.2.

Under the assumption of normality, $RSS/\sigma^2 = \boldsymbol{\varepsilon}'\mathbf{R}\boldsymbol{\varepsilon}/\sigma^2$ is distributed as χ^2_{n-p} by Theorem 1.10 in Sect. 1.9, as $\mathbf{R}$ is idempotent of rank (and trace) $n-p$. When we don't have normality and the ε_i have kurtosis $\gamma_2 = (\mu_4 - 3\sigma^4)/\sigma^4$, then we have from Theorem 1.9(ii) and $\text{trace}[\mathbf{R}^2] = n - p$

$$\text{var}[s^2] = \frac{2\sigma^4}{n-p}\left\{1 + \frac{\gamma_2}{2}\frac{\mathbf{d}'\mathbf{d}}{n-p}\right\}, \tag{3.8}$$

where $\mathbf{d}'\mathbf{d} = \sum_{i=1}^n r_{ii}^2$ and $\mathbf{R} = (r_{ij})$. As $\mathbf{R}$ is idempotent

$$r_{ii} = r_{ii}^2 + \sum_i \sum_{j, j \neq i} r_{ij}^2$$

so that $r_{ii} > r_{ii}^2$ and $0 \leq r_{ii} < 1$. If $\mathbf{P}_\Omega = (p_{ij})$ then

$$\begin{aligned}\sum_i r_{ii}^2 &= \sum_i (1 - p_{ii})^2 \\ &= n - 2\,\text{trace}[\mathbf{P}_\Omega] + \sum_i p_{ii}^2 \\ &= n - 2p + \sum_i p_{ii}^2 \\ &> n - 2p.\end{aligned}$$

Hence

$$\text{var}[s^2] > \frac{2\sigma^4}{n-p}\left\{1 + \frac{\gamma_2}{2}\frac{n-2p}{n-p}\right\}.$$

Clearly inferences for σ^2 based on s^2 will be strongly affected by long-tailed distributions ($\gamma_2 > 0$). If we have quadratic balance, then the p_{ii} will all be equal to p/n so that the r_{ii} are $(n-p)/n$. Hence $\mathbf{d}'\mathbf{d} = n(n-p)^2/n^2$ and from (3.8)

$$\text{var}[s^2] = \frac{2\sigma^4}{n-p}\left\{1 + \frac{\gamma_2}{2}\frac{n-p}{n}\right\}.$$

If $(n-p)/n$ is small, inferences about σ^2 can be robust to non-normality. This won't be the case for a single sample as then $p = 1$.

3.7 Assumptions and Residuals

Before we have confidence in a particular linear model, we need to check on its underlying assumptions, which are: (i) the ε_i have mean zero, (ii), the ε_i all have the same variance, (iii) the ε_i are mutually independent, and (iv) the ε_i are normally distributed. An estimate of $\varepsilon_i = y_i - \theta_i$ is given by $r_i = y_i - \hat{\theta}_i = y_i - \hat{y}_i$, say, where the $\hat{y}_i$ are called the *fitted values*. The properties of the ε_i will be reflected in the properties of the r_i, so that the latter can be used to investigate the assumptions. The elements r_i of the vector

$$\begin{aligned}\mathbf{r} &= \mathbf{y} - \hat{\boldsymbol{\theta}} \\ &= (\mathbf{I}_n - \mathbf{P}_\Omega)\mathbf{y} \\ &= (\mathbf{I}_n - \mathbf{P}_\Omega)\boldsymbol{\varepsilon},\end{aligned}$$

(since $(\mathbf{I}_n - \mathbf{P}_\Omega)\mathbf{X} = \mathbf{0}$), are called the *residuals*. The sum of squares $(\mathbf{y} - \hat{\boldsymbol{\theta}})'(\mathbf{y} - \hat{\boldsymbol{\theta}})$ is called the *residual sum of squares* or *RSS*. There is an extensive literature on how the r_i and scaled versions of them can be used to investigate the underlying assumptions (e.g., Seber and Lee 2003, chapter 10) which we shall not consider here apart from a few properties. We note that $\hat{\mathbf{y}} = \mathbf{P}_\Omega \mathbf{y}$, where the projection matrix $\mathbf{P}_\Omega$ is usually referred to as the *hat matrix*. We summarize the following properties without assuming normality of $\boldsymbol{\varepsilon}$:

$$\begin{aligned}\mathrm{E}[\mathbf{r}] &= (\mathbf{I}_n - \mathbf{P}_\Omega)\mathrm{E}[\boldsymbol{\varepsilon}] = \mathbf{0}, \\ \mathrm{Var}[\mathbf{r}] &= (\mathbf{I}_n - \mathbf{P}_\Omega)\mathrm{Var}[\mathbf{y}](\mathbf{I}_n - \mathbf{P}_\Omega)' \\ &= \sigma^2(\mathbf{I}_n - \mathbf{P}_\Omega)(\mathbf{I}_n - \mathbf{P}_\Omega)' \\ &= \sigma^2(\mathbf{I}_n - \mathbf{P}_\Omega), \\ \mathrm{E}[\hat{\mathbf{y}}] &= \mathbf{P}_\Omega\boldsymbol{\theta} = \boldsymbol{\theta}, \\ \mathrm{Var}[\hat{\mathbf{y}}] &= \mathbf{P}_\Omega \mathrm{Var}[\mathbf{y}]\mathbf{P}'_\Omega \\ &= \sigma^2\mathbf{P}_\Omega,\end{aligned}$$

and from Theorem 1.5(ii),

$$\mathrm{Cov}[\mathbf{r}, \hat{\mathbf{y}}] = \mathrm{Cov}[(\mathbf{I}_n - \mathbf{P}_\Omega)\mathbf{y}, \mathbf{P}_\Omega\mathbf{y}] = \sigma^2(\mathbf{I}_n - \mathbf{P}_\Omega)\mathbf{P}_\Omega = \mathbf{0}.$$

If we now assume a normal distribution for $\mathbf{y}$, the last result implies that $\mathbf{r}$ and $\hat{\mathbf{y}}$ are statistically independent (by Theorem 1.6(v) in Sect. 1.7). Also from the above equations, $\mathbf{r} \sim N_n[\mathbf{0}, \sigma^2(\mathbf{I}_n - \mathbf{P}_\Omega)]$, a singular distribution as $\mathbf{I}_n - \mathbf{P}_\Omega (= \mathbf{P}_{\Omega\perp})$ has rank $n - \dim[\Omega] < n$, and is therefore a singular matrix. If $\mathbf{P}_\Omega = (p_{ij})$, the diagonal elements p_{ii} are called the *hat matrix diagonals*. By suitably scaling the r_i, these scaled residuals can be used for checking on the normality of their distribution and constant variance, for looking for outliers, and for checking on the linearity of the model. For example, if $s^2 = \mathbf{r}'\mathbf{r}/(n-p)$, we can use the so-called *internally Studentized residual*

$$r_i* = \frac{r_i}{s(1-p_{ii})^{1/2}},$$

which can be shown to have the property that the $r_i^{*2}/(n-p)$ are identically distributed with the Beta $[\frac{1}{2}, \frac{1}{2}(n-p-1)]$ distribution (Cook and Weisberg (1982, 18)). Regression computing packages automatically produce various residuals and their plots.

Systematic bias can sometimes be a problem in linear models so that assumption (i) at the beginning of the section may not hold. In the case of a regression model, there may be systematic bias because of an incorrect specification of the model. The effect of this is discussed in detail in Seber and Lee (2003, section 9.2). In the case of analysis of variance models, careful experimentation using randomization in the experimental design will usually minimize any bias.

We can also consider the effect of serial correlation on the ε_i by assuming that the ε_is have a first-order autoregressive model AR(1), namely

$$\varepsilon_i = \phi\varepsilon_{i-1} + a_i \quad |\phi| < 1,$$

where the a_i $(i = 0, \pm 1, \pm 2, \ldots$ are independent with $\text{E}[a_i] = 0$ and $\text{var}[a_i] = \sigma_a^2$ $(i = 1, 2, \ldots)$. From Seber and Wild (1989, 275–276) we have that the correlation between ε_i and $\varepsilon_{i\pm k}$ is $\rho_k = \rho_1^k$. Hence the correlation matrix $\mathbf{V} = (v_{ij})$ is given by $v_{ij} = \rho_1^{|i-j|}$ and $\text{Var}[\varepsilon] = \sigma^2\mathbf{V}$. In terms of the AR(1) model, $\rho_1 = \phi$ and $\sigma_a^2 = (1-\phi^2)\sigma^2$. We consider the simple regression model of sampling from a single population, namely $y_i = \mu + \varepsilon_i$, where $\hat{\mu} = \bar{y}$ and $(n-1)s^2 = \sum_i (y_i - \bar{y})^2 = \mathbf{y}'\mathbf{A}\mathbf{y}$. Now

$$\text{var}[\bar{y}] = \text{var}[\mathbf{1}_n'\mathbf{y}/n] = \sigma^2\mathbf{1}_n'\mathbf{V}\mathbf{1}_n/n^2 = \sigma^2\sum_i\sum_j v_{ij}/n^2,$$

where, after some algebra,

$$\sum_i\sum_j v_{ij} = n + \frac{2n\rho - 2\rho^n}{1-\rho} - \frac{2\rho(1-\rho^{n-1})}{(1-\rho)^2} \approx n(1 + \frac{2\rho}{1-\rho}).$$

This is larger than it should bc ($> n$) for large n as without autocorrelation we have $\text{var}[\bar{y}] = \sigma^2/n$. Also, from Theorem 1.8(i) in Sect. 1.9,

$$\begin{aligned} E[(n-1)s^2] &= \sigma^2\,\text{trace}[\mathbf{AV}] + \mu\mathbf{1}_n'\mathbf{A}\mathbf{1}_n\mu \\ &= \sigma^2\,\text{trace}[\mathbf{AV}] \end{aligned}$$

$$
\begin{aligned}
&= \sigma^2 \sum_i \sum_j a_{ij} v_{ij} \\
&= \sigma^2 \sum_i \sum_j (\delta_{ij} - n^{-1}) v_{ij} \\
&= \sigma^2 (n - n^{-1} \sum_i \sum_j v_{ij}) \\
&\approx \sigma^2 [n - (1 + \frac{2\rho}{1-\rho})],
\end{aligned}
$$

so that

$$
E[s^2] \approx \sigma^2 \left[1 - \frac{2\rho}{(n-1)(1-\rho)} \right]
$$

and s^2 is an approximately unbiased estimate of σ^2 for large n and small ρ. Now the usual t-statistic for testing $\mu = 0$ assuming $\rho = 0$ is based on

$$
t = \frac{\sqrt{n}\overline{y}}{s} = \frac{\overline{y}}{\sqrt{\widehat{\text{var}}[\overline{y}]}}, \tag{3.9}
$$

where $\widehat{\text{var}}[\overline{y}]$ is $\text{var}[\overline{y}]$ with σ^2 replaced by s^2. When $\rho \neq 0$, the denominator is underestimated and t is larger than it should be so that it can give a significant result when it is not actually significant.

We now consider the effect of autocorrelation on a more general regression model for which $\text{Var}[\varepsilon] = \sigma^2 \mathbf{V}$, where $\mathbf{V}$ is the same as before, $\text{E}[\mathbf{y}] = \mathbf{X}\boldsymbol{\beta}$, and $\mathbf{X}$ is $n \times p$ of rank p. The ordinary least squares estimate of $\boldsymbol{\beta}$ is $\hat{\boldsymbol{\beta}} = (\mathbf{X}'\mathbf{X})^{-1}\mathbf{X}'\mathbf{y}$, which is still unbiased, with variance matrix $\sigma^2(\mathbf{X}'\mathbf{X})^{-1}\mathbf{X}'\mathbf{V}\mathbf{X}(\mathbf{X}'\mathbf{X})^{-1}$ that in general will not be equal to $\sigma^2(\mathbf{X}'\mathbf{X})^{-1}$. Suppose we wish to test $H : \mathbf{a}'\boldsymbol{\beta} = 0$, then when $\mathbf{V} = \mathbf{I}_n$ we would use the t-statistic

$$
t = \frac{\mathbf{a}'\hat{\boldsymbol{\beta}}}{\sqrt{\hat{v}}},
$$

where $\hat{v} = s^2 \mathbf{a}'(\mathbf{X}'\mathbf{X})^{-1}\mathbf{a}$ will normally be a biased estimate of

$$
\text{var}[\mathbf{a}'\hat{\boldsymbol{\beta}}] = \sigma^2 \mathbf{a}'(\mathbf{X}'\mathbf{X})^{-1}\mathbf{X}'\mathbf{V}\mathbf{X}(\mathbf{X}'\mathbf{X})^{-1}\mathbf{a}.
$$

Also from Theorem 1.8(i)) in Sect. 1.9

$$
\begin{aligned}
E[s^2] &= \frac{1}{n-p} E[\varepsilon'(\mathbf{I}_n - \mathbf{P}_\Omega)\varepsilon] \\
&= \frac{\sigma^2}{n-p} \text{trace}[\mathbf{V}(\mathbf{I}_n - \mathbf{P}_\Omega)],
\end{aligned}
$$

where $\mathbf{P}_\Omega = \mathbf{X}(\mathbf{X}'\mathbf{X})^{-1}\mathbf{X}'$. Then, if

$$E[\hat{v}] = \text{var}[\mathbf{a}'\hat{\beta}] + b,$$

Swindel (1968, 315) showed that

$$\frac{c_1 - d_1}{d_1} \leq \frac{b}{\text{var}[\mathbf{a}'\hat{\beta}]} \leq \frac{c_2 - d_2}{d_2}, \tag{3.10}$$

where c_1 is the mean of the $(n - p)$ smallest eigenvalues of $\mathbf{V}$, c_2 is the mean of the $(n - p)$ largest eigenvalues, d_1 is the largest eigenvalue value of $\mathbf{V}$, and d_2 the smallest eigenvalue; the bounds are attainable. Although we can't find the eigenvalues explicitly, we can use some approximations. We begin by considering the inverse of $\mathbf{V}$, namely (Seber 2008, 8.114b)

$$\mathbf{V}^{-1} = \frac{1}{(1-\rho^2)} \begin{pmatrix} 1 & -\rho & 0 & \cdots & 0 & 0 \\ -\rho & 1+\rho^2 & -\rho & \cdots & 0 & 0 \\ 0 & -\rho & 1+\rho^2 & \cdots & 0 & 0 \\ \cdot & \cdot & \cdot & \cdots & \cdot & \cdot \\ 0 & 0 & 0 & \cdots & -\rho & 1 \end{pmatrix}.$$

Assuming that ρ^2 is small we can approximate $\mathbf{V}^{-1}$ by $\mathbf{W}/(1-\rho^2)$, where we have $w_{11} = w_{nn} = 1 + \rho^2$ and the other elements are unchanged, thus giving us a symmetric tridiagonal matrix whose eigenvalues are known, namely (Seber 2008, 8.110(b))

$$\lambda_j = f + 2\sqrt{g^2}\cos(j\pi/(n+1)) \quad (j = 1, 2, \ldots, n),$$

where $f = 1+\rho^2$ and $g = -\rho$. As $\lambda_{\min} > f - 2|g| = (1-|\rho|)^2 > 0$, the eigenvalues of $\mathbf{W}^{-1}$ are λ_j^{-1} which, apart from the scale factor $(1-\rho^2)$ (which cancels out of the above ratios in (3.10)), are approximations for the eigenvalues of $\mathbf{V}$. The cosine terms can be readily computed and, given an estimate of ρ, we can obtain some idea of the bounds on b as a multiple of $\text{var}[\mathbf{a}'\hat{\beta}]$.

There are a number of methods for testing for serial correlation that can arise if the y_i observations are collected serially in time. A plot of r_i versus time order, which is often a plot of r_i versus i, may show up the presence of any correlation between time consecutive ε_i. For positively correlated errors a residual tends to have the same sign as its predecessor giving a slow up and down effect, while for negatively correlated errors (which is much less common), the signs of the residuals tend to alternate giving a saw-toothed plot. Another plot is to divide time-ordered residuals into consecutive pairs and plot one member of the pair against the other. Serially correlated data shows up as a linear trend in the plot. A useful graphical method is the correlogram or autocorrelation plot of the sample autocorrelations of the residuals r_h from the regression versus the time lags h. We would want the sample autocorrelations to be close to zero. In addition, the Durbin-Watson statistic (cf. Seber and Lee 2003, 292–294) provides a test for significant residual autocorrelation at lag 1. I won't proceed any further into the time series literature.

We finally consider briefly the effect of unequal variances by looking once again at sampling from a single population, i.e. $y_i = \mu + \varepsilon_i$. We assume that

$$\mathrm{Var}[\boldsymbol{\varepsilon}] = \boldsymbol{\Sigma} = \mathrm{diag}(\sigma_1^2, \sigma_2^2, \ldots, \sigma_n^2) \quad \text{where} \quad \sigma_1^2 \leq \sigma_2^2 \leq \cdots \leq \sigma_n^2.$$

Then since the y_i are independent,

$$\mathrm{var}[\overline{y}] = \sum_i \sigma_i^2/n^2 = \overline{\sigma^2}/n,$$

where $\overline{\sigma^2}$ is the mean of the σ_i^2. Also

$$E[(n-1)s^2] = \mathrm{trace}[\mathbf{A}\boldsymbol{\Sigma}] = \sum_i a_{ii}\sigma_i^2 = \frac{n-1}{n}\sum_i \sigma_i^2$$

so that $E[s^2] = \overline{\sigma^2}$. We find then that for large n, t of (3.9) will be approximately $N[0, 1]$ and generally insensitive to unequal variances for large n. In the case of multiple regression, the eigenvalues of $\boldsymbol{\Sigma}$ are simply the diagonal elements of $\boldsymbol{\Sigma}$. Hence, from (3.10), with the σ_i^2 being the eigenvalues,

$$\frac{\frac{1}{n-p}\sum_{i=1}^{n-p}\sigma_i^2}{\sigma_n^2} - 1 \leq \frac{b}{\mathrm{var}[\mathbf{a}'\hat{\boldsymbol{\beta}}]} \leq \frac{\frac{1}{n-p}\sum_{i=p+1}^{n}\sigma_i^2}{\sigma_1^2} - 1.$$

3.8 Multiple Correlation Coefficient

A helpful assessment tool in regression analysis is the multiple correlation coefficient defined to be

$$R = \frac{\sum_i (y_i - \overline{y})(\hat{y}_i - \overline{\hat{y}}_i)}{\left\{\sum_i (y_i - \overline{y})^2 \sum_i (\hat{y}_i - \overline{\hat{y}}_i)^2\right\}^{1/2}},$$

the simple correlation between the two vectors $\mathbf{y}$ and the fitted values $\hat{\mathbf{y}}$. This is applied to regression models with a constant term β_0 so that the regression matrix $\mathbf{X}$ has its first column all ones and, from $\mathbf{P}_\Omega\mathbf{X} = \mathbf{X}$, we have $\mathbf{P}_\Omega\mathbf{1}_n = \mathbf{1}_n$. We now show how an alternative expression for R^2 can be derived using projection matrices.

We begin with

$$\sum_i (y_i - \hat{y}_i) = \mathbf{1}_n'(\mathbf{I}_n - \mathbf{P}_\Omega)\mathbf{y} = 0$$

so that $\overline{y} = \overline{\hat{y}}_i$. Also, from $(\mathbf{I}_n - \mathbf{P}_\Omega)\mathbf{P}_\Omega = \mathbf{0}$ and the above equation,

$$(\mathbf{I}_n - \mathbf{P}_\Omega)(\mathbf{P}_\Omega - n^{-1}\mathbf{1}_n\mathbf{1}_n') = \mathbf{0}.$$

Hence

$$\begin{aligned}\sum_i (y_i - \overline{y})^2 &= \sum_i y_i^2 - n\overline{y}^2 \\ &= \mathbf{y}'(\mathbf{I}_n - n^{-1}\mathbf{1}_n\mathbf{1}_n')\mathbf{y} \\ &= \mathbf{y}'(\mathbf{I}_n - \mathbf{P}_\Omega + \mathbf{P}_\Omega - n^{-1}\mathbf{1}_n\mathbf{1}_n')\mathbf{y} \\ &= \mathbf{y}'(\mathbf{I}_n - \mathbf{P}_\Omega)\mathbf{y} + \mathbf{y}'(\mathbf{P}_\Omega - n^{-1}\mathbf{1}_n\mathbf{1}_n')\mathbf{y} \\ &= \| (\mathbf{I}_n - \mathbf{P}_\Omega)\mathbf{y} \|^2 + \| (\mathbf{P}_\Omega - n^{-1}\mathbf{1}_n\mathbf{1}_n')\mathbf{y} \|^2 \\ &= \sum_i (y_i - \hat{y}_i)^2 + \sum_i (\hat{y}_i - \overline{y})^2, \end{aligned} \tag{3.11}$$

since $\mathbf{I}_n - \mathbf{P}_\Omega$ and $\mathbf{P}_\Omega - n^{-1}\mathbf{1}_n\mathbf{1}_n'$ are both idempotent. Now

$$\begin{aligned}\sum_i (y_i - \overline{y})(\hat{y}_i - \overline{\hat{y}}_i) &= \sum_i (y_i - \overline{y})(\hat{y}_i - \overline{y}) \\ &= \mathbf{y}'(\mathbf{I}_n - n^{-1}\mathbf{1}_n\mathbf{1}_n')(\mathbf{P}_\Omega - n^{-1}\mathbf{1}_n\mathbf{1}_n')\mathbf{y} \\ &= \mathbf{y}'(\mathbf{P}_\Omega - n^{-1}\mathbf{1}_n\mathbf{1}_n')\mathbf{y} \\ &= \sum_i (\hat{y}_i - \overline{y})^2. \end{aligned}$$

so that using (3.11),

$$\begin{aligned} R^2 &= \frac{\sum_i (\hat{y}_i - \overline{y})^2}{\sum_i (y_i - \overline{y})^2} \\ &= 1 - \frac{\sum_i (y_i - \hat{y}_i)^2}{\sum_i (y_i - \overline{y}_i)^2} \\ &= 1 - \frac{RSS}{\sum_i (y_i - \overline{y})^2}. \end{aligned}$$

3.9 Maximum Likelihood Estimation

If $\mathbf{y} \sim N_n[\boldsymbol{\theta}, v\mathbf{I}_n]$, where $v = \sigma^2$, then the likelihood function of $\mathbf{y}$ is

$$\ell(\boldsymbol{\theta}, v) = (2\pi v)^{-n/2} \exp\left\{-\frac{1}{2v} \parallel \mathbf{y} - \boldsymbol{\theta} \parallel^2\right\}, \tag{3.12}$$

and the log likelihood function (ignoring constants) is

$$L(\boldsymbol{\theta}, v) = -\frac{n}{2} \log v - \frac{1}{2v} \parallel \mathbf{y} - \boldsymbol{\theta} \parallel^2 .$$

To find the maximum likelihood estimates of v and $\boldsymbol{\theta}$ subject to $\boldsymbol{\theta} \in \Omega$ we wish to maximize $L(\boldsymbol{\theta}, v)$ subject to the constraints on $\boldsymbol{\theta}$ and v. Clearly, for any $v > 0$ this is maximized by minimizing $\parallel \mathbf{y} - \boldsymbol{\theta} \parallel^2$ using the least squares estimate $\hat{\boldsymbol{\theta}}$. Hence $L(\hat{\boldsymbol{\theta}}, v) \leq L(\boldsymbol{\theta}, v)$ for all $v > 0$. We now wish to maximize $L(\hat{\boldsymbol{\theta}}, v)$ with respect to v. Setting $\partial L/\partial v = 0$, we get a stationary value of $\hat{v} = \parallel \mathbf{y} - \hat{\boldsymbol{\theta}} \parallel^2 /n$. Then

$$L(\hat{\boldsymbol{\theta}}, \hat{v}) - L(\hat{\boldsymbol{\theta}}, v) = -\frac{n}{2}\left[\log\left(\frac{\hat{v}}{v}\right) + 1 - \frac{\hat{v}}{v}\right] \geq 0,$$

since $x \leq e^{x-1}$ and therefore $\log x \leq x - 1$ for all $x > 0$ (with equality when $x = 1$). Hence

$$L(\boldsymbol{\theta}, v) \leq L(\hat{\boldsymbol{\theta}}, v) \leq L(\hat{\boldsymbol{\theta}}, \hat{v}) \quad \text{for all } v > 0$$

with equality if and only if $\boldsymbol{\theta} = \hat{\boldsymbol{\theta}}$ and $v = \hat{v}$. Thus $\hat{\boldsymbol{\theta}}$ and $\hat{v} = \hat{\sigma}^2$ are the maximum likelihood estimates of $\boldsymbol{\theta}$ and v. Also, for future use,

$$\ell(\hat{\boldsymbol{\theta}}, \hat{\sigma}^2) = (2\pi\hat{\sigma}^2)^{-(n/2)} e^{-n/2}. \tag{3.13}$$

Another method is to use a Lagrange multiplier $\boldsymbol{\lambda}$ (cf. Sect. 1.10) and minimize $L(\boldsymbol{\theta}, v)$ subject to $(\mathbf{I}_n - \mathbf{P}_\Omega)\boldsymbol{\theta} = \mathbf{0}$. Differentiating $L(\boldsymbol{\theta}, v)$ with respect to v and $\boldsymbol{\theta}$ gives us the equations (cf. A.20)

$$-\frac{n}{2v} + \frac{1}{2v^2}(\mathbf{y} - \boldsymbol{\theta})'(\mathbf{y} - \boldsymbol{\theta}) = 0$$

$$-\mathbf{y} + \boldsymbol{\theta} + (\mathbf{I}_n - \mathbf{P}_\Omega)'\boldsymbol{\lambda} v = \mathbf{0}, \quad \text{and}$$

$$(\mathbf{I}_n - \mathbf{P}_\Omega)\boldsymbol{\theta} = \mathbf{0}.$$

Multiplying the second equation by $\mathbf{P}_\Omega$ we get $\boldsymbol{\theta} = \mathbf{P}_\Omega\boldsymbol{\theta} = \mathbf{P}_\Omega\mathbf{y} = \hat{\boldsymbol{\theta}}$ and

$$v = \frac{1}{n}(\mathbf{y} - \hat{\boldsymbol{\theta}})'(\mathbf{y} - \hat{\boldsymbol{\theta}}) = \hat{v},$$

giving us the same estimates, as expected.

References

Atiqullah, M. (1962). The estimation of residual variance in quadratically balanced least squares problems and the robustness of the *F* test. *Biometrika, 49*, 83–91.
Cook, R. D., & Weisberg, S. (1982). *Residuals and influence in regression.* New York: Chapman & Hall.
Rao, C. R. (1952). Some theorems on minimum variance estimation. *Sankhyā, 12*, 27–42.
Seber, G. A. F. (2008). *A matrix handbook for statisticians*. New York: Wiley.
Seber, G. A. F., & Lee, A. J. (2003). *Linear regression analysis* (2nd ed.). New York: Wiley.
Seber, G. A. F., & Wild, C. J. (1989). *Nonlinear regression.* New York: Wiley. Also reproduced in paperback by Wiley in (2004).
Swindel, B. F. (1968). On the bias of some least-squares estimators of variance in a general linear model. *Biometrika, 55*, 313–316.

Chapter 4
Hypothesis Testing

4.1 The Likelihood Ratio Test

Given the model $\mathbf{y} \sim N_n(\boldsymbol{\theta}, \sigma^2\mathbf{I}_n)$ and assumption G that $\boldsymbol{\theta} \in \Omega$, a p-dimensional subspace of $\mathbb{R}^n$, we wish to test the linear hypothesis $H : \boldsymbol{\theta} \in \omega$, where ω is a $p - q$ dimensional subspace of Ω. If $v = \sigma^2$, and $\ell(\boldsymbol{\theta}, v)$ is the normal probability density function for $\mathbf{y}$ (given by (3.12)), the usual test statistic for H is based on the likelihood ratio test $\Lambda[H|G]$, where

$$\Lambda[H|G] = \frac{\sup_{\boldsymbol{\theta}\in\omega,v>0} \ell(\boldsymbol{\theta}, v)}{\sup_{\boldsymbol{\theta}\in\Omega,v>0} \ell(\boldsymbol{\theta}, v)}$$
$$= \frac{\max_{\boldsymbol{\theta}\in\omega,v>0} \ell(\boldsymbol{\theta}, v)}{\max_{\boldsymbol{\theta}\in\Omega,v>0} \ell(\boldsymbol{\theta}, v)}.$$

We accept H if $\Lambda[H|G]$ is "near" enough to unity. Any monotone function of $\Lambda[H|G]$ would also be suitable as a test statistic, and for reasons we shall see later we choose

$$F = (\{\Lambda[H|G]\}^{-2/n} - 1)(n - p)/q.$$

We would now accept H if F is "small" enough.

Let $\hat{\sigma}_H^2$ and $\hat{\boldsymbol{\theta}}_H$ be the maximum likelihood estimates for $\boldsymbol{\theta} \in \omega$. Then $\hat{\boldsymbol{\theta}}_H = \mathbf{P}_\omega\mathbf{y}$ and $n\hat{\sigma}_H^2 = \| \mathbf{y} - \hat{\boldsymbol{\theta}}_H \|^2 = RSS_H$, say, where $\mathbf{P}_\omega$ is the symmetric idempotent matrix representing the orthogonal projection of $\mathbb{R}^n$ onto ω. Then from (3.13),

$$\Lambda(H|G] = (\hat{\sigma}^2/\hat{\sigma}_H^2)^{n/2}$$

G.A.F. Seber, *The Linear Model and Hypothesis*, Springer Series in Statistics,
DOI 10.1007/978-3-319-21930-1_4

and

$$\begin{aligned}
F &= \frac{(n-p)}{q}\frac{(\hat{\sigma}_H^2-\hat{\sigma}^2)}{\hat{\sigma}^2} \\
&= \frac{(n-p)}{q}\frac{\mathbf{y}'(\mathbf{P}_\Omega-\mathbf{P}_\omega)\mathbf{y}}{\mathbf{y}'(\mathbf{I}_n-\mathbf{P}_\Omega)\mathbf{y}} \qquad (4.1)\\
&= \frac{(n-p)}{q}\frac{(RSS_H-RSS)}{RSS} \qquad (4.2)\\
&= \frac{(n-p)}{q}\frac{(Q_H-Q)}{Q}, \text{ say} \qquad (4.3)\\
&= (Q_H-Q)/(qs^2). \qquad (4.4)
\end{aligned}$$

To find the distribution of F we shall need the following theorem.

Theorem 4.1 *Let* $\mathbf{y} \sim N_n(\boldsymbol{\theta}, \sigma^2\mathbf{I}_n)$ *and let* $\mathbf{A}_i$, $i = 1, 2, \ldots, m$ *be a sequence of* $n\times n$ *symmetric matrices with ranks* r_i *such that* $\sum_{i=1}^m \mathbf{A}_i = \mathbf{I}_n$. *If one (and therefore all, by Theorem A.12) of the following conditions hold, namely*

(i) $\sum_{i=1}^m r_i = n$, *where* $r_i = \text{rank}[\mathbf{A}_i]$,
(ii) $\mathbf{A}_i\mathbf{A}_j = \mathbf{0}$ *for all* $i, j, i \neq j$,
(iii) $\mathbf{A}_i^2 = \mathbf{A}_i$ *for* $i = 1, 2, \ldots, m$,

then the quadratics $\mathbf{y}'\mathbf{A}_i\mathbf{y}$ *are independently distributed as non-central chi-square with* r_i *degrees of freedom and non-centrality parameters* $\boldsymbol{\theta}'\mathbf{A}_i\boldsymbol{\theta}/\sigma^2$.

Proof Since $\mathbf{A}_i$ is symmetric and $\mathbf{A}_i\mathbf{A}_j = \mathbf{0}$ we have (Theorem 1.1 in Sect. 1.2)

$$\mathcal{C}[\mathbf{A}_j] \subset \mathcal{N}[\mathbf{A}_i] = \{\mathcal{C}[\mathbf{A}_i]\}^\perp.$$

Hence the $\mathcal{C}[\mathbf{A}_i]$ are mutually orthogonal vector spaces and, as $\mathbf{I}_n\mathbf{y} = \sum_i \mathbf{A}_i\mathbf{x}$ for every $\mathbf{y}$, their direct sum is $\mathbb{R}^n$. We can therefore construct an orthonormal basis $\mathbf{t}_1, \mathbf{t}_2, \ldots, \mathbf{t}_n$ of $\mathbb{R}^n$ such that $\mathbf{t}_1, , \mathbf{t}_2, \ldots, \mathbf{t}_{r_1}$ form a basis of $\mathcal{C}[\mathbf{A}_1]$; $\mathbf{t}_{r_1+1}, \ldots, \mathbf{t}_{r_1+r_2}$ a basis for $\mathcal{C}[\mathbf{A}_2]$, and so forth. Let $\mathbf{T} = (\mathbf{t}_1, \mathbf{t}_2, \ldots, \mathbf{t}_n)$, then $\mathbf{T}'\mathbf{T} = \mathbf{I}_n$. Now as $\mathbf{A}_1$ is symmetric and idempotent, it represents an orthogonal projection of $\mathbb{R}^n$ onto $\mathcal{C}[\mathbf{A}_1]$ (Theorem 3.3 in Sect. 3.2). Hence

$$\mathbf{A}_1\mathbf{T} = (\mathbf{t}_1, \mathbf{t}_2, \ldots, \mathbf{t}_{r_1}, \mathbf{0}, \ldots \mathbf{0})$$

and

$$\mathbf{T}'\mathbf{A}_1\mathbf{T} = \begin{pmatrix} \mathbf{I}_{r_1} & \mathbf{0} \\ \mathbf{0} & \mathbf{0} \end{pmatrix}.$$

Also

$$\mathbf{T}'\mathbf{A}_2\mathbf{T} = \begin{pmatrix} \mathbf{0} & \mathbf{0} & \mathbf{0} \\ \mathbf{0} & \mathbf{I}_{r_2} & \mathbf{0} \\ \mathbf{0} & \mathbf{0} & \mathbf{0} \end{pmatrix},$$

and similar expressions are given for the other quadratics. Transforming $\mathbf{z} = \mathbf{T}'\mathbf{y}$ or $\mathbf{y} = \mathbf{T}\mathbf{z}$ give us

$$\mathbf{y}'\mathbf{A}_1\mathbf{y} = \mathbf{z}'\mathbf{T}'\mathbf{A}_1\mathbf{T}\mathbf{z} = z_1^2 + z_2^2 + \cdots + z_{r_1}^2,$$

$$\mathbf{y}'\mathbf{A}_2\mathbf{y} = z_{r_1+1}^2 + z_{r_1+2}^2 + \cdots + z_{r_1+r_2}^2,$$

and so forth, where the z_i are independently distributed as $N(\mathbf{t}_i'\boldsymbol{\theta}, \sigma^2)$ (by Theorem 1.6(ii) in Sect. 1.7). Hence, by Theorem 1.10 in Sect. 1.9, $\sum_{i=1}^{r_1} z_i^2/\sigma^2$ is non-central chi-square with r_1 degrees of freedom and non-centrality parameter $\delta_1 = \boldsymbol{\theta}'\mathbf{A}_1\boldsymbol{\theta}/\sigma^2$. Similarly $\mathbf{x}'\mathbf{A}_2\mathbf{x}$ is independently distributed as non-central chi-square with r_2 degrees of freedom and non-centrality parameter $\delta_2 = \boldsymbol{\theta}'\mathbf{A}_2\boldsymbol{\theta}/\sigma^2$, and so forth. This completes the proof.

We now use the above theorem to find the distribution of F given by (4.1). Consider the identity

$$\mathbf{I}_n = (\mathbf{I}_n - \mathbf{P}_\Omega) + (\mathbf{P}_\Omega - \mathbf{P}_\omega) + \mathbf{P}_\omega. \tag{4.5}$$

Since $\mathcal{C}[\mathbf{P}_\omega] = \omega \subset \Omega$ (by Theorem 3.4), $\mathbf{P}_\Omega\mathbf{P}_\omega = \mathbf{P}_\omega (= \mathbf{P}_\omega' = \mathbf{P}_\omega\mathbf{P}_\Omega)$ and $(\mathbf{P}_\Omega - \mathbf{P}_\omega)^2 = \mathbf{P}_\Omega - \mathbf{P}_\Omega\mathbf{P}_\omega - \mathbf{P}_\omega\mathbf{P}_\Omega + \mathbf{P}_\omega = \mathbf{P}_\Omega - \mathbf{P}_\omega$. As $\mathbf{I}_n - \mathbf{P}_\Omega$ is idempotent with trace and rank $n-p$ (by (3.4)), and $\mathbf{P}_\omega$ is idempotent of rank $p-q$, the conditions of Theorem 4.1 hold so that by (i) of the theorem,

$$n = n - p + \text{rank}[\mathbf{P}_\Omega - \mathbf{P}_\omega] + p - q$$

and $\text{rank}[\mathbf{P}_\Omega - \mathbf{P}_\omega] = q$. It follows from the theorem that the quadratic $Q/\sigma^2 = \mathbf{y}'(\mathbf{I}_n - \mathbf{P}_\Omega)\mathbf{y}/\sigma^2$ is χ^2_{n-p} (as the non-centrality parameter, namely $\boldsymbol{\theta}'(\mathbf{I}_n - \mathbf{P}_\Omega)\boldsymbol{\theta}/\sigma^2$, is zero as $\mathbf{P}_\Omega\boldsymbol{\theta} = \boldsymbol{\theta}$) and $(Q_H - Q)/\sigma^2 = \mathbf{y}'(\mathbf{P}_\Omega - \mathbf{P}_\omega)\mathbf{y}/\sigma^2$ is independently distributed as non-central $\chi^2_q(\delta)$, where the non-centrality parameter δ is given by $\delta = \boldsymbol{\theta}'(\mathbf{P}_\Omega - \mathbf{P}_\omega)\boldsymbol{\theta}/\sigma^2$. Note that $E[Q_H - Q] = \sigma^2(q + \delta)$, from Theorem 1.7(ii) in Sect. 1.8. When H is true, $\delta = 0$ as $\mathbf{P}_\omega\boldsymbol{\theta} = \boldsymbol{\theta}$ and F has the $F_{q,n-p}$ distribution, while if $\mathbf{H}$ is false, F has the non-central F-distribution $F_{q,n-p}(\delta)$ (cf. Sect. 1.8)

The computations for calculating F are usually set out in the form of an analysis of variance (ANOVA) table as given below (Table 4.1). There *df* is the degrees of freedom and $MSS = SS/df$ is the Mean Sum of Squares. The difference $Q_H - Q$ is sometimes referred to as the *hypothesis sum of squares*. Looking at the ratio of the two *MSS* we see that F is roughly $1 + \delta/q$, and so we would reject H if F is much greater than unity and accept H if $F \approx 1$. In fact we reject H at the $100\alpha\%$ level of

Table 4.1 ANOVA table

Source	SS	df	MSS	E[MSS]
H	$Q_H - Q$	q	$(Q_H - Q)/q$	$\sigma^2 + \delta\sigma^2/q$
Residual	Q	$n-p$	$Q/(n-p)$	σ^2
Total	Q_H	$n-p+q$		

significance if $F > F_\alpha$, where F_α is determined by $\Pr[F_{q,n-p} \leq F_\alpha] = 1 - \alpha$. We note that the hypothesis sum of squares is given by

$$Q_H - Q = \| (\mathbf{P}_\Omega - \mathbf{P}_\omega)\mathbf{y} \|^2 = \| \hat{\boldsymbol{\theta}} - \hat{\boldsymbol{\theta}}_H \|^2 . \tag{4.6}$$

4.2 The Hypothesis Sum of Squares

In this section we look more closely at the matrix $\mathbf{P}_\Omega - \mathbf{P}_\omega$ from the hypothesis sum of squares. We shall show that the F-test is not only based on the likelihood ratio principle but it is also the test statistic obtained by applying a general principle due to Wald. We shall require the following theorems.

Theorem 4.2 *$\mathbf{P}_\Omega - \mathbf{P}_\omega$ represents the orthogonal projection of $\mathbb{R}^n$ onto $\omega^\perp \cap \Omega$, that is $\mathbf{P}_\Omega - \mathbf{P}_\omega = \mathbf{P}_{\omega^\perp \cap \Omega}$.*

Proof In the previous section we saw that $\mathbf{P}_\omega \mathbf{P}_\Omega = \mathbf{P}_\Omega \mathbf{P}_\omega = \mathbf{P}_\omega$, and $\mathbf{P}_\Omega - \mathbf{P}_\omega$ being symmetric and idempotent is a projection matrix. If $\boldsymbol{\theta} \in \omega^\perp \cap \Omega$, then $\mathbf{P}_\omega \boldsymbol{\theta} = \mathbf{0}$ and $\boldsymbol{\theta} = \mathbf{P}_\Omega \boldsymbol{\theta} = (\mathbf{P}_\Omega - \mathbf{P}_\omega)\boldsymbol{\theta} \in \mathcal{C}[\mathbf{P}_\Omega - \mathbf{P}_\omega]$. Conversely, if $\boldsymbol{\theta} = (\mathbf{P}_\Omega - \mathbf{P}_\omega)\mathbf{a}$ then $\mathbf{P}_\Omega \boldsymbol{\theta} = \boldsymbol{\theta}$ and $\mathbf{P}_\omega \boldsymbol{\theta} = \mathbf{0}$, so that $\boldsymbol{\theta} \in \omega^\perp \cap \Omega$. Thus $\omega^\perp \cap \Omega = \mathcal{C}[\mathbf{P}_\Omega - \mathbf{P}_\omega]$, and the result follows from Theorem 3.3 in Sect. 3.2. [We note in passing that any vector $\boldsymbol{\theta} \in \Omega$ takes the form $\boldsymbol{\theta} = \mathbf{P}_\Omega \mathbf{b} = (\mathbf{P}_\Omega - \mathbf{P}_\omega)\mathbf{b} + \mathbf{P}_\omega \mathbf{b}$ for some $\mathbf{b}$. Thus we see intuitively that we have an orthogonal decomposition corresponding to $\Omega = (\omega^\perp \cap \Omega) \oplus \omega$.]

Theorem 4.3 *If $\mathbf{A}$ is any matrix such that $\omega = \Omega \cap \mathcal{N}[\mathbf{A}]$, then*

$$\omega^\perp \cap \Omega = \mathcal{C}[\mathbf{P}_\Omega \mathbf{A}'].$$

Proof By Theorems 1.2 and 1.1,

$$\begin{aligned} \omega^\perp \cap \Omega &= (\Omega \cap \mathcal{N}[\mathbf{A}])^\perp \cap \Omega \\ &= (\Omega^\perp + \mathcal{C}[\mathbf{A}']) \cap \Omega. \end{aligned}$$

If $\boldsymbol{\theta}$ belongs to the right-hand side of the above equation, then $\mathbf{P}_\Omega \boldsymbol{\theta} = \boldsymbol{\theta}$ and $\boldsymbol{\theta} = (\mathbf{I}_n - \mathbf{P}_\Omega)\mathbf{a} + \mathbf{A}'\mathbf{b}$ for some $\mathbf{a}$ and $\mathbf{b}$, which together implies $\boldsymbol{\theta} = \mathbf{P}_\Omega \mathbf{A}'\mathbf{b} \in \mathcal{C}[\mathbf{P}_\Omega \mathbf{A}']$.

Conversely, suppose $\boldsymbol{\theta} = \mathbf{P}_\Omega \mathbf{A}'\mathbf{b}$, then since $\omega \in \mathcal{N}[\mathbf{A}]$ we have $\omega \perp \mathcal{C}[\mathbf{A}']$. Hence

$$\mathbf{P}_\omega \boldsymbol{\theta} = \mathbf{P}_\omega \mathbf{P}_\Omega \mathbf{A}'\mathbf{b} = \mathbf{P}_\omega \mathbf{A}'\mathbf{b} = \mathbf{0}$$

and $\mathbf{P}_\Omega \boldsymbol{\theta} = \boldsymbol{\theta}$ so that $\boldsymbol{\theta} \in \omega^\perp \cap \Omega$. Thus

$$(\Omega^\perp + \mathcal{C}[\mathbf{A}']) \cap \Omega = \mathcal{C}[\mathbf{P}_\Omega \mathbf{A}']$$

and the theorem is proved.

Theorem 4.4 *If* $\mathbf{A}$ *is a* $q \times n$ *matrix of rank* q, *then* $\text{rank}[\mathbf{P}_\Omega \mathbf{A}'] = q$ *if and only if* $\mathcal{C}[\mathbf{A}'] \cap \Omega^\perp = \mathbf{0}$.

Proof Let the rows of $\mathbf{A}$ be $\mathbf{a}_i'$ $(i = 1, 2, \ldots, q)$. If $\text{rank}[\mathbf{P}_\Omega \mathbf{A}'] \neq q$, then the columns of $\mathbf{P}_\Omega \mathbf{A}'$ are linearly dependent, that is there exist $c_1, c_2, \ldots, c_q$ not all zero such that $\sum_i c_i \mathbf{P}_\Omega \mathbf{a}_i = \mathbf{0}$. This implies there exists a vector $\sum_i c_i \mathbf{a}_i \in \mathcal{C}[\mathbf{A}']$ which is perpendicular to Ω and therefore $\mathcal{C}[\mathbf{A}'] \cap \Omega^\perp \neq \mathbf{0}$. We have established a contradiction and the theorem is proved.

4.3 Wald Test

We now use the results of the previous section to consider an alternative form of the likelihood ratio test. Let $\mathbf{y} = \mathbf{X}\boldsymbol{\beta} + \boldsymbol{\varepsilon}$, where $\mathbf{X}$ is $n \times p$ of rank p, $\boldsymbol{\beta} = (\beta_0, \beta_1, \ldots, \beta_{p-1})'$, $\boldsymbol{\theta} = \mathbf{X}\boldsymbol{\beta}$, and $\boldsymbol{\varepsilon}$ is $N_n[\mathbf{0}, \sigma^2 \mathbf{I}_n]$. Let $\mathbf{A}_1$ be a $q \times n$ matrix of rank q such that $\mathcal{N}[\mathbf{A}_1] = \omega \oplus \Omega^\perp$. Then

$$\Omega \cap \mathcal{N}[\mathbf{A}_1] = \Omega \cap (w \oplus \Omega^\perp) = \omega$$

by Theorem 1.3 with $\mathcal{V}_0 = \Omega$ and $\mathcal{V}_1 = \omega$. Also, from Theorem 4.3, $\mathbf{P}_\Omega - \mathbf{P}_\omega$ represents the orthogonal projection onto $\mathcal{C}[\mathbf{P}_\Omega \mathbf{A}_1']$. Now, by Theorem 1.1 in Sect. 1.2,

$$\begin{aligned}
\mathcal{C}[\mathbf{A}_1'] \cap \Omega^\perp &= \mathcal{N}[\mathbf{A}_1]^\perp \cap \Omega^\perp \\
&= (\omega \oplus \Omega^\perp)^\perp \cap \Omega^\perp \\
&= \omega^\perp \cap \Omega \cap \Omega^\perp \\
&= \mathbf{0}.
\end{aligned}$$

Hence, by Theorem 4.4, the $n \times q$ matrix $\mathbf{P}_\Omega \mathbf{A}_1'$ has rank q, $\mathbf{A}_1 \mathbf{P}_\Omega \mathbf{A}_1'$ is $q \times q$ of rank q (by A.4(ii) and $\mathbf{P}_\Omega^2 = \mathbf{P}_\Omega$), and is therefore nonsingular. From Example 3.1

in Sect. 3.3 with $\mathbf{X} = \mathbf{P}_\Omega \mathbf{A}_1'$, and $\hat{\boldsymbol{\theta}}$ the least squares estimate of $\boldsymbol{\theta}$,

$$\begin{aligned}\mathbf{y}'(\mathbf{P}_\Omega - \mathbf{P}_\omega)\mathbf{y} &= \mathbf{y}'\mathbf{X}(\mathbf{X}'\mathbf{X})^{-1}\mathbf{X}'\mathbf{y} \\ &= \mathbf{y}'\mathbf{P}_\Omega \mathbf{A}_1'(\mathbf{A}_1\mathbf{P}_\Omega\mathbf{A}_1')^{-1}\mathbf{A}_1\mathbf{P}_\Omega\mathbf{y} \qquad (4.7)\\ &= (\mathbf{A}_1\hat{\boldsymbol{\theta}})'(\mathbf{A}_1\mathbf{P}_\Omega\mathbf{A}_1')^{-1}\mathbf{A}_1\hat{\boldsymbol{\theta}}. \qquad (4.8)\end{aligned}$$

The variance matrix of $\mathbf{A}_1\hat{\boldsymbol{\theta}} = \mathbf{A}_1\mathbf{P}_\Omega\mathbf{y}$ is (by Theorem 1.5(iii) in Sect. 1.6(iii))

$$\mathrm{var}[\mathbf{A}_1\hat{\boldsymbol{\theta}}] = \mathbf{A}_1\mathbf{P}_\Omega\sigma^2\mathbf{I}_n\mathbf{P}_\Omega\mathbf{A}_1' = \sigma^2\mathbf{A}_1\mathbf{P}_\Omega\mathbf{A}_1',$$

and if $\hat{\mathbf{D}}$ is its value for $\sigma^2 = \hat{\sigma}^2 = \mathbf{y}'(\mathbf{I}_n - \mathbf{P}_\Omega)\mathbf{y}/n$, the maximum likelihood estimate of σ^2, then we find that

$$\begin{aligned}F &= \frac{(n-p)}{q}\frac{\mathbf{y}'(\mathbf{P}_\Omega - \mathbf{P}_\omega)\mathbf{y}}{\mathbf{y}'(\mathbf{I}_n - \mathbf{P}_\Omega)\mathbf{y}} \\ &= \frac{n-p}{nq}(\mathbf{A}_1\hat{\boldsymbol{\theta}})'\hat{\mathbf{D}}^{-1}(\mathbf{A}_1\hat{\boldsymbol{\theta}}).\end{aligned}$$

Thus to test $H : \mathbf{A}_1\boldsymbol{\theta} = \mathbf{0}$, we replace $\boldsymbol{\theta}$ by its maximum likelihood estimate and see if $\mathbf{A}_1\hat{\boldsymbol{\theta}}$ is "near enough" to zero by calculating F, a simple positive-semidefinite quadratic function of $\mathbf{A}_1\hat{\boldsymbol{\theta}}$. This simple test principle, due to Wald (1943), is discussed again later.

Example 4.1 Suppose we consider the regression model discussed above and we wish to test $H : \beta_r = \beta_{r+1} = \ldots = \beta_{p-1} = 0$. Then $\omega = \mathcal{C}[\mathbf{X}_r]$, where $\mathbf{X}_r$ consists of the first r columns of $\mathbf{X}$ so that

$$\mathbf{P}_\Omega = \mathbf{X}(\mathbf{X}'\mathbf{X})^{-1}\mathbf{X}', \quad \mathbf{P}_\omega = \mathbf{X}_r(\mathbf{X}_r'\mathbf{X}_r)^{-1}\mathbf{X}_r',$$

and we can immediately write down our F-statistic. However, using the Wald principle, we can express $\mathbf{P}_\Omega - \mathbf{P}_\omega$ as a single matrix as follows. We first of all show that if $\mathbf{X} = (\mathbf{X}_r, \mathbf{X}_{p-r})$, then

$$\omega = \Omega \cap \mathcal{N}[\mathbf{X}_{p-r}'(\mathbf{I}_n - \mathbf{P}_\omega)]. \qquad (4.9)$$

If $\boldsymbol{\theta} \in \omega$, then $\boldsymbol{\theta} \in \Omega$, $(\mathbf{I}_n - \mathbf{P}_\omega)\boldsymbol{\theta} = \mathbf{0}$ and $\boldsymbol{\theta}$ belongs to the right-hand side of the above equation. Conversely, if $\boldsymbol{\theta}$ belongs to the right-hand side, then $\boldsymbol{\theta} = \mathbf{X}\boldsymbol{\beta} = \mathbf{X}_r\boldsymbol{\beta}_r + \mathbf{X}_{p-r}\boldsymbol{\beta}_{p-r}$ and $\mathbf{X}_{p-r}'(\mathbf{I}_n - \mathbf{P}_\omega)\boldsymbol{\theta} = \mathbf{0}$. Thus, since $(\mathbf{I}_n - \mathbf{P}_\omega)\mathbf{X}_r = \mathbf{0}$,

$$\begin{aligned}\mathbf{0} &= \mathbf{X}_{p-r}'(\mathbf{I}_n - \mathbf{P}_\omega)(\mathbf{X}_r\boldsymbol{\beta}_r + \mathbf{X}_{p-r}\boldsymbol{\beta}_{p-r}) \\ &= \mathbf{X}_{p-r}'(\mathbf{I}_n - \mathbf{P}_\omega)\mathbf{X}_{p-r}\boldsymbol{\beta}_{p-r}.\end{aligned}$$

By Theorem 4.4 with $\Omega^\perp$ replaced by ω and $\mathbf{A} = \mathbf{X}'_{p-r}, (\mathbf{I}_n - \mathbf{P}_\omega)\mathbf{X}_{p-r}$ with rank $p-r$ as $\mathcal{C}[\mathbf{X}_{p-r}] \cap \omega = \mathbf{0}$ then $\mathbf{X}'_{p-r}(\mathbf{I}_n - \mathbf{P}_\omega)\mathbf{X}_{p-r}$ is non-singular. Hence $\boldsymbol{\beta}_{p-r} = \mathbf{0}$, $\boldsymbol{\theta} \in \mathcal{C}[\mathbf{X}_r] = \omega$, and (4.9) is established. Since $\mathbf{P}_\Omega \mathbf{X}_{p-r} = \mathbf{X}_{p-r}$,

$$\mathbf{P}_\Omega(\mathbf{I}_n - \mathbf{P}_\omega)\mathbf{X}_{p-r} = (\mathbf{I}_n - \mathbf{P}_\omega)\mathbf{X}_{p-r}$$

and it follows from Eq. (4.7), with $\mathbf{A}_1 = \mathbf{X}'_{r-p}(\mathbf{I}_n - \mathbf{P}_\omega)$, that

$$\mathbf{P}_\Omega - \mathbf{P}_\omega = (\mathbf{I}_n - \mathbf{P}_\omega)\mathbf{X}_{p-r}[\mathbf{X}'_{p-r}(\mathbf{I}_n - \mathbf{P}_\omega)\mathbf{X}_{p-r}]^{-1}\mathbf{X}'_{p-r}(\mathbf{I}_n - \mathbf{P}_\omega), \tag{4.10}$$

which can be used for a Wald test.

Example 4.2 Suppose $\mathbf{X}$ is defined as in Example 4.1, and we wish to test $\mathbf{A}\boldsymbol{\beta} = \mathbf{b}$, where $\mathbf{A}$ is $q \times p$ of rank q. Let $\boldsymbol{\beta}_0$ be any solution of $\mathbf{A}\boldsymbol{\beta} = \mathbf{b}$, put $\mathbf{z} = \mathbf{y} - \mathbf{X}\boldsymbol{\beta}_0$ and let $\boldsymbol{\gamma} = \boldsymbol{\beta} - \boldsymbol{\beta}_0$. Then our original model and hypothesis are equivalent to $\mathbf{z} = \mathbf{X}\boldsymbol{\gamma} + \boldsymbol{\varepsilon}$, where $\boldsymbol{\varepsilon}$ is $N_n[\mathbf{0}, \sigma^2\mathbf{I}_n]$, and $\omega : \mathbf{A}\boldsymbol{\gamma} = \mathbf{0}$. If $\boldsymbol{\phi} = \mathbf{X}\boldsymbol{\gamma}$, then since $\boldsymbol{\gamma} = (\mathbf{X}'\mathbf{X})^{-1}\mathbf{X}'\boldsymbol{\phi}$, $\omega = \Omega \cap \mathcal{N}[\mathbf{A}_1]$, where $\mathbf{A}_1 = \mathbf{A}(\mathbf{X}'\mathbf{X})^{-1}\mathbf{X}'$. Now $\mathbf{A}_1'\mathbf{c} = \mathbf{0}$ implies that $\mathbf{X}(\mathbf{X}'\mathbf{X})^{-1}\mathbf{A}'\mathbf{c} = \mathbf{0}$, which pre-multiplying by $\mathbf{X}'$ give us $\mathbf{A}'\mathbf{c} = \mathbf{0}$ or $\mathbf{c} = \mathbf{0}$, as the q columns of $\mathbf{A}'$ are linearly independent. Hence the rows of $\mathbf{A}_1$ are linearly independent and $\mathbf{A}_1$ has rank q. Also,

$$\mathcal{C}[\mathbf{A}_1'] \cap \Omega^\perp \subset \mathcal{C}[\mathbf{X}] \cap \Omega^\perp = \mathbf{0}.$$

Thus (4.7) applies with $\mathbf{A}_1\mathbf{P}_\Omega = \mathbf{A}_1$ (since $\mathbf{P}_\Omega\mathbf{X} = \mathbf{X}$). Substituting for $\mathbf{A}_1$, and using

$$\hat{\boldsymbol{\beta}} = (\mathbf{X}'\mathbf{X})^{-1}\mathbf{X}'\mathbf{z} = (\mathbf{X}'\mathbf{X})^{-1}\mathbf{X}'(\mathbf{y} - \mathbf{X}\boldsymbol{\beta}_0) = \hat{\boldsymbol{\beta}} - \boldsymbol{\beta}_0$$

and $\mathbf{A}\boldsymbol{\beta}_0 = \mathbf{b}$, we get

$$\begin{aligned}
\mathbf{z}'(\mathbf{P}_\Omega - \mathbf{P}_\omega)\mathbf{z} &= (\mathbf{y} - \mathbf{X}\boldsymbol{\beta}_0)'\mathbf{A}_1'(\mathbf{A}_1\mathbf{A}_1')^{-1}\mathbf{A}_1(\mathbf{y} - \mathbf{X}\boldsymbol{\beta}_0) \\
&= (\mathbf{y} - \mathbf{X}\boldsymbol{\beta}_0)'\mathbf{X}(\mathbf{X}'\mathbf{X})^{-1}\mathbf{A}'[\mathbf{A}(\mathbf{X}'\mathbf{X})^{-1}\mathbf{A}']^{-1}\mathbf{A}(\mathbf{X}'\mathbf{X})^{-1}\mathbf{X}'(\mathbf{y} - \mathbf{X}\boldsymbol{\beta}_0) \\
&= (\hat{\boldsymbol{\beta}} - \boldsymbol{\beta}_0)'\mathbf{A}'[\mathbf{A}(\mathbf{X}'\mathbf{X})^{-1}\mathbf{A}']^{-1}\mathbf{A}(\hat{\boldsymbol{\beta}} - \boldsymbol{\beta}_0) \\
&= (\mathbf{A}\hat{\boldsymbol{\beta}} - \mathbf{b})'[\mathbf{A}(\mathbf{X}'\mathbf{X})^{-1}\mathbf{A}']^{-1}(\mathbf{A}\hat{\boldsymbol{\beta}} - \mathbf{b}) \\
&= (\mathbf{A}\hat{\boldsymbol{\beta}} - \mathbf{b})'\{\text{Var}[\mathbf{A}\hat{\boldsymbol{\beta}}]\}^{-1}(\mathbf{A}\hat{\boldsymbol{\beta}} - \mathbf{b})'\sigma^2.
\end{aligned}$$

The above equation can be used for a Wald test.

Example 4.3 We now consider a theoretical model that we shall use in later asymptotic theory. Let $\mathbf{z} = \boldsymbol{\phi} + \boldsymbol{\eta}$ where $\boldsymbol{\eta}$ is $N_n[\mathbf{0}, \mathbf{I}_n]$. We assume that $G : \Omega = \mathbb{R}^n$ and $H : \omega = \mathcal{N}[\mathbf{C}]$, where the rows of $\mathbf{C}$ are linearly independent. Then $\mathbf{P}_\Omega = \mathbf{I}_n$ so that $\hat{\boldsymbol{\phi}} = \mathbf{z}$ and, since $\mathcal{N}[\mathbf{C}] = \mathcal{C}[\mathbf{C}']^\perp$ (by Theorem 1.1 in Sect. 1.2), we have

$\mathbf{P}_\omega = \mathbf{I}_n - \mathbf{P} = \mathbf{I}_n - \mathbf{C}'(\mathbf{C}\mathbf{C}')^{-1}\mathbf{C}$, and $\hat{\phi}_H = \mathbf{P}_\omega \mathbf{z}$. Because of the usefulness of this model we express our result as a general theorem.

Theorem 4.5 *Let* $\mathbf{z} = \phi + \eta$, *where* $\eta \sim N_p[\mathbf{0}, \mathbf{I}_p]$, *and consider* $H : \mathbf{C}\phi = \mathbf{0}$, *where* $\mathbf{C}$ *is* $q \times p$ *of rank* q. *Let* $\tilde{\boldsymbol{\lambda}}$ *and* $\tilde{\phi}$ *be the restricted least-squares solutions under* H *of*

$$\mathbf{z} - \tilde{\phi} + \mathbf{C}'\tilde{\boldsymbol{\lambda}} = \mathbf{0} \tag{4.11}$$

and

$$\mathbf{C}\tilde{\phi} = \mathbf{0}. \tag{4.12}$$

Then H *can be tested using*

$$\begin{aligned} t &= (\mathbf{z} - \tilde{\phi})'(\mathbf{z} - \tilde{\phi}) & (4.13)\\ &= \mathbf{z}'\mathbf{C}'(\mathbf{C}\mathbf{C}')^{-1}\mathbf{C}\mathbf{z} & (4.14)\\ &= \tilde{\boldsymbol{\lambda}}\mathbf{C}\mathbf{C}'\tilde{\boldsymbol{\lambda}}, & (4.15) \end{aligned}$$

where $t \sim \chi_q^2(\delta)$, *the non-central chi-square distribution with non-centrality parameter* $\delta = \phi'\mathbf{C}'(\mathbf{C}\mathbf{C}')^{-1}\mathbf{C}\phi$. *The test statistic* t *is also the likelihood-ratio test.*

Proof To find $\tilde{\phi}$ we can use A.20 to differentiate (cf. Sect. 1.10)

$$\frac{1}{2} \parallel \mathbf{z} - \phi \parallel^2 + \lambda'\mathbf{C}\phi = \frac{1}{2}(\mathbf{z}'\mathbf{z} - 2\phi'\mathbf{z} + \phi'\mathbf{z}'\mathbf{z}\phi) + \phi'\mathbf{C}'\boldsymbol{\lambda},$$

to obtain (4.11). Multiplying (4.11) by $\mathbf{C}$ and using (4.12) we obtain $\tilde{\boldsymbol{\lambda}} = -(\mathbf{C}\mathbf{C}')^{-1}\mathbf{C}\mathbf{z}$. Substituting in (4.11) give us $\tilde{\phi} = (\mathbf{I}_p - \mathbf{P})\mathbf{z}$, where $\mathbf{P} = \mathbf{C}'(\mathbf{C}\mathbf{C}')^{-1}\mathbf{C}$, a projection matrix of rank q. Thus $(\mathbf{z} - \tilde{\phi})'(\mathbf{z} - \tilde{\phi}) = \mathbf{z}'\mathbf{P}\mathbf{z}$, and has a non-central chi-square distribution $\chi_q^2(\delta)$ with $\delta = \phi'\mathbf{P}\phi$ by Theorem 1.10 in Sect. 1.9. Putting the above results together we obtain our three expressions for t. Finally, the log of the likelihood function for the multivariate normal with $\sigma^2 = 1$ is $L(\phi) = -(\mathbf{z} - \phi)'(\mathbf{z} - \phi)$ so that the likelihood ratio test is

$$2[L(\hat{\phi}) - L(\tilde{\phi})] = (\mathbf{z} - \tilde{\phi})'(\mathbf{z} - \tilde{\phi}),$$

as $L(\hat{\phi}) = L(\mathbf{z}) = 0$. This completes the proof.

Example 4.4 (Less than full rank) Suppose $\mathbf{y} = \mathbf{X}\boldsymbol{\beta} + \boldsymbol{\varepsilon}$, where $\mathbf{X}$ is $n \times p$ of rank r $(r < p)$. Instead of introducing identifiability constraints we can focus on what linear restrictions $H : \mathbf{a}_i'\boldsymbol{\beta} = 0$ $(i = 1, 2, \ldots, q)$ we might be able to test, or in matrix terms $\mathbf{A}\boldsymbol{\beta} = \mathbf{0}$ where rank$[\mathbf{A}] = q$. A natural assumption is that the constraints are all estimable, which implies $\mathbf{a}_i' = \mathbf{m}_i'\mathbf{X}$ (by end of Sect. 3.4) for some $\mathbf{m}_i$, or $\mathbf{A} = \mathbf{M}\mathbf{X}$, where $\mathbf{M}$ is $q \times n$ of rank q (as $q = \text{rank}[\mathbf{A}] \le \text{rank}[\mathbf{M}]$ by A.2).

Since $\mathbf{A}\boldsymbol{\beta} = \mathbf{M}\mathbf{X}\boldsymbol{\beta} = \mathbf{M}\boldsymbol{\theta}$ we can therefore find the least squares estimate of $\boldsymbol{\theta}$ under H by minimizing $\| \mathbf{y} - \boldsymbol{\theta} \|^2$ subject to $\boldsymbol{\theta} \in \mathcal{C}[\mathbf{X}] = \Omega$ and $\mathbf{M}\mathbf{X}\boldsymbol{\beta} = \mathbf{M}\boldsymbol{\theta} = \mathbf{0}$, that is subject to $\boldsymbol{\theta} \in \Omega \cap \mathcal{N}[\mathbf{M}]$ $(= \omega)$. Now using Theorem 4.3 in Sect. 4.2, $\omega^\perp \cap \Omega = \mathcal{C}[\mathbf{P}_\Omega \mathbf{M}']$, where by A.15(ii)

$$\mathbf{P}_\Omega \mathbf{M}' = \mathbf{X}(\mathbf{X}'\mathbf{X})^- \mathbf{X}'\mathbf{M}' = \mathbf{X}(\mathbf{X}'\mathbf{X})^- \mathbf{A}',$$

is $n \times q$ of rank q (cf. Theorem 4.4 in Sect. 4.2) and $(\mathbf{X}'\mathbf{X})^-$ is a weak inverse of $\mathbf{X}'\mathbf{X}$. Hence

$$\begin{aligned}
\mathbf{y}'(\mathbf{P}_\Omega - \mathbf{P}_\omega)\mathbf{y} &= \mathbf{y}'\mathbf{P}_{\omega^\perp \cap \Omega}\mathbf{y} \\
&= \mathbf{y}'(\mathbf{P}_\Omega \mathbf{M}')[\mathbf{M}\mathbf{P}_\Omega \mathbf{M}']^{-1}(\mathbf{P}_\Omega \mathbf{M}')'\mathbf{y} \\
&= \mathbf{y}'\mathbf{X}(\mathbf{X}'\mathbf{X})^- \mathbf{A}'[\mathbf{A}(\mathbf{X}'\mathbf{X})^- \mathbf{A}']^{-1}\mathbf{A}(\mathbf{X}'\mathbf{X})^- \mathbf{X}'\mathbf{y} \\
&= (\mathbf{A}\hat{\boldsymbol{\beta}})'[\mathbf{A}(\mathbf{X}'\mathbf{X})^- \mathbf{A}']^{-1}\mathbf{A}\hat{\boldsymbol{\beta}}.
\end{aligned}$$

4.4 Contrasts

A contrast of the vector $\boldsymbol{\theta}$ is any linear function $\mathbf{c}'\boldsymbol{\theta}$ such that $\sum_i c_i = 0$. Two contrasts $\mathbf{c}'\boldsymbol{\theta}$ and $\mathbf{d}'\boldsymbol{\theta}$ are said to be *orthogonal* if $\mathbf{c}'\mathbf{d} = 0$. For example, $\theta_1 - \theta_2$ and $\theta_1 + \theta_2 + \theta_3 - 3\theta_4$ are two orthogonal contrasts.

Example 4.5 The situation that we often meet in factorial experiments is that we are given a set of independent contrasts $\mathbf{a}_i'\boldsymbol{\theta}$ $(i = 1, 2, \ldots, n - p)$ equal to zero and we wish to test whether a further set of q orthogonal contrasts $\mathbf{a}_{1i}'\boldsymbol{\theta}$ $(i = 1, 2, \ldots, q)$, which are orthogonal to the previous set, are also zero. If $\mathbf{A} = (\mathbf{a}_1, \mathbf{a}_2, \ldots, \mathbf{a}_{n-p})'$ and $\mathbf{A}_1 = (\mathbf{a}_{11}, \mathbf{a}_{12}, \ldots, \mathbf{a}_{1q})'$, then G is $\mathbf{A}\boldsymbol{\theta} = \mathbf{0}$ and H is $\mathbf{A}\boldsymbol{\theta} = \mathbf{0}$, $\mathbf{A}_1\boldsymbol{\theta} = \mathbf{0}$, where $\mathbf{A}_1\mathbf{A}' = \mathbf{0}$ and $\mathbf{A}_1\mathbf{A}_1'$ is diagonal, namely $\mathbf{D} = \operatorname{diag}(\mathbf{a}_{11}'\mathbf{a}_{11}, \ldots, \mathbf{a}_{1q}'\mathbf{a}_{1q})$. Define $d_i(\boldsymbol{\theta}) = \| \mathbf{a}_{1i} \|^{-1} \mathbf{a}_{1i}'\boldsymbol{\theta}$ $(i = 1, 2, \ldots, q)$, a set of orthonormal contrasts. Since, by Example 3.2 of Sect. 3.3, $\mathbf{P}_\Omega = \mathbf{I}_n - \mathbf{A}'(\mathbf{A}\mathbf{A}')^{-1}\mathbf{A}$, we have $\mathbf{A}_1\mathbf{P}_\Omega\mathbf{A}_1' = \mathbf{A}_1\mathbf{A}_1'$ and, from (4.8),

$$\begin{aligned}
F &= \frac{n-p}{q} \frac{(\mathbf{A}_1\hat{\boldsymbol{\theta}})'\mathbf{D}^{-1}(\mathbf{A}_1\hat{\boldsymbol{\theta}})}{\mathbf{y}'(\mathbf{I}_n - \mathbf{P}_\Omega)\mathbf{y}} \\
&= \frac{\sum_{i=1}^q d_i^2(\hat{\boldsymbol{\theta}})}{qs^2}.
\end{aligned}$$

Example 4.6 Consider a factorial experiment with two factors a and b, each at two levels a_1, a_2 and b_1, b_2. Then the yields from the four treatment combinations may be represented symbolically by a_2b_2, a_2b_1, a_1b_2, and a_1b_1. We can now define the following quantities

$$\text{effect of } a \text{ at level } b_1 = a_2b_1 - a_1b_1,$$

$$\text{effect of } a \text{ at level } b_2 = a_2b_2 - a_1b_2,$$

and the average effect is

$$A = \frac{1}{2}(a_2b_1 - a_1b_1 + a_2b_2 - a_1b_2).$$

If the two factors were acting independently we would expect the two effects at levels b_1 and b_2 to be equal, but in general they will be different, and their difference is a measure of the extent to which the factors interact. We define the interaction AB to be the difference between these two effects, namely

$$AB = \frac{1}{2}(a_2b_2 - a_1b_2 - a_2b_1 + a_1b_1).$$

In a similar manner we can define the average effect of b as

$$B = \frac{1}{2}(a_2b_2 - a_2b_1 + a_1b_2 - a_1b_1),$$

and BA as half of the difference of the two effects of b. However, $AB = BA$ and the concept of interaction is a symmetrical one, as we would expect. We note that A, B, and AB are three orthonormal contrasts of the four treatments, and denoting the mean yield by M we have the orthogonal transformation

$$\begin{pmatrix} 2M \\ A \\ B \\ AB \end{pmatrix} = \frac{1}{2}\begin{pmatrix} 1 & 1 & 1 & 1 \\ 1 & 1 & -1 & -1 \\ 1 & -1 & 1 & -1 \\ 1 & -1 & -1 & 1 \end{pmatrix}\begin{pmatrix} a_2b_2 \\ a_2b_1 \\ a_1b_2 \\ a_1b_1 \end{pmatrix},$$

or

$$\boldsymbol{\delta} = \mathbf{T}\boldsymbol{\mu},$$

say, where $\mathbf{T}$ is orthogonal. We have denoted the four combinations a_ib_j by μ_i ($i = 1, 2, 3, 4$), and suppose we have t observations on each combination μ_i. Then the hypothesis of interest, H, is that the four population means are equal. If y_{ij} is the jth observation on the ith mean ($i = 1, 2, \ldots 4; j = 1, 2, \ldots, t$) then we assume the model $y_{ij} = \theta_{ij} + \varepsilon_{ij}$ and $G : \theta_{ij} = \mu_i$ for all i, j. If $\boldsymbol{\theta}$ is the vector of elements θ_{ij}, then G states that certain contrasts $\theta_{ij} - \overline{\theta}_{i\cdot}$ of $\boldsymbol{\theta}$ are zero, and H is equivalent to

$A = B = AB = 0$, where A, B, and AB are orthogonal contrasts in $\boldsymbol{\theta}$. For example

$$\begin{aligned} 2A(\boldsymbol{\theta}) &= a_2 b_2 + a_2 b_1 - a_1 b_2 - a_1 b_1 \\ &= \mu_1 + \mu_2 - \mu_3 - \mu_4 \\ &= \overline{\theta}_{1\cdot} + \overline{\theta}_{2\cdot} - \overline{\theta}_{3\cdot} - \overline{\theta}_{4\cdot}, \end{aligned}$$

and this is a contrast in $\boldsymbol{\theta}$ since $\theta_{1j} + \theta_{2j} - \theta_{3}j - \theta_{4j}$ is a contrast and a sum (and average) of contrasts is still a contrast. Also the two sets of contrasts for G and H are orthogonal, and therefore the general theory described in the previous example, Example 4.5, can be applied to this example. All we require is $\hat{\boldsymbol{\theta}}$, and our F-statistic for testing H is

$$F = \frac{4(t-1)t}{3} \frac{A(\hat{\boldsymbol{\theta}})^2 + B(\hat{\boldsymbol{\theta}})^2 + AB(\hat{\boldsymbol{\theta}})^2}{\| \mathbf{y} - \hat{\boldsymbol{\theta}} \|^2}.$$

Minimizing $\sum_i \sum_j (y_{ij} - \mu_i)^2$ gives us $\hat{\theta}_{ij} = \hat{\mu}_i = \overline{y}_{i\cdot}$, the least squares estimate of μ_i. Hence

$$2A(\hat{\boldsymbol{\theta}}) = \overline{y}_{1\cdot} + \overline{y}_{2\cdot} - \overline{y}_{3\cdot} - \overline{y}_{4\cdot} \quad \text{etc.},$$

$$4M(\hat{\boldsymbol{\theta}}) = \overline{y}_{1\cdot} + \overline{y}_{2\cdot} + \overline{y}_{3\cdot} + \overline{y}_{4\cdot} = 4\overline{y}_{\cdot\cdot},$$

and

$$\| \mathbf{y} - \hat{\boldsymbol{\theta}} \|^2 = \sum_{i=1}^{4} \sum_{j=1}^{t} (y_{ij} - \overline{y}_{i\cdot})^2.$$

Since $\hat{\boldsymbol{\delta}}'\hat{\boldsymbol{\delta}} = \hat{\boldsymbol{\mu}}'\mathbf{T}'\mathbf{T}\hat{\boldsymbol{\mu}} = \hat{\boldsymbol{\mu}}'\hat{\boldsymbol{\mu}}$, we see that

$$A(\hat{\boldsymbol{\theta}})^2 + B(\hat{\boldsymbol{\theta}})^2 + AB(\hat{\boldsymbol{\theta}})^2 = \sum_{i=1}^{4} \overline{y}_{i\cdot}^2 - 4M(\hat{\boldsymbol{\theta}})^2 = \sum_{i=1}^{4} (\overline{y}_{i\cdot} - \overline{y}_{\cdot\cdot})^2.$$

4.5 Confidence Regions and Intervals

In most practical applications of linear hypothesis theory our prime interest is not just in significance tests but also in the finding of confidence regions and confidence intervals for the unknown parameters. Suppose we are given $G : \boldsymbol{\theta} \in \Omega$ and we wish to test the hypothesis $\mathbf{A}_1\boldsymbol{\theta} = \mathbf{0}$, where the q rows of $\mathbf{A}_1$ are linearly independent. Let $\mathbf{B}_1 = \mathbf{A}_1\mathbf{P}_\Omega\mathbf{A}_1'$, where $\sigma^2\mathbf{B}_1$ is the variance-covariance matrix of $\mathbf{A}_1\hat{\boldsymbol{\theta}} = \mathbf{A}_1\mathbf{P}_\Omega\mathbf{y}$. Since $\mathbf{P}_\Omega\boldsymbol{\theta} = \boldsymbol{\theta}$,

$$\begin{aligned}(\mathbf{A}_1\hat{\boldsymbol{\theta}} - \mathbf{A}_1\boldsymbol{\theta})'\mathbf{B}_1^{-1}(\mathbf{A}_1\hat{\boldsymbol{\theta}} - \mathbf{A}_1\boldsymbol{\theta}) &= (\mathbf{y} - \boldsymbol{\theta})'\mathbf{P}_\Omega\mathbf{A}_1'\mathbf{B}_1^{-1}\mathbf{A}_1\mathbf{P}_\Omega(\mathbf{y} - \boldsymbol{\theta}) \\ &= \boldsymbol{\varepsilon}'(\mathbf{P}_\Omega - \mathbf{P}_\omega)\boldsymbol{\varepsilon},\end{aligned}$$

by Eq. (4.7) in Sect. 4.3, and $\mathbf{y}'(\mathbf{I}_n - \mathbf{P}_\Omega)\mathbf{y} = \boldsymbol{\varepsilon}'(\mathbf{I}_n - \mathbf{P}_\Omega)\boldsymbol{\varepsilon}$. Hence from (4.1),

$$F = (\mathbf{A}_1\hat{\boldsymbol{\theta}} - \mathbf{A}_1\boldsymbol{\theta})'\mathbf{B}_1^{-1}(\mathbf{A}_1\hat{\boldsymbol{\theta}} - \mathbf{A}_1\boldsymbol{\theta})/qs^2 \tag{4.16}$$

$$= \frac{(n-p)}{q}\frac{\boldsymbol{\varepsilon}'(\mathbf{P}_\Omega - \mathbf{P}_\omega)\boldsymbol{\varepsilon}}{\boldsymbol{\varepsilon}'(\mathbf{I}_n - \mathbf{P}_\Omega)\boldsymbol{\varepsilon}} \tag{4.17}$$

has the (central) $F_{q,n-p}$ distribution. Thus if

$$\Pr[F_{q,n-p} \leq F_{q,n-p}(\alpha)] = 1 - \alpha,$$

then a $100(1-\alpha)$ per cent confidence region for $\boldsymbol{\theta}$ is given by

$$(\mathbf{A}_1\hat{\boldsymbol{\theta}} - \mathbf{A}_1\boldsymbol{\theta})'\mathbf{B}_1^{-1}(\mathbf{A}_1\hat{\boldsymbol{\theta}} - \mathbf{A}_1\boldsymbol{\theta}) \leq qs^2F_{q,n-p}(\alpha). \tag{4.18}$$

If we wanted to obtain a confidence interval for a single constraint $\psi = \mathbf{a}'\boldsymbol{\theta}$, then by the Gauss-Markov Theorem 3.6 in Sect. 3.5

$$\hat{\psi} = \mathbf{a}'\hat{\boldsymbol{\theta}} = \mathbf{a}'\mathbf{P}_\Omega\mathbf{y}$$

has minimum variance, and the confidence interval for ψ is

$$\frac{(\hat{\psi} - \psi)^2}{(\mathbf{a}'\mathbf{P}_\Omega\mathbf{a})} \leq s^2F_{1,n-p}(\alpha).$$

As $F_{1,n-p}$ is t_{n-p}^2, where t_{n-p} is the t-distribution with $n-p$ degrees of freedom, this confidence interval can also be expressed in the form

$$\hat{\psi} - s(\mathbf{a}'\mathbf{P}_\Omega\mathbf{a})^{1/2}t_{n-p}(\alpha/2) \leq \psi \leq \hat{\psi} + s(\mathbf{a}'\mathbf{P}_\Omega\mathbf{a})^{1/2}t_{n-p}(\alpha/2).$$

We can also obtain simultaneous confidence intervals using Scheffé's so-called S-method (Scheffé 1959, 68) as follows. Let $\boldsymbol{\phi} = \mathbf{A}_1\boldsymbol{\theta}$ and $\hat{\boldsymbol{\phi}} = \mathbf{A}_1\hat{\boldsymbol{\theta}}$. Then, from (4.18),

$$\begin{aligned}1 - \alpha &= \Pr[F_{q,n-p} \leq F_{q,n-p}(\alpha)] \\ &= \Pr[(\hat{\boldsymbol{\phi}} - \boldsymbol{\phi})'\mathbf{B}_1^{-1}(\hat{\boldsymbol{\phi}} - \boldsymbol{\phi}) \leq qs^2F_{q,n-p}(\alpha)]\end{aligned}$$

$$
\begin{aligned}
&= \Pr[\mathbf{b}'\mathbf{B}_1^{-1}\mathbf{b} \le m], \quad \text{say} \\
&= \Pr\left[\sup_{\mathbf{h},\mathbf{h}\neq\mathbf{0}} \left\{\frac{(\mathbf{h}'\mathbf{b})^2}{\mathbf{h}'\mathbf{B}_1\mathbf{h}}\right\} \le m\right] \quad \text{by A.21(i)} \\
&= \Pr\left[\frac{(\mathbf{h}'\mathbf{b})^2}{\mathbf{h}'\mathbf{B}_1\mathbf{h}} \le m, \quad \text{for all } \mathbf{h}\ (\mathbf{h}\neq\mathbf{0})\right] \\
&= \Pr\left[\frac{|\mathbf{h}'\hat{\boldsymbol{\phi}} - \mathbf{h}'\boldsymbol{\phi}|}{s(\mathbf{h}'\mathbf{B}_1\mathbf{h})^{1/2}} \le (qF_{q,n-p}(\alpha))^{1/2}, \quad \text{for all } \mathbf{h}\ (\mathbf{h}\neq\mathbf{0})\right].
\end{aligned}
$$

We can therefore construct a confidence interval for any linear function $\mathbf{h}'\boldsymbol{\phi}$, namely

$$
\mathbf{h}'\hat{\boldsymbol{\phi}} \pm (qF_{q,n-p}(\alpha))^{1/2} s(\mathbf{h}'\mathbf{B}_1\mathbf{h})^{1/2}, \tag{4.19}
$$

and the overall probability for the entire class of such intervals is exactly $1-\alpha$. We note that the term $s^2\mathbf{h}'\mathbf{B}_1\mathbf{h}$ involved in calculating (4.19) is simply an unbiased estimate of var$[\mathbf{h}'\hat{\boldsymbol{\phi}}]$ that can often be found directly. We can therefore write (4.19) in the form

$$
\mathbf{h}'\hat{\boldsymbol{\phi}} \pm (qF_{q,n-p}(\alpha))^{1/2}\hat{\sigma}_{\mathbf{h}'\hat{\boldsymbol{\phi}}}.
$$

Suppose $\boldsymbol{\theta} = \mathbf{X}\boldsymbol{\beta}$, where $\mathbf{X}$ is $n \times p$ of rank p and $\boldsymbol{\beta} = (\beta_0, \beta_1, \ldots, \beta_{p-1})'$, and we wish to use the test of $\mathbf{A}\boldsymbol{\beta} = \mathbf{0}$, where $\mathbf{A}$ is $q \times p$ of rank q, to obtain a set of confidence intervals. Then $\mathbf{A}\boldsymbol{\beta} = \mathbf{A}(\mathbf{X}'\mathbf{X})^{-1}\mathbf{X}'\boldsymbol{\theta} = \mathbf{A}_1\boldsymbol{\theta}$, say, and $\mathbf{A}\hat{\boldsymbol{\beta}} = \mathbf{A}(\mathbf{X}'\mathbf{X})^{-1}\mathbf{X}'\hat{\boldsymbol{\theta}} = \mathbf{A}_1\hat{\boldsymbol{\theta}}$. Also Var$[\mathbf{A}\hat{\boldsymbol{\beta}}] = \sigma^2\mathbf{A}(\mathbf{X}'\mathbf{X})^{-1}\mathbf{A}' = \sigma^2\mathbf{B}$, say. Hence, from (4.16)

$$
(\mathbf{A}\hat{\boldsymbol{\beta}} - \mathbf{A}\boldsymbol{\beta})'\mathbf{B}^{-1}(\mathbf{A}\hat{\boldsymbol{\beta}} - \mathbf{A}\boldsymbol{\beta})/(qs^2) \sim F_{q,n-p}.
$$

Setting $\hat{\boldsymbol{\eta}} = \mathbf{A}\hat{\boldsymbol{\beta}}$ and $\boldsymbol{\eta} = \mathbf{A}\boldsymbol{\beta}$, we have

$$
\begin{aligned}
1-\alpha &= \Pr[(\hat{\boldsymbol{\eta}} - \boldsymbol{\eta})'\mathbf{B}^{-1}(\hat{\boldsymbol{\eta}} - \boldsymbol{\eta}) \le m] \\
&= \Pr\left[\frac{|\mathbf{h}'\hat{\boldsymbol{\eta}} - \mathbf{h}'\boldsymbol{\eta}|}{s(\mathbf{h}'\mathbf{B}\mathbf{h})^{1/2}} \le (qF_{q,n-p}(\alpha))^{1/2}, \quad \text{for all } \mathbf{h}\ (\mathbf{h}\neq\mathbf{0})\right],
\end{aligned}
$$

and we end up with a confidence interval for $\mathbf{h}'\boldsymbol{\eta}$, namely

$$
\mathbf{h}'\hat{\boldsymbol{\eta}} \pm (qF_{q,n-p}(\alpha))^{1/2} s(\mathbf{h}'\mathbf{B}\mathbf{h})^{1/2}.
$$

If we set $\mathbf{h}'\boldsymbol{\eta} = \eta_j$ we include intervals for every $\eta_j = \mathbf{a}_j'\boldsymbol{\beta}$, where $\mathbf{a}_j'$ is the jth row of $\mathbf{A}$, namely

$$
\mathbf{a}_j'\hat{\boldsymbol{\beta}} \pm (F_{q,n-p}(\alpha))^{1/2}\hat{\sigma}_{\mathbf{a}_j'\hat{\boldsymbol{\beta}}}.
$$

If we set $\mathbf{A} = \mathbf{I}_p$, $\mathbf{a}_j'\boldsymbol{\beta}$ is the jth element of $\boldsymbol{\beta}$, thus giving us a set of confidence intervals for the β_j, $(j = 0, 1, 2, \ldots, p-1)$.

Other confidence intervals can also be obtained. For example, we can use the p Bonferroni intervals $\widehat{\beta}_j \pm s t_{n-p}(\alpha/(2p)) d_{jj}$, where d_{jj} is the $(j+1)$th diagonal element of $(\mathbf{X}'\mathbf{X})^{-1}$. We can also use maximum-modulus t-intervals. For further details see Seber and Lee (2003, chapter 5).

References

Scheffé, H. (1959). *The analysis of variance*. New York: Wiley.

Seber, G. A. F., & Lee, A. J. (2003). *Linear regression analysis* (2nd ed.). New York: Wiley.

Wald, A. (1943). Tests of statistical hypotheses concerning several parameters when the number of observations is large. *Transactions of the American Mathematical Society, 54*, 426–482.

Chapter 5
Inference Properties

5.1 Power of the F-Test

We assume the model $\mathbf{y} = \boldsymbol{\theta} + \boldsymbol{\varepsilon}$, $G : \boldsymbol{\theta} \in \Omega$, a p-dimensional vector space in $\mathbb{R}^n$, and $H : \boldsymbol{\theta} \in \omega$, a $p - q$ dimensional subspace of Ω; $\boldsymbol{\varepsilon}$ is $N_n[\mathbf{0}, \sigma^2 \mathbf{I}_n]$. To test H we choose a region W called the *critical region* and we reject H if and only if $\mathbf{y} \in W$. The *power* of the test $\beta(W, \boldsymbol{\theta})$ is defined to be probability of rejecting H when $\boldsymbol{\theta}$ is the true value of E[$\mathbf{y}$]. Thus,

$$\beta(W, \boldsymbol{\theta}) = \Pr[\mathbf{y} \in W | \boldsymbol{\theta}]$$

and is a function of W and $\boldsymbol{\theta}$. The *size* of a critical region W is $\sup_{\boldsymbol{\theta} \in W} \beta(W, \boldsymbol{\theta})$, and if $\beta(W, \boldsymbol{\theta}) = \alpha$ for all $\boldsymbol{\theta} \in \omega$, then W is said to be a *similar* region of size α. If W is of size α and $\beta(W, \boldsymbol{\theta}) \geq \alpha$ for every $\boldsymbol{\theta} \in \Omega - \omega$ (the set of all points in Ω which are not in ω), then W is said to be *unbiased*. In particular, if we have the strict inequality $\beta(W, \boldsymbol{\theta}) > \alpha$ for $\boldsymbol{\theta} \in \Omega - \omega$, then W is said to be *consistent*. Finally we define W to be a uniformly most powerful (UMP) critical region of a given class C if $W \in C$ and if, for any $W' \in C$ and all $\boldsymbol{\theta} \in \Omega - \omega$,

$$\beta(W, \boldsymbol{\theta}) \geq \beta(W', \boldsymbol{\theta}).$$

Obviously a wide choice of W is possible for testing H, and so we would endeavor to choose a critical region which has some, or if possible, all of the desired properties mentioned above, namely similarity, unbiasedness or consistency, and providing a UMP test for certain reasonable classes of critical regions. Other criteria such as invariance are also used (Lehmann and Romano 2005). The F-test for H, given by

$$F = \frac{f_2}{f_1} \frac{\mathbf{y}'(\mathbf{P}_\Omega - \mathbf{P}_\omega)\mathbf{y}}{\mathbf{y}'(\mathbf{I}_n - \mathbf{P}_\Omega)\mathbf{y}},$$

G.A.F. Seber, *The Linear Model and Hypothesis*, Springer Series in Statistics,
DOI 10.1007/978-3-319-21930-1_5

where $f_1 = q$ and $f_2 = n - p$, provides such a critical region W_0, say, and we now consider some properties of W_0.

We note first of all that W_0 is the set of vectors $\mathbf{y}$ such that $F > F_\alpha$, where

$$\Pr[F > F_\alpha | \boldsymbol{\theta} \in \omega] = \alpha,$$

so that W_0 is a similar region of size α. The similarity property holds because F is distributed as F_{f_1,f_2} when H is true, and it therefore doesn't depend on $\boldsymbol{\theta}$ when $\boldsymbol{\theta} \in \omega$. The power of W_0 depends on $\boldsymbol{\theta}$ through the non-centrality parameter $\delta = \boldsymbol{\theta}'(\mathbf{P}_\Omega - \mathbf{P}_\omega)\boldsymbol{\theta}/\sigma^2$ and is therefore a function of δ and W_0, say $\beta(W_0, \delta)$. Also

$$\beta(W_0, \delta) = \Pr[v > v_\alpha = f_1 F_\alpha / (f_1 F_\alpha + f_2)],$$

where $v = f_1 F/(f_1 F + f_2)$ has a non-central Beta distribution (cf. Sect. 1.8). It is known that $\beta(W_0, \delta)$ can be increased by (a) decreasing f_1 keeping f_2 and δ fixed, (b) increasing f_2 keeping f_1 and δ fixed, or (c) increasing δ keeping f_1 and f_2 fixed. Now since $\delta = 0$ if and only if $\boldsymbol{\theta} \in \omega$, and $\beta(W_0, \delta)$ is a monotonic strictly increasing function of δ, then $\beta(W_0, \delta) > \beta(W_0, 0) = \alpha$ when $\boldsymbol{\theta} \in \Omega - \omega$, and W_0 is consistent.

It is known that W_0 has a number of optimal properties. However we shall only consider one due to Saw (1964) as it demonstrates the geometric approach used in this book. The result is stated as a theorem.

Theorem 5.1 *W_0 is UMP among the class C of all consistent, variance-ratio type tests. (A variance ratio test is a test of the form $s\mathbf{y}'\mathbf{A}_r\mathbf{y}/r\mathbf{y}'\mathbf{B}_s\mathbf{y}$ where the numerator and denominator sums of squares (SS) are independently distributed as σ^2 times a non-central chi-square distribution with r and s degrees of freedom respectively. Also the non-centrality parameter $\boldsymbol{\theta}'\mathbf{B}_s\boldsymbol{\theta}/\sigma^2$ for the denominator SS, $\mathbf{y}'\mathbf{B}_s\mathbf{y}$, is zero when $\boldsymbol{\theta} \in \Omega$ and the non-centrality parameter $\boldsymbol{\theta}'\mathbf{A}_r\boldsymbol{\theta}/\sigma^2$ for the numerator SS is zero when $\boldsymbol{\theta} \in \omega$.)*

Proof For a consistent test we must have $\boldsymbol{\theta}'\mathbf{A}_r\boldsymbol{\theta} > 0$ when $\boldsymbol{\theta} \in \Omega - \omega$. The quadratic $\mathbf{y}'\mathbf{B}_s\mathbf{y}$ is distributed as $\sigma^2\chi^2_s$ if and only if $\mathbf{B}_s$ is symmetric and idempotent (Theorems 1.10 and 1.11 in Sect. 1.9), and therefore $\mathbf{B}_s$ represents an orthogonal projection of $\mathbb{R}^n$ on some vector space $\mathcal{B}_s$ of dimension s. The non-centrality parameter $\boldsymbol{\theta}'\mathbf{B}_s\boldsymbol{\theta}/\sigma^2 = (\mathbf{B}_s\boldsymbol{\theta})'(\mathbf{B}_s\boldsymbol{\theta})/\sigma^2$ is, in units of $1/\sigma^2$, the square of the distance from the origin to the projection of $\boldsymbol{\theta}$ on $\mathcal{B}_s$. If this is to be zero for every $\boldsymbol{\theta} \in \Omega$, then $\mathcal{B}_s \perp \Omega$ so that $s \leq n-p$. When $s = n-p$ there is a unique vector space $\mathcal{B}_s = \Omega^\perp$, so that $\mathbf{y}'(\mathbf{I}_n - \mathbf{P}_\Omega)\mathbf{y}$ is the unique (error) SS with maximum degrees of freedom $n-p$. If $\mathbf{y}'\mathbf{A}_r\mathbf{y}$ is $\sigma^2\chi^2_r$, then $\mathbf{A}_r$ represents an orthogonal projection on some vector space $\mathcal{A}_r$ of dimension r. Since $\mathbf{y}'\mathbf{A}_r\mathbf{y}$ and $\mathbf{y}'\mathbf{B}_s\mathbf{y}$ are statistically independent then $\mathbf{A}_r\mathbf{B}_s = \mathbf{0}$ (Theorem 1.12) or geometrically $\mathcal{A}_r \perp \mathcal{B}_s$. As $\mathbf{y}'\mathbf{A}_r\mathbf{y}$ has zero non-centrality parameter when $\boldsymbol{\theta} \in \omega$, then $\mathcal{A}_r \perp \omega$, that is, $\mathcal{A}_r \subset \omega^\perp \cap \mathcal{B}_s^\perp$. Now, by Theorem 1.3 in Sect. 1.4 with $\mathcal{V}_0 = \omega^\perp$, $\mathcal{V}_1 = \Omega^\perp$, and $\mathcal{V}_2 = \Omega$,

$$\omega^\perp = (\omega^\perp \cap \Omega) \oplus \Omega^\perp.$$

Therefore using $\Omega \subset \mathcal{B}_s^{\perp}$ and Theorem 1.3 again with $\mathcal{V}_0 = \mathcal{B}_s^{\perp}$, $\mathcal{V}_1 = \omega^{\perp} \cap \Omega$, and $\mathcal{V}_2 = \Omega^{\perp}$ we have

$$\begin{aligned}\omega^{\perp} \cap \mathcal{B}_s^{\perp} &= [(\omega^{\perp} \cap \Omega) \oplus \Omega^{\perp}] \cap \mathcal{B}_s^{\perp} \\ &= (\omega^{\perp} \cap \Omega) \oplus (\Omega^{\perp} \cap \mathcal{B}_s^{\perp}) = \mathcal{D},\end{aligned}$$

where $\mathcal{D}$ is the sum of two orthogonal vector spaces. We now show that for a consistent test, $r \geq \dim[\omega^{\perp} \cap \Omega] = q$.

Let $\mathcal{A}_r^*$ be the orthogonal projection of $\mathcal{A}_r$ onto $\omega^{\perp} \cap \Omega$ so that since $\mathcal{A}_r \subset \omega^{\perp} \cap \mathcal{B}_s^{\perp} = \mathcal{D}$, we have $\mathcal{A}_r \subset \mathcal{A}_r^* \oplus (\Omega^{\perp} \cap \mathcal{B}_s^{\perp})$. Suppose $r < q$ so that $\dim[\mathcal{A}_r^*] < q$ and $\mathcal{A}_r^*$ is a proper subset of $\omega^{\perp} \cap \Omega$. Then there exists $\boldsymbol{\theta} \in \omega^{\perp} \cap \Omega$ such that $\boldsymbol{\theta} \perp \mathcal{A}_r^*$. Since $\boldsymbol{\theta} \perp \Omega^{\perp} \cap \mathcal{B}_s^{\perp}$, $\boldsymbol{\theta} \perp \mathcal{A}_r$, and $\mathbf{A}_r\boldsymbol{\theta} = \mathbf{0}$. Hence there exists $\boldsymbol{\theta} \in \Omega - \omega$ such that $\boldsymbol{\theta}'\mathbf{A}_r\boldsymbol{\theta} = 0$. This contradicts the requirement of consistency, so that $r \geq q$.

If $\mathcal{B}_s = \Omega^{\perp}$, then for a consistent test we must have $\mathcal{A}_r = \omega \cap \Omega^{\perp}$, and $\mathbf{y}'\mathbf{A}_r\mathbf{y}$ $(= \mathbf{y}'(\mathbf{P}_\Omega - \mathbf{P}_\omega)\mathbf{y})$ is the unique hypothesis sum of squares with minimum degrees of freedom q. For a general variance-ratio test, however, when $\mathcal{B}_s \subset \Omega^{\perp}$, we have $\mathcal{A}_r \neq \omega^{\perp} \cap \Omega$.

We now focus our attention on $\sigma^2\delta$, where δ is the non-centrality parameter. Since $\mathcal{A}_r \subset \mathcal{D}$, $\mathbf{P}_{\mathcal{D}}\mathbf{A}_r = \mathbf{A}_r$, where $\mathbf{P}_{\mathcal{D}}$ is the projection matrix on $\mathcal{D}$. Suppose $\boldsymbol{\theta} \in \Omega - \omega$, then $\boldsymbol{\theta} = \boldsymbol{\theta}_1 + \boldsymbol{\theta}_2$, where $\boldsymbol{\theta}_1 \in \omega$, $\boldsymbol{\theta}_2 \in \omega^{\perp} \cap \Omega$, and $\boldsymbol{\theta}_2 \neq \mathbf{0}$. Now $\boldsymbol{\theta}_1 \perp \mathcal{D}$ and, since $\mathbf{0} \in \mathcal{B}_s^{\perp} \cap \Omega^{\perp}$, $\boldsymbol{\theta}_2 + \mathbf{0} \in \mathcal{D}$. Hence by Theorem 4.2 in Sect. 4.2, $(\mathbf{P}_\Omega - \mathbf{P}_\omega)\boldsymbol{\theta} = (\mathbf{P}_\Omega - \mathbf{P}_\omega)\boldsymbol{\theta}_2 = \boldsymbol{\theta}_2 = \mathbf{P}_{\mathcal{D}}\boldsymbol{\theta}_2 = \mathbf{P}_{\mathcal{D}}\boldsymbol{\theta}$. Thus

$$\begin{aligned}\| (\mathbf{P}_\Omega - \mathbf{P}_\omega)\boldsymbol{\theta} \|^2 &= \| \mathbf{P}_{\mathcal{D}}\boldsymbol{\theta} \|^2 \\ &= \| \mathbf{P}_{\mathcal{D}}\mathbf{A}_r\boldsymbol{\theta} \|^2 + \| \mathbf{P}_{\mathcal{D}}(\mathbf{I}_n - \mathbf{A}_r)\boldsymbol{\theta} \|^2 \\ &\geq \| \mathbf{P}_{\mathcal{D}}\mathbf{A}_r\boldsymbol{\theta} \|^2 = \| \mathbf{A}_r\boldsymbol{\theta} \|^2,\end{aligned}$$

or

$$\boldsymbol{\theta}'(\mathbf{P}_\Omega - \mathbf{P}_\omega)\boldsymbol{\theta} \geq \boldsymbol{\theta}'\mathbf{A}_r\boldsymbol{\theta}$$

with strict equality occurring for every $\boldsymbol{\theta} \in \Omega - \omega$ if and only if $\mathcal{A}_r \supset \omega^{\perp} \cap \Omega$ (since $\mathbf{A}_r\boldsymbol{\theta}_2 = \boldsymbol{\theta}_2$).

It has been shown that (1) $r \geq q$ (ii) $n - p \geq s$, (iii) the F-test is the unique consistent variance-ratio test with $r = q$ and $s = n - p$, and (iv) the F-test has a non-centrality parameter as large as that of any other variance-ratio test and that if there is a different test with the same non-centrality parameter, then $\mathcal{A}_r \supset \omega^{\perp} \cap \Omega$ (with strict inclusion) and $r > q$. By virtue of the remarks made above prior to the theorem statement about the power being monotonic increasing with respect to δ and s and monotonic decreasing with respect to r, it follows that $\beta(W_0, \boldsymbol{\theta}) \geq \beta(W, \boldsymbol{\theta})$ for every $\boldsymbol{\theta} \in \Omega - \omega$ and every $W \subset C$, with equality if and only if $W = W_0$.

5.2 Robustness of the F-Test and Non-normality

Although optimality properties of the F-test are of theoretical interest, what is important is the degree of robustness the test has with regard to departures from the underlying assumptions of the test. These assumptions are spelt out in Sect. 3.7 along with mention of some diagnostic tools for detecting departures from them. We now examine the effect of various departures on the validity of the F-test. We first begin with the assumption of normality and the effect of some departures from it are described in the following two theorems from Atiqullah (1962) that make the following assumptions. Let the y_i be independent random variables with means θ_i $(i = 1, 2, \ldots, n)$, with common variance σ^2, and common third and fourth moments μ_3 and μ_4 respectively about their means. Let $\gamma_2 = (\mu_4 - 3\sigma^4)/\sigma^4$ be their common kurtosis.

Theorem 5.2 *Let* $\mathbf{P}_i$ $(i = 1, 2)$ *be a symmetric idempotent matrix of rank* f_i *such that* $E[\mathbf{y}'\mathbf{P}_i\mathbf{y}] = \sigma^2 f_i$*, and let* $\mathbf{P}_1\mathbf{P}_2 = \mathbf{0}$*. If* $\mathbf{p}_i$ *is the column vector of the diagonal elements of* $\mathbf{P}_i$*, then:*

(i) $\text{var}[\mathbf{y}'\mathbf{P}_i\mathbf{y}] = 2\sigma^4(f_i + \frac{1}{2}\gamma_2\mathbf{p}_i'\mathbf{p}_i)$.
(ii) $\text{cov}[\mathbf{y}'\mathbf{P}_1\mathbf{y}, \mathbf{y}'\mathbf{P}_2\mathbf{y}] = \sigma^4\gamma_2\mathbf{p}_1'\mathbf{p}_2$.

Proof

(i) Since $\mathbf{P}_i$ is symmetric and idempotent, $\text{trace}[\mathbf{P}_i] = \text{rank}[\mathbf{P}_i] = f_i$ (by Theorem 1.4 in Sect. 1.5). Also, by Theorem 1.8(iii) in Sect. 1.9,

$$E[\mathbf{y}'\mathbf{P}_i\mathbf{y}] = \sigma^2\,\text{trace}[\mathbf{P}_i] + \boldsymbol{\theta}'\mathbf{P}_i\boldsymbol{\theta} = \sigma^2 f_i,$$

so that $\boldsymbol{\theta}'\mathbf{P}_i^2\boldsymbol{\theta} = \boldsymbol{\theta}'\mathbf{P}_i\boldsymbol{\theta} = 0$ for all $\boldsymbol{\theta}$; that is $\mathbf{P}_i\boldsymbol{\theta} = \mathbf{0}$ for all $\boldsymbol{\theta}$. Therefore substituting $\mathbf{A} = \mathbf{P}_i$ and $\boldsymbol{\mu} = \boldsymbol{\theta}$ in Theorem 1.9(ii) of Sect. 1.9, we have

$$\begin{aligned}\text{var}[\mathbf{y}'\mathbf{P}_i\mathbf{y}] &= 2\sigma^4\,\text{trace}[\mathbf{P}_i^2] + (\mu_4 - 3\sigma^4)\mathbf{p}_i'\mathbf{p}_i \\ &= 2\sigma^4\{\text{trace}[\mathbf{P}_i] + \frac{1}{2}\gamma_2\mathbf{p}_i'\mathbf{p}_i\} \\ &= 2\sigma^4(f_i + \frac{1}{2}\gamma_2\mathbf{p}_i'\mathbf{p}_i).\end{aligned}$$

(ii) Given $\mathbf{P}_1\mathbf{P}_2 = \mathbf{0}$, we have

$$\begin{aligned}(\mathbf{P}_1 + \mathbf{P}_2)^2 &= \mathbf{P}_1^2 + \mathbf{P}_1\mathbf{P}_2 + \mathbf{P}_2\mathbf{P}_1 + \mathbf{P}_2^2 \\ &= \mathbf{P}_1 + \mathbf{P}_1\mathbf{P}_2 + (\mathbf{P}_1\mathbf{P}_2)' + \mathbf{P}_2 \\ &= \mathbf{P}_1 + \mathbf{P}_2.\end{aligned}$$

Therefore $\mathbf{P}_1 + \mathbf{P}_2$ is idempotent and, by (i),

$$\begin{aligned}
\text{var}[\mathbf{y}'\mathbf{P}_1\mathbf{y} + \mathbf{y}'\mathbf{P}_2\mathbf{y}] &= \text{var}[\mathbf{y}'(\mathbf{P}_1 + \mathbf{P}_2)\mathbf{y}] \\
&= 2\sigma^4\{\text{trace}[\mathbf{P}_1 + \mathbf{P}_2] + \frac{1}{2}\gamma_2(\mathbf{p}_1 + \mathbf{p}_2)'(\mathbf{p}_1 + \mathbf{p}_2)\} \\
&= 2\sigma^4\{f_1 + f_2 + \frac{1}{2}\gamma_2(\mathbf{p}_1'\mathbf{p}_1 + 2\mathbf{p}_1'\mathbf{p}_2 + \mathbf{p}_2'\mathbf{p}_2)\} \\
&= \text{var}[\mathbf{y}'\mathbf{P}_1\mathbf{y}] + \text{var}[\mathbf{y}'\mathbf{P}_2\mathbf{y}] + 2\sigma^4\gamma_2\mathbf{p}_1'\mathbf{p}_2.
\end{aligned}$$

Hence

$$\begin{aligned}
\text{cov}[\mathbf{y}'\mathbf{P}_1\mathbf{y}, \mathbf{y}'\mathbf{P}_2\mathbf{y}] &= \frac{1}{2}\{\text{var}[\mathbf{y}'(\mathbf{P}_1 + \mathbf{P}_2)\mathbf{y}] - \text{var}[\mathbf{y}'\mathbf{P}_1\mathbf{y}] - \text{var}[\mathbf{y}'\mathbf{P}_2\mathbf{y}]\} \\
&= \sigma^4\gamma_2\mathbf{p}_1'\mathbf{p}_2.
\end{aligned}$$

Theorem 5.3 *Suppose that* $\mathbf{P}_1$ *and* $\mathbf{P}_2$ *satisfy the conditions of Theorem 5.2 above. Let* $Z = \frac{1}{2}\log F$, *where*

$$F = \frac{\mathbf{y}'\mathbf{P}_1\mathbf{y}/f_1}{\mathbf{y}'\mathbf{P}_2\mathbf{y}/f_2} \quad \left(= \frac{s_1^2}{s_2^2}, \text{ say} \right),$$

Then, for large f_1 *and* f_2 *we have asymptotically*

$$\begin{aligned}
E[Z] &\approx \tfrac{1}{2}\,(f_2^{-1} - f_1^{-1}) \\
&\quad \times [1 + \frac{1}{2}\gamma_2(f_1\mathbf{p}_2 - f_2\mathbf{p}_1)'(f_1\mathbf{p}_2 + f_2\mathbf{p}_1)\{f_1 f_2(f_1 - f_2)\}^{-1}], \qquad (5.1)
\end{aligned}$$

and

$$\text{var}[Z] \approx \frac{1}{2}(f_2^{-1} + f_1^{-1})[1 + \frac{1}{2}\gamma_2(f_1\mathbf{p}_2 - f_2\mathbf{p}_1)'(f_1\mathbf{p}_2 - f_2\mathbf{p}_1)\{f_1 f_2(f_1 + f_2)\}^{-1}]. \qquad (5.2)$$

Proof Using a Taylor expansion of $\log s_i^2$ about $\log\sigma^2$, we have

$$\log s_i^2 \approx \log\sigma^2 + \frac{s_i^2 - \sigma^2}{\sigma^2} - \frac{(s_i^2 - \sigma^2)^2}{2\sigma^4}. \qquad (5.3)$$

Taking expected values, and using $E[s_i^2] = \sigma^2$, we have

$$E[\log s_i^2] \approx \log\sigma^2 - \frac{1}{2\sigma^4}\text{var}[s_i^2],$$

where, from Theorem 5.2,

$$\mathrm{var}[s_i^2] = \frac{\mathrm{var}[\mathbf{y}'\mathbf{P}_i\mathbf{y}]}{f_i^2} = 2\sigma^4(f_i^{-1} + \frac{1}{2}\gamma_2 f_i^{-2}\mathbf{p}_i'\mathbf{p}_i).$$

Substituting in

$$E[Z] = \frac{1}{2}(E[\log s_1^2] - E[\log s_2^2])$$

leads to (5.1).

To find an asymptotic expression for var[Z], we first note that

$$\mathrm{var}[Z] = \frac{1}{4}\{\mathrm{var}[\log s_1^2] + \mathrm{var}[\log s_2^2] - 2\mathrm{cov}[\log s_1^2, \log s_2^2]\}. \tag{5.4}$$

Then ignoring the third term of (5.3), we have $E[\log s_i^s] \sim \log \sigma^2$ and

$$\begin{aligned}\mathrm{var}[\log s_i^2] &\approx E[(\log s_i^2 - \log\sigma^2)^2] \\ &\approx \frac{E[(s_i^2 - \sigma^2)^2]}{\sigma^4} \\ &= \frac{\mathrm{var}[s_i^2]}{\sigma^4}.\end{aligned}$$

Similarly,

$$\begin{aligned}\mathrm{cov}[\log s_1^2, \log s_2^2] &\approx E[(\log s_1^2 - \log\sigma^2)(\log s_2^2 - \log\sigma^2)] \\ &\approx \frac{E[(s_1^2 - \sigma^2)(s_2^2 - \sigma^2)]}{\sigma^4} \\ &= \frac{\mathrm{cov}[s_1^2, s_2^2]}{\sigma^4}.\end{aligned}$$

Finally, substituting in

$$\mathrm{var}[Z] \approx \frac{1}{4\sigma^4}(\mathrm{var}[s_1^2] + \mathrm{var}[s_2^2] - 2\mathrm{cov}[s_1^2, s_2^2])$$

and using Theorem 5.2 leads to Eq. (5.2). This completes the proof.

We now apply the above theory to the F-test of $H : \boldsymbol{\theta} \in \omega$ given $G : \boldsymbol{\theta} \in \Omega$. This test is given by

$$F = \frac{\mathbf{y}'(\mathbf{P}_\Omega - \mathbf{P}_\omega)\mathbf{y}/q}{\mathbf{y}'(\mathbf{I}_n - \mathbf{P}_\Omega)\mathbf{y}/(n-p)}$$

$$= \frac{\mathbf{y}'\mathbf{P}_1\mathbf{y}/f_1}{\mathbf{y}'\mathbf{P}_2\mathbf{y}/f_2}$$

$$= \frac{s_1^2}{s_2^2} \text{ say,}$$

where $\mathbf{P}_1\mathbf{P}_2 = (\mathbf{P}_\Omega - \mathbf{P}_\omega)(\mathbf{I}_n - \mathbf{P}_\Omega) = -\mathbf{P}_\omega + \mathbf{P}_\omega\mathbf{P}_\Omega = \mathbf{0}$. We now relax the assumptions underlying F and assume only that the $\varepsilon_i = y_i - \theta_i$ are independently and identically distributed with mean zero and variance σ^2; i.e., $\mathrm{E}[\boldsymbol{\varepsilon}] = \mathbf{0}$ and $\mathrm{Var}[\boldsymbol{\varepsilon}] = \sigma^2\mathbf{I}_n$. We note that $E[s_2^2] = \sigma^2$ (from (3.4) in Sect. 3.6) and, from Theorem 1.8(i) in Sect. 1.9,

$$\begin{aligned} E[qs_1^2] &= E[\mathbf{y}'(\mathbf{P}_\Omega - \mathbf{P}_\omega)\mathbf{y}] \\ &= \sigma^2(\mathrm{trace}[\mathbf{P}_\Omega] - \mathrm{trace}[\mathbf{P}_\omega]) + \boldsymbol{\theta}'(\mathbf{P}_\Omega - \mathbf{P}_\omega)\boldsymbol{\theta} \\ &= \sigma^2[p - (p-q)] = \sigma^2 q, \end{aligned}$$

when H is true as $\mathbf{P}_\Omega\boldsymbol{\theta} - \mathbf{P}_\omega\theta = \boldsymbol{\theta} - \boldsymbol{\theta} = \mathbf{0}$. Thus the conditions of Theorem 5.3 are satisfied. It is known that when the ε_i are normally distributed and f_1 and f_2 are large, $Z = \frac{1}{2}\log F$ is approximately normally distributed with mean and variance given by (5.1) and (5.2), but with $\gamma_2 = 0$. This approximation is evidently quite good even when f_1 and f_2 are as small as four so that it is not unreasonable to accept the proposition that Z is still approximately normal for a moderate amount of non-normality with mean and variance given approximately by (5.1) and (5.2). On this assumption, Z and therefore F will be approximately independent of γ_2 if the coefficient of γ_2 in (5.1) and (5.2) is zero; that is if

$$f_1\mathbf{p}_2 = f_2\mathbf{p}_1. \tag{5.5}$$

Now using Atiqullah's (1962) terminology, we say that F is quadratically balanced if the diagonal elements of $\mathbf{P}_i$ ($i = 1, 2$) are equal, that is if the diagonal elements of $\mathbf{P}_\omega$ are equal and those of $\mathbf{P}_\Omega$ are equal; most of the usual F-tests for balanced experimental designs belong to this category. In this case, since $\mathrm{trace}[\mathbf{P}_i] = f_i$, we have

$$\mathbf{p}_i = \frac{f_i}{n}\mathbf{1}_n \quad \text{and} \quad f_1\mathbf{p}_2 = \frac{f_1f_2}{n}\mathbf{1}_n = f_2\mathbf{p}_1.$$

Thus a sufficient condition for (5.5) to hold is that F is quadratically balanced.

Example 5.1 We revisit Example 2.3 in Sect. 2.2 where we compare I Normal populations. Let y_{ij} be the jth observation ($j = 1, 2, \ldots, J_i$) on the ith population $N[\mu_i, \sigma^2]$ ($i = 1, 2, \ldots, I$), and let $n = \sum_i J_i$. This gives us the model

$$y_{ij} = \theta_{ij} + \varepsilon_{ij} \quad (i = 1, 2, \ldots, I; j = 1, 2, \ldots, J_i)$$

with $G : \theta_{ij} = \mu_i$ for all i, j. Then setting

$$\mathbf{y} = (y_{11}, y_{12} \ldots, y_{1J_1}, y_{21}, y_{22}, \ldots, y_{2J_2}, \ldots, y_{I1}, y_{I2}, \ldots, y_{IJ_I})'$$

with $\boldsymbol{\theta}$ and $\boldsymbol{\varepsilon}$ similarly defined, we have $\mathbf{y} = \boldsymbol{\theta} + \boldsymbol{\varepsilon}$, where $\boldsymbol{\varepsilon} \sim N_n[\mathbf{0}, \sigma^2 \mathbf{I}_n]$. We have $G : \theta_{ij} = \mu_i$ and we wish to test $H : \mu_1 = \mu_2 = \cdots = \mu_I$ (= μ, say). The least squares estimate $\hat{\mu}_i$ of μ_i is obtained by minimizing $\sum_i \sum_j (y_{ij} - \mu_i)^2$ with respect to μ_i, namely $\hat{\mu}_i = \sum_{j=1}^{J_i} y_{ij}/J_i = \overline{y}_{i\cdot}$ say, and the residual sum of squares $(\mathbf{y}'(\mathbf{I}_n - \mathbf{P}_\Omega)\mathbf{y})$ is

$$RSS = \sum_i \sum_j (y_{ij} - \hat{\mu}_i)^2 = \sum_i \sum_j (y_{ij} - \overline{y}_{i\cdot})^2. \tag{5.6}$$

Similarly, under H we minimize $\sum_i \sum_j (y_{ij} - \mu)^2$ with respect to μ giving

$$\hat{\mu}_H = \sum_i \sum_j y_{ij}/n = \overline{y}_{\cdot\cdot}$$

and

$$RSS_H = \sum_i \sum_j (y_{ij} - \hat{\mu}_H)^2 = \sum_i \sum_j (y_{ij} - \overline{y}_{\cdot\cdot})^2.$$

Using the matrix approach, we have from (2.1) that RSS has $n - I$ degrees of freedom (since rank$[\mathbf{X}] = I$), and $RSS_H = \mathbf{y}'(\mathbf{I}_n - \mathbf{P}_\omega)\mathbf{y}$ has $n - 1$ degrees of freedom (since rank$[\mathbf{X}_H]$ = rank$[\mathbf{1}_n] = 1$). The F-test of H is now

$$F = \frac{(RSS_H - RSS)/(I - 1)}{RSS/(n - I)},$$

which has an $F_{I-1,n-I}$ distribution when H is true. Alternative parameterizations have been used for this model and the reader is referred to Seber and Lee (2003, section 8.2.1). We note from (4.6) that

$$Q_H - Q = RSS_H - RSS = \| \hat{\boldsymbol{\theta}} - \hat{\boldsymbol{\theta}}_H \|^2$$

which for the above example gives us

$$\sum_i \sum_j (\hat{\mu}_i - \hat{\mu}_H)^2 = \sum_i J_i (\overline{y}_{i\cdot} - \overline{y}_{\cdot\cdot})^2. \tag{5.7}$$

In order for F to be quadratically balanced we require the coefficients of the y_{ij}^2 terms to be all equal for each of RSS_H and RSS. Now expanding RSS,

$$RSS = \sum_I \sum_j y_{ij}^2 - \sum_i \frac{y_{i\cdot}^2}{J_i},$$

where $y_{i\cdot} = \sum_j y_{ij}$ so that for quadratic balance we must have all the J_i equal (to J, say). In this case

$$Q_H = \sum_i \sum_j (y_{ij} - \bar{y}_{\cdot\cdot})^2 = \sum_i \sum_j y_{ij}^2 - (\sum_i \sum_j y_{ij})^2/IJ,$$

which will also have equal diagonal elements for $\mathbf{P}_\omega$.

5.3 Unequal Variances

One of our assumptions is that the y_i all have the same variance. We now allow the variances to vary and consider by way of illustration Example 5.1 in the previous section. We assume that for each $i = 1, 2, \ldots, I$, $y_{i1}, y_{i2}, \ldots, y_{iJ_i}$ is a random sample from a population with mean μ_i, variance σ_i^2 and kurtosis γ_{i2} $(= \mu_{i4}/\sigma_i^4 - 3)$. Assuming normality of the observations and equal variances, the F-ratio that we can use for testing the hypothesis H that the μ_i's are all equal can be expressed in the form

$$F = \frac{n-p}{q} \cdot \frac{\boldsymbol{\varepsilon}'(\mathbf{P}_\Omega - \mathbf{P}_\omega)\boldsymbol{\varepsilon}}{\boldsymbol{\varepsilon}'(\mathbf{I}_n - \mathbf{P}_\Omega)\boldsymbol{\varepsilon}}, \tag{5.8}$$

which has an $F_{q,n-p}$ distribution with $q = I - 1$ and $n - p = \sum_i (J_i - 1)$. To actually carry out the test we replace $\boldsymbol{\varepsilon}$ by $\mathbf{y}$ and then F has an F-distribution when H is true. However (5.8) is useful for examining the effects of non-normality and unequal variances. We note that because $\mathbf{P}_\Omega \boldsymbol{\theta} = \boldsymbol{\theta}$, the denominator of (5.8) is $(n-p)s^2 = \mathbf{y}'(\mathbf{I}_n - \mathbf{P}_\Omega)\mathbf{y}$. Referring to Example 5.1 and using (5.6) and (5.7) we have from (5.8)

$$F = \sum_{i=1}^{I} J_i (v_i - \bar{v})^2 / [(I-1)s^2], \tag{5.9}$$

where, replacing y_{ij} by $y_{ij} - \mu_i$,

$$v_i = \bar{y}_{i\cdot} - \mu_i, \quad \bar{v} = \sum_i J_i v_i / n,$$

$$s^2 = \sum_i (J_i - 1)s_i^2/(n - I), \quad \text{and}$$

$$s_i^2 = \sum_{j=1}^{J_i} (y_{ij} - \bar{y}_{i\cdot})^2/(J_i - 1).$$

Following Scheffé (1959, 341–342) we now allow the J_i and n to go off to infinity in such a way that J_i/n is fixed. Then $s_i^2 \to \sigma_i^2$, and approximating $J_i - 1$ by J_i and $n - I$ by n in s^2, F is approximately distributed as

$$F_1 = \frac{1}{(I-1)\sigma_w^2} \sum_i J_i(v_i - \overline{v})^2,$$

where $\sigma_w^2 = \sum_i J_i\sigma_i^2/n$, a weighted average of the variances σ_i^2, and

$$\sum_i J_i(v_i - \overline{v})^2 = \sum_i J_i v_i^2 - n\overline{v}^2 = \mathbf{v}'\mathbf{A}\mathbf{v},$$

where $\mathbf{A} = (a_{ij})$ and $a_{ij} = \delta_{ij}J_i - J_iJ_j/n$. The next step is to find the mean and variance of F_1. Now v_i is approximately normal with mean zero and variance σ_i^2/J_i so that

$$\begin{aligned} E[F_1] &= \frac{1}{(I-1)\sigma_w^2}\left\{\sum_i J_i \frac{\sigma_i^2}{J_i} - n\text{var}[\overline{v}]\right\} \\ &= \frac{1}{(I-1)\sigma_w^2}\left\{\sum_i \sigma_i^2 - \sum_i J_i\sigma_i^2/n\right\} \\ &= \frac{1}{(I-1)\sigma_w^2}(I\sigma_u^2 - \sigma_w^2), \end{aligned} \tag{5.10}$$

where σ_u^2 is the unweighted average of the σ_i^2. We can find the variance of F_1 using Theorem 1.9(iii) for Normal data, namely

$$\text{var}[\mathbf{v}'\mathbf{A}\mathbf{v}] = 2\sum_i\sum_j a_{ij}^2 \mu_{2i}\mu_{2j},$$

where $\mu_{2i} = \sigma_i^2/J_i$ and $a_{ij} = \delta_{ij}J_i - J_iJ_j/n$. Hence

$$\begin{aligned} \text{var}[F_1] &= \frac{2}{(I-1)^2(\sigma_w^2)^2}\left\{\sum_i J_i^2\left(1 - \frac{J_i}{n}\right)^2 \frac{\sigma_i^4}{J_i^2} + \sum_i\sum_{j:j\neq i} \frac{J_i^2 J_j^2 \sigma_i^2\sigma_j^2}{n^2 J_i J_j}\right\} \\ &= \frac{2}{(I-1)^2(\sigma_w^2)^2}\left\{\sum_i \sigma_i^4 - 2\sum_i J_i\sigma_i^4/n + (\sum_i J_i\sigma_i^2/n)^2\right\} \\ &= \frac{2}{(I-1)^2(\sigma_w^2)^2}\left\{\sum_i \sigma_i^4 - 2\sum_i J_i\sigma_i^4/n + (\sigma_w^2)^2\right\}. \end{aligned}$$

Now for large n, $F_{I-1,n-I} \rightarrow (I-1)^{-1}\chi^2_{I-1}$, and given normality and equal variances F_1 will be distributed as $(I-1)^{-1}\chi^2_{I-1}$. In general it therefore has an expected value of 1 and variance $2/(I-1)$. We see from (5.10) that the expected value of F_1 will be unity only if $\sigma_u^2 = \sigma_w^2$, that is if all the $\{J_i\}$ are equal. When this happens

$$\text{var}[F_1] = \frac{2}{(I-1)^2(\sigma_u^2)^2}\left\{\sum_i \sigma_i^4 - 2\sum_i \sigma_i^4/I + (\sigma_u^2)^2\right\}.$$

Using $\sum_i \sigma_i^4 - I(\sigma_u^2)^2 = \sum_i(\sigma_i^2 - \sigma_u^2)^2$, we can readily prove that the above expression is equal to

$$\text{var}[F_1] = \frac{2}{I-1}\left(1 + V_u\frac{I-2}{I-1}\right), \tag{5.11}$$

where

$$V_u = \frac{1}{I(\sigma_u^2)^2}\sum_i(\sigma_i^2 - \sigma_u^2)^2.$$

The result (5.11) was proved by Scheffé (1959, 342) using a different approach. He noted that if $I = 2$ or if the $\{\sigma_i^2\}$ are all equal so that $V_u = 0$ in (5.11), then F_1 has the correct variance of $2/(I-1)$; otherwise it is inflated. We conclude that except for comparing just two populations, the F-test can be seriously affected by unequal variances.

References

Atiqullah, M. (1962). The estimation of residual variance in quadratically balanced least squares problems and the robustness of the F test. *Biometrika, 49*, 83–91.

Lehmann, E. L., & Romano, J. P. (2005). *Testing statistical hypotheses* (3rd ed.). New York: Springer.

Saw, J. G. (1964). Some notes on variance-ratio tests of the general linear hypothesis. *Biometrika, 51*, 511–518.

Scheffé, H. (1959). *The analysis of variance*. New York: Wiley.

Seber, G. A. F., & Lee, A. J. (2003). *Linear regression analysis* (2nd ed.). New York: Wiley.

Chapter 6
Testing Several Hypotheses

6.1 The Nested Procedure

Let $\boldsymbol{\theta}$ be an unknown vector parameter, let G be the hypothesis that $\boldsymbol{\theta} \in \Omega$, a p-dimensional vector space in $\mathbb{R}^n$, and assume that $\mathbf{y} \sim N_n[\boldsymbol{\theta}, \sigma^2\mathbf{I}_n]$. Let H_i ($i = 1, 2, \ldots, k$) be the hypothesis that $\boldsymbol{\theta} \in \omega_i$, a $p - q_i$-dimensional subspace of Ω, and denote the joint hypotheses $\boldsymbol{\theta} \in \omega_1 \cap \omega_2$, $\boldsymbol{\theta} \in \omega_1 \cap \omega_2 \cap \omega_3$ etc., by H_{12}, H_{123}, etc. Suppose we wish to test the hypothesis $H_{12\ldots k}$ versus G. Obviously we could test this hypothesis directly, but if it was rejected we would not know why it was rejected and which of the H_i were responsible. What we want is a sequence of tests that tell us how much of $H_{12\ldots k}$ we can accept. One such method is the nested test procedure where we accept $H_{12\ldots k}$ only if the tests of H_1 versus G, H_{12} versus H_1, H_{123} versus $H_{12}, \ldots$, $H_{12\ldots k}$ versus $H_{12\ldots k-1}$ are not significant. The question immediately arises: is such a procedure reasonable, and what sort of power does it have as a test method? If we use the likelihood ratio as our test criterion, we have (Sect. 4.1)

$$\Lambda[H_{12\ldots k}|G] = \Lambda[H_1|G]\,\Lambda[H_{12}|H_1]\cdots\Lambda[H_{12\ldots k}|H_{12\ldots k-1}].$$

Thus if each of the likelihood ratio statistics on the right-hand side is "near" unity then the left-hand side will also be "near" unity. This implies that if each of the nested test statistics is well below its significance level then this nested procedure is "nearly" equivalent to a direct likelihood ratio test of $H_{12\ldots k}$ versus G. As the F-test—and therefore the likelihood ratio test—has good power, this procedure will also have good power. If the nested method led to an "acceptance" of $H_{12\ldots k}$, we could make a final check and carry out a dircct F-test of $H_{12\ldots k}$ versus G.

G.A.F. Seber, *The Linear Model and Hypothesis*, Springer Series in Statistics,
DOI 10.1007/978-3-319-21930-1_6

The appropriate distribution theory for the nested method follows from the orthogonal decomposition

$$\begin{aligned}\mathbf{y} &= (\mathbf{y} - \hat{\boldsymbol{\theta}}_{\Omega}) + (\hat{\boldsymbol{\theta}}_{\Omega} - \hat{\boldsymbol{\theta}}_1) + (\hat{\boldsymbol{\theta}}_1 - \hat{\boldsymbol{\theta}}_{12}) + \cdots + \hat{\boldsymbol{\theta}}_{12\ldots k}\\ &= (\mathbf{I}_n - \mathbf{P}_{\Omega})\mathbf{y} + (\mathbf{P}_{\Omega} - \mathbf{P}_1)\mathbf{y} + (\mathbf{P}_1 - \mathbf{P}_{12})\mathbf{y} + \cdots + \mathbf{P}_{12\ldots k},\end{aligned}$$

where $\mathbf{P}_{12\ldots i} = \mathbf{P}_{\omega_1 \cap \omega_2 \cap \ldots \cap \omega_i}$ and $\hat{\boldsymbol{\theta}}_{12\ldots i} = \mathbf{P}_{12\ldots i}\mathbf{y}$ is the least squares estimate of $\boldsymbol{\theta} \in \omega_1 \cap \omega_2 \cap \ldots \cap \omega_i$ $(i = 1, 2, \ldots, k)$. The orthogonality follows by multiplying the appropriate projection matrices together and using the fact that the product of two projection matrices is equal to the projection matrix projecting onto the smaller subset vector space. For example

$$(\mathbf{P}_{\Omega} - \mathbf{P}_1)(\mathbf{P}_{12} - \mathbf{P}_1) = \mathbf{P}_{\Omega}\mathbf{P}_{12} - \mathbf{P}_{\Omega}\mathbf{P}_1 - \mathbf{P}_1\mathbf{P}_{12} + \mathbf{P}_1^2 = \mathbf{P}_{12} - \mathbf{P}_1 - \mathbf{P}_{12} + \mathbf{P}_1 = \mathbf{0}.$$

Since from the orthogonal decomposition of $\mathbf{y}$ above,

$$\mathbf{I}_n = (\mathbf{I}_n - \mathbf{P}_{\Omega}) + (\mathbf{P}_{\Omega} - \mathbf{P}_1) + (\mathbf{P}_1 - \mathbf{P}_{12}) + \cdots + (\mathbf{P}_{12\ldots k-1} - \mathbf{P}_{12\ldots k}) + \mathbf{P}_{12\ldots k}$$

and the projection matrices in parenthesis are each idempotent representing orthogonal projections onto the mutually orthogonal subspaces $\Omega^{\perp}, \omega_1^{\perp} \cap \Omega, (\omega_1 \cap \omega_2)^{\perp} \cap \omega_1, \ldots,$ and $\omega_{12\ldots k}$, then Cochran's Theorem 4.1 in Sect. 4.1 applies. Hence the quadratics obtained by multiplying the bracketed terms in right-hand side of the above equation on the left by $\mathbf{y}'$, on the right by $\mathbf{y}$, and dividing by σ^2, namely

$$Q/\sigma^2, (Q_1 - Q)/\sigma^2, (Q_{12} - Q_1)/\sigma^2, \ldots, (Q_{12\ldots k} - Q_{12\ldots k-1})/\sigma^2,$$

are all distributed independently as chi-square with $n-p$ and $r_{i-1}-r_i$ $(i = 1, 2, \ldots k)$ degrees of freedom respectively, where $r_i = \dim[\omega_1 \cap \omega_2 \cap \ldots \cap \omega_i]$ and $r_0 = \dim[\Omega] = p$. The distributions are central or non-central depending on which of the H_i are true. Thus the test statistics for the nested method

$$\frac{(n-p)(Q_1 - Q)}{(p - r_1)Q}, \frac{(n - r_1)(Q_{12} - Q_1)}{(r_1 - r_2)Q_1}, \frac{(n - r_2)(Q_{123} - Q_{12})}{(r_2 - r_3)Q_{12}}, \text{ and so forth,}$$

all have F-distributions and the nesting procedure is continued until a significant test is obtained. We notice that the denominator or residual sum of squares (SS) of each test is obtained by pooling the previous numerator and residual SS. For this reason the nested method is essentially one of "pooling non-significant sums of squares."

Example 6.1 The nested procedure can be applied to a set of hypotheses in which there is a natural ordering of the hypotheses. An example of this is found in polynomial regression where our basic underlying model is

$$y_i = \beta_0 + \beta_1 x_i + \beta_2 x_i^2 + \cdots + \beta_{p-1} x_i^{p-1} + \varepsilon_i \quad (i = 1, 2, \ldots, n),$$

and the problem is to estimate p. The first step would be to decide what is the highest value of k necessary for a polynomial approximation of the form

$$y_i = \beta_0 + \beta_1 x_i + \beta_2 x_i^2 + \cdots + \beta_k x_i^k + \varepsilon_i \tag{6.1}$$

to represent an adequate fit to the observations $\mathbf{y}$. We could then apply the nested procedure to the following sequence of hypotheses $H_1 : \beta_k = 0$, $H_{12} : \beta_k = \beta_{k-1} = 0$ etc. and carry on until a significant test is obtained. If the test of $\beta_j = 0$ given that $\beta_k = \beta_{k-1} = \cdots = \beta_{j+1} = 0$, is the first significant test, then j is our estimate of p. Before leaving this example it should be noted that polynomial fitting has some problems. It is known from the Weierstrass approximation theorem that any continuous function on a finite interval can be approximated arbitrarily closely by a polynomial (Davis 1975, chapter VI). We would therefore be tempted to fit a low degree polynomial to a well-behaved curved trend in a scatter plot for the pairs (x_i, y_i). Although the approximation could be improved by increasing the order of the polynomial, the cost is an increase in the number of the β_i and some oscillation between data points. Although it is possible to fit a polynomial of degree up to $n-1$, there are a number of practical difficulties when k is large. In particular, for k greater than about 6, we find that the regression matrix $\mathbf{X}$ associated with (6.1) becomes ill-conditioned, that is becomes close to being less than full rank. For further details about the problem see Seber and Lee (2003, Section 7.1).

When there is no natural ordering of a set of hypotheses, the most thorough procedure would be to test all possible combinations of hypotheses using special computer selection methods. This problem arises in multiple regression where we are given the model

$$y_i = \beta_0 + \beta_1 x_{i1} + \beta_2 x_{i2} + \cdots + \beta_{p-1} x_{i,p-1} + \varepsilon_i$$

and we wish to find out which of the β's can be put equal to zero without giving a significant increase in the residual SS. Obviously the subset of β's selected will not be unique, especially when there are high correlations among the x-variables, and what we require is some criterion for choosing the best subset of β's from the class of admissible subsets. Various methods are available, and these are discussed in detail in Seber and Lee (2003, Chapter 12) for example.

In many situations, especially in analysis of variance applied to experimental designs, the order of nesting is immaterial because of a certain property of the system of hypotheses known as "orthogonality," and a simpler procedure that we describe below is available.

6.2 Orthogonal Hypotheses

One method for testing $H_{12...k}$ versus G would be to accept the hypothesis if we accepted each of the k hypotheses $H_i : \boldsymbol{\theta} \in \omega_i$ $(i = 1, 2, \ldots, k)$ versus G separately. As a first step, we assume that $\sigma^2 = 1$, which arises in large sample tests considered in later chapters. Now the individual test for H_i is (cf. Sect. 4.1 with $v = 1$) $-2\log L[H_i|G] = \mathbf{y}'(\mathbf{P}_\Omega - \mathbf{P}_i)\mathbf{y}$ and the corresponding test statistic for $H_{12...k}$ is $\mathbf{y}'(\mathbf{P}_\Omega - \mathbf{P}_{12...k})\mathbf{y}$. Following Darroch and Silvey (1963), a useful requirement would be to have the individual test statistics independent of one another, and we ask what constraints must be put on the vector spaces $\Omega, \omega_1, \omega_2, \ldots, \omega_k$ to achieve this. Now a reasonable criterion for independence would be

$$\Lambda[H_{12...k}|G] = \prod_{i=1}^{k} \Lambda[H_i|G],$$

and taking logarithms this is true if and only if

$$\mathbf{P}_\Omega - \mathbf{P}_{12...k} = \sum_{i=1}^{k} (\mathbf{P}_\Omega - \mathbf{P}_i),$$

where $\mathbf{P}_\Omega - \mathbf{P}_i$ represents the orthogonal projection onto $\omega_i^p = \omega_i^\perp \cap \Omega$ (Theorem 4.2). We therefore have from the above equation (cf. the special case of (4.5) in Sect. 4.1)

$$\mathbf{I}_n = \mathbf{I}_n - \mathbf{P}_\Omega + \sum_{1=1}^{k} (\mathbf{P}_\Omega - \mathbf{P}_i) + \mathbf{P}_{12...k},$$

where all the matrices are symmetric and idempotent. Hence by Theorem 4.1 in Sect. 4.1, the subspaces $\Omega^\perp$, $\omega_1^p, \omega_2^p, \ldots, \omega_k^p$, and $\omega_{12...k}$ are mutually orthogonal and the test statistics for the H_i are mutually independent. We are thus led to the following definition due to Darroch and Silvey (1963, 564). An experimental design is *orthogonal* relative to a general linear model G and linear hypotheses H_1, $H_2, \ldots, H_k$ if, with this design, the subspaces $\Omega, \omega_1, \omega_2, \ldots, \omega_k$ satisfy the conditions $\omega_i^p \perp \omega_j^p$ for all $i, j, i \neq j$. Since $\omega_i^p \perp \omega_j^p$ if and only if $\omega_i \oplus \Omega^\perp = (\omega_i^\perp \cap \Omega)^\perp \supset \omega_j^p$ (Theorem 1.2(ii) in Sect.1.4) if and only if $\omega_i \supset \omega_j^p$ (since $\Omega \supset \omega_j^p$ and $(\omega_i \oplus \Omega^\perp) \cap \Omega = \omega_i$), we have an equivalent definition of orthogonality, namely $\omega_i \supset \omega_j^p$ for all $i, j, i \neq j$. Because we have symmetry between i and j in the original definition, we see that $\omega_i \supset \omega_j^p$ if and only if $\omega_j \supset \omega_i^p$.

If σ^2 is unknown and ω_i has dimension $p - q_i$, then the F-statistics for testing the individual hypotheses are

$$\frac{(n-p)}{q_i}\frac{\mathbf{y}'(\mathbf{P}_\Omega - \mathbf{P}_i)\mathbf{y}}{\mathbf{y}'(\mathbf{I}_n - \mathbf{P}_\Omega)\mathbf{y}} = \frac{(n-p)}{q_i}\frac{Q_i - Q}{Q} \quad (i = 1, 2, \ldots, k).$$

An advantage of having the above property of orthogonality is given by the following theorem.

Theorem 6.1 *The sums of squares $Q_1 - Q$, $Q_{12} - Q_1$, $\ldots Q_{12\ldots k} - Q_{12\ldots k-1}$ are the same independent of the order of nesting of the hypotheses if and only if $\omega_i^p \perp \omega_j^p$, $i \neq j$, that is, the hypotheses are orthogonal.*

Proof (Sufficiency) Suppose $\omega_i^p \perp \omega_j^p$, $i \neq j$. The matrix $\mathbf{P}_{12\ldots i-1} - \mathbf{P}_{12\ldots i}$ represents an orthogonal projection onto

$$\begin{aligned} W_i &= \omega_1 \cap \omega_2 \cap \ldots \cap \omega_{i-1} \cap (\omega_1 \cap \omega_2 \cap \ldots \cap \omega_i)^\perp \\ &= \omega_1 \cap \ldots \cap \omega_{i-1} \cap (\omega_1^\perp + \cdots + \omega_i^\perp) \quad \text{(by Theorem 1.2)} \\ &= \omega_i^p. \end{aligned}$$

Justification for this last step is as follows: if $\boldsymbol{\theta} \in \omega_i^p$ then $\boldsymbol{\theta} \in \omega_1 \cap \ldots \cap \omega_{i-1}$ (by the alternative definition of orthogonality), $\boldsymbol{\theta} \in \omega_i^\perp$, and $\boldsymbol{\theta} \in W_i$. Conversely, if $\boldsymbol{\theta} \in W_i$, then $\boldsymbol{\theta} \in \omega_1 \cap \omega_2 \ldots \cap \omega_{i-1}$ and $\boldsymbol{\theta} = \mathbf{P}_{12\ldots i-1}\boldsymbol{\theta}$. Also $\boldsymbol{\theta} \in (\omega_1^\perp + \cdots + \omega_i^\perp)$ so that for some $\boldsymbol{\alpha}_1, \boldsymbol{\alpha}_2, \ldots, \boldsymbol{\alpha}_i$, $\boldsymbol{\theta} = \sum_{j=1}^i (\mathbf{I}_n - \mathbf{P}_j)\boldsymbol{\alpha}_j$ and

$$\begin{aligned} \boldsymbol{\theta} &= \mathbf{P}_{12\ldots i-1}\sum_{j=1}^{i}(\mathbf{I}_n - \mathbf{P}_j)\boldsymbol{\alpha}_j \\ &= \mathbf{P}_{12\ldots i-1}(\mathbf{I}_n - \mathbf{P}_i)\boldsymbol{\alpha}_i \\ &= \mathbf{P}_{12\ldots i-1}(\mathbf{P}_\Omega - \mathbf{P}_i)\boldsymbol{\alpha}_i \\ &= \mathbf{P}_{12\ldots i-1}\mathbf{P}_{\omega_i^p}\boldsymbol{\alpha}_i \\ &= \mathbf{P}_{\omega_i^p}\boldsymbol{\alpha}_i, \end{aligned}$$

since $\omega_i^p \subset \omega_1 \cap \omega_2 \cap \ldots, \omega_{i-1}$. Hence $\boldsymbol{\theta} \in \omega_i^p$ and $W_i = \omega_i^p$. Equating the projection matrices on these two subspaces gives us

$$\mathbf{P}_{12\ldots i-1} - \mathbf{P}_{12\ldots i} = \mathbf{P}_\Omega - \mathbf{P}_i \tag{6.2}$$

and the sums of squares are equal to $Q_i - Q$ for $i = 1, 2, \ldots, k$, which are independent of the order of nesting.

(Necessity). Given the sums of squares independent of the order of nesting, we can choose ω_i to be the first in the sequence so that (6.2) must hold, that is $\omega_i^p = W_i \subset \omega_j$

$(i = 1, 2, \ldots, k; j = 1, 2, \ldots, i-1)$. Hence $\omega_i^p \perp \omega_j^p$ for all $i, j, i \neq j$. This completes the proof of the theorem.

Having established the definition of orthogonality, we can now demonstrate using the above theorem that the separate test method for testing $H_{12\ldots k}$ versus G is a reasonable one when we have hypothesis orthogonality. The following justification is due to Darroch and Silvey (1963). From Sect. 4.1 we have, with orthogonality,

$$\begin{aligned}
\{\Lambda[H_{12\ldots k}|G]\}^{-2/n} - 1 &= \frac{Q_{12\ldots k} - Q}{Q} \\
&= \frac{(Q_1 - Q) + (Q_{12} - Q_1) + \cdots + (Q_{12\ldots k} - Q_{12\ldots k-1})}{Q} \\
&= \sum_{i=1}^{k} \frac{Q_i - Q}{Q} \text{ (by Theorem 6.1)} \\
&= \sum_{i=1}^{k} (\{\Lambda[H_i|G]\}^{-2/n} - 1).
\end{aligned}$$

If each $\Lambda[H_i|G]$ is "near" unity, then $\Lambda[H_{12\ldots k}|G]$ is "near" unity, and by the same argument applied to the nested procedure we see that the separate test method will also have good power.

6.3 Orthogonal Hypotheses in Regression Models

In this section we shall show that the idea of hypothesis orthogonality is usually associated with those experimental designs in which least squares estimates of certain parameters are uncorrelated.

Example 6.2 Suppose Ω takes the form $\boldsymbol{\theta} = \mathbf{X}\boldsymbol{\beta}$, where $\mathbf{X}$ is an $n \times p$ matrix of rank r $(r < p)$ and $\mathbf{H}\boldsymbol{\beta} = \mathbf{0}$ are suitable identifiability conditions. Let $\mathbf{X}$ be partitioned into $k + 1$ submatrices $(\mathbf{X}_0, \mathbf{X}_1, \mathbf{X}_2, \ldots, \mathbf{X}_k)$ with a corresponding partition of $\boldsymbol{\beta} = (\boldsymbol{\beta}_0', \boldsymbol{\beta}_1', \ldots, \boldsymbol{\beta}_k')'$ and of $\mathbf{H}$. We are interested in testing the hypotheses $H_i : \boldsymbol{\beta}_i = \mathbf{0}$ $(i = 1, 2, \ldots, k)$. Thus $\omega_i = \{\boldsymbol{\theta} = \mathbf{X}_i^* \boldsymbol{\beta}_i^*, \mathbf{H}_i^* \boldsymbol{\beta}_i^* = \mathbf{0}\}$, where $\mathbf{X}_i^*$ is the matrix $\mathbf{X}$ with the submatrix $\mathbf{X}_i$ deleted; $\mathbf{H}_i^*$ and $\boldsymbol{\beta}_i^*$ are similarly defined. We shall now prove that the least squares estimates $\hat{\boldsymbol{\beta}}_i$ of $\boldsymbol{\beta}_i$ $(i = 1, 2, \ldots, k)$ for $\boldsymbol{\theta} \in \Omega$ are uncorrelated if and only if we have orthogonality. The proof rests on the following Theorem.

Theorem 6.2 *Let Ω be a vector space and let $\mathbf{A}_i$ $(i = 1, 2, \ldots, k)$ be any matrix such that $\omega_i = \mathcal{N}[\mathbf{A}_i] \cap \Omega$ is a proper subspace of Ω. Then $\omega_i^p \perp \omega_j^p$ if and only if $\mathbf{A}_i \mathbf{P}_\Omega \mathbf{A}_j' = \mathbf{0}$. Furthermore, if $\Omega = \mathcal{N}[\mathbf{A}]$ and $\mathbf{A}\mathbf{A}_i' = \mathbf{0}$ for $i = 1, 2, \ldots, k$, then $\omega_i^p \perp \omega_j^p$ if and only if $\mathbf{A}_i \mathbf{A}_j' = \mathbf{0}$ (for all $i, j, i \neq j$).*

Proof From Theorem 4.3 in Sect. 4.2,

$$\omega_i^p = \omega_i^{\perp} \cap \Omega = \mathcal{C}[\mathbf{P}_\Omega \mathbf{A}_i']$$

and therefore $\omega_i^p \perp \omega_j^p$ if and only if

$$(\mathbf{P}_\Omega \mathbf{A}_i')'\mathbf{P}_\Omega \mathbf{A}_j' = \mathbf{A}_i \mathbf{P}_\Omega^2 \mathbf{A}_j' = \mathbf{A}_i \mathbf{P}_\Omega \mathbf{A}_j' = \mathbf{0}.$$

If $\Omega = \mathcal{N}[\mathbf{A}] = \mathcal{C}[\mathbf{A}']^{\perp}$ and $\mathbf{AA}_i' = \mathbf{0}$, then

$$\mathbf{A}_i \mathbf{P}_\Omega \mathbf{A}_j' = \mathbf{A}_i[\mathbf{I}_n - \mathbf{A}'(\mathbf{AA}')^{-}\mathbf{A}]\mathbf{A}_j' = \mathbf{A}_i \mathbf{A}_j',$$

where $(\mathbf{AA}')^{-}$ is a weak inverse (cf. A.15(ii)), and the result follows.

We are now in the position to prove the following theorem for our regression example given above.

Theorem 6.3 *The vectors $\hat{\beta}_i$ and $\hat{\beta}_j$ are uncorrelated if and only if $\omega_i^p \perp \omega_j^p$.*

Proof We note from Theorem 3.5(i) in Sect. 3.4 that $\hat{\beta} = (\mathbf{G}'\mathbf{G})^{-1}\mathbf{X}'\mathbf{y}$. Since the constraints $\mathbf{H}\beta = \mathbf{0}$ are suitable for identifiability, we have from Theorem 3.5 in Sect. 3.4

$$\mathbf{P}_\Omega = \mathbf{X}(\mathbf{G}'\mathbf{G})^{-1}\mathbf{X}' \quad \text{and} \quad \mathbf{H}(\mathbf{G}'\mathbf{G})^{-1}\mathbf{X}' = \mathbf{0}, \tag{6.3}$$

where $\mathbf{G}'\mathbf{G} = \mathbf{X}'\mathbf{X} + \mathbf{H}'\mathbf{H}$. Now $\beta = (\mathbf{G}'\mathbf{G})^{-1}\mathbf{X}'\theta$, and the hypothesis $\beta_i = \mathbf{0}$ is equivalent to $\mathbf{B}_i\beta = \mathbf{0}$, where $\mathbf{B}_i$, if partitioned in the same way as $\mathbf{X}$, has the identity matrix in the $(i+1)$th partition and zero matrices elsewhere. Thus testing the hypothesis H_i is equivalent to testing $\mathbf{B}_i(\mathbf{G}'\mathbf{G})^{-1}\mathbf{X}'\theta = \mathbf{A}_i\theta = \mathbf{0}$, and from Theorem 6.2 above, $\omega_i^p \perp \omega_j^p$ if and only if

$$\mathbf{A}_i \mathbf{P}_\Omega \mathbf{A}_j' = \mathbf{B}_i(\mathbf{G}'\mathbf{G})^{-1}\mathbf{X}'\mathbf{X}(\mathbf{G}'\mathbf{G})^{-1}\mathbf{X}'\mathbf{X}(\mathbf{G}'\mathbf{G})^{-1}\mathbf{B}_j' = \mathbf{0}.$$

or using (6.3) with $\mathbf{P}_\Omega \mathbf{X} = \mathbf{X}$ and adding $\mathbf{H}'\mathbf{H}$ to $\mathbf{X}'\mathbf{X}$,

$$\mathbf{B}_i(\mathbf{G}'\mathbf{G})^{-1}\mathbf{X}'\mathbf{X}(\mathbf{G}'\mathbf{G})^{-1}\mathbf{B}_j' = \mathbf{0}.$$

But from Theorem 1.5(iii) in Sect. 1.6,

$$\begin{aligned}
\mathrm{Cov}[\hat{\beta}_i, \hat{\beta}_j] &= \mathrm{Cov}[\mathbf{B}_i\hat{\beta}, \mathbf{B}_j\hat{\beta}] \\
&= \mathbf{B}_i \mathrm{Var}[\hat{\beta}]\mathbf{B}_j' \\
&= \sigma^2 \mathbf{B}_i(\mathbf{G}'\mathbf{G})^{-1}\mathbf{X}'\mathbf{X}(\mathbf{G}'\mathbf{G})^{-1}\mathbf{B}_j',
\end{aligned}$$

and we therefore have orthogonality if and only if $\mathrm{Cov}[\hat{\beta}_i, \hat{\beta}_j] = \mathbf{0}$.

Example 6.3 We return to Example 6.1 on polynomial regression where we considered the model (6.1). In applying the nested method of hypothesis testing we run into the problem that the least squares estimates of the β_i have to be recalculated at each stage of the nesting. However, the algebra would be much simpler if each of the hypotheses $H_i : \beta_i = 0, i = 0, 1, 2, \ldots, k$ was orthogonal, for then the least squares estimates $\hat{\beta}_i$ would be uncorrelated, that is $\text{cov}[\hat{\beta}_i, \hat{\beta}_j] = 0$ for $i \neq j$, and they would be the same irrespective of whether or not some of the β_i were made zero. This means we would not have to recalculate these estimates at each stage. One method of achieving this desired simplification is by the use of orthogonal polynomials. Our model then becomes

$$y_i = \gamma_0\phi_0(x_i) + \gamma_1\phi_1(x_i) + \cdots + \gamma_k\phi_k(x_i) + \varepsilon_i,$$

where $\phi_0(x_i) = 1$, $\phi_r(x_i)$ is a polynomial of degree r, and $\sum_i \phi_r(x_i)\phi_s(x_i) = 0$ for all $r, s = 0, 1, 2, \ldots, k$, $r \neq s$. if $\boldsymbol{\gamma} = (\gamma_0, \gamma_1, \ldots, \gamma_k)'$, then $\text{E}[\mathbf{y}] = \mathbf{W}\boldsymbol{\gamma}$, where $\mathbf{W} = (\phi_j(x_i))$ has mutually orthogonal columns. Let $\hat{\boldsymbol{\gamma}}$ be the least-squares estimate of $\boldsymbol{\gamma}$, then $\text{Var}[\hat{\boldsymbol{\gamma}}] = \sigma^2(\mathbf{W}'\mathbf{W})^{-1}$, which is diagonal as $\mathbf{W}'\mathbf{W}$ is diagonal, and the $\hat{\gamma}_i$ are uncorrelated. Hence, by Theorem 6.3, the hypotheses $H_i : \gamma_i = 0$ are orthogonal; also $\beta_k = \beta_{k-1} = \cdots = \beta_i = 0$ if and only $\gamma_k = \gamma_{k-1} = \ldots = \gamma_i = 0$. For further details concerning orthogonal polynomials see Seber and Lee (2003, Chapter 7). This example can be generalized in the following theorem.

Theorem 6.4 *Suppose* $\mathbf{X} = (\mathbf{X}_0, \mathbf{X}_1, \ldots, \mathbf{X}_k)$ *with linearly independent columns, with a corresponding partition of* $\boldsymbol{\beta} = (\boldsymbol{\beta}_0', \boldsymbol{\beta}_1', \ldots, \boldsymbol{\beta}_k')'$. *We wish to test the hypotheses* $H_i : \boldsymbol{\beta}_i = \mathbf{0}$ $(i = 1, 2, \ldots, k)$. *Thus* $\omega_i = \mathcal{C}[\mathbf{X}_i^*]$, *where* $\mathbf{X}_i^*$ *is the matrix* $\mathbf{X}$ *with the submatrix* $\mathbf{X}_i$ *deleted. Then* $\omega_i^p \perp \omega_j^p$ *for all* $i, j = 1, 2, \ldots k, i \neq j$ *if and only if* $\mathbf{X}_j'(\mathbf{I}_n - \mathbf{P}_0)\mathbf{X}_i = \mathbf{0}$, *where* $\mathbf{P}_0$ *is the orthogonal projection onto* $\mathcal{C}[\mathbf{X}_0]$.

Proof We shall use the results that (i) $\mathbf{P}_\omega\mathbf{P}_\Omega = \mathbf{P}_\Omega\mathbf{P}_\omega = \mathbf{P}_\omega$ for $\omega \in \Omega$, and (ii) $\mathcal{C}[\mathbf{V}] = \mathcal{C}[\mathbf{P}_\mathbf{V}]$, where $\mathbf{P}_\mathbf{V}$ represents the orthogonal projection onto $\mathcal{C}[\mathbf{V}]$. Also let $\mathbf{P}_i = \mathbf{P}_{\omega_i}$.

We first show that

$$(\mathcal{C}[\mathbf{X}_i^*])^\perp \cap \mathcal{C}[\mathbf{X}] = \mathcal{C}[(\mathbf{I}_n - \mathbf{P}_i)\mathbf{X}],$$

where the left-hand side (LHS) of the above equation is ω_i^p. If $\boldsymbol{\theta} \in$ LHS, then $(\mathbf{I}_n - \mathbf{P}_i)\boldsymbol{\theta} = \boldsymbol{\theta}$ and $\boldsymbol{\theta} = \mathbf{X}\boldsymbol{\beta}$ for some $\boldsymbol{\beta}$, that is $\boldsymbol{\theta} = (\mathbf{I}_n - \mathbf{P}_i)\mathbf{X}\boldsymbol{\beta}$ and $\boldsymbol{\theta} \in$ RHS. Conversely, if $\boldsymbol{\theta} \in$ RHS, then $\boldsymbol{\theta} = (\mathbf{I}_n - \mathbf{P}_i)\mathbf{X}\boldsymbol{\beta}$ for some $\boldsymbol{\beta}$ and $\boldsymbol{\theta} \in (\mathcal{C}[\mathbf{X}_i^*])^\perp$. Now

$$\begin{aligned}
(\mathbf{I}_n - \mathbf{P}_\Omega)\boldsymbol{\theta} &= (\mathbf{I}_n - \mathbf{P}_\Omega)(\mathbf{I}_n - \mathbf{P}_i)\mathbf{X}\boldsymbol{\beta} \\
&= (\mathbf{I}_n - \mathbf{P}_\Omega - \mathbf{P}_i + \mathbf{P}_i)\mathbf{X}\boldsymbol{\beta} \quad (\text{as } \mathbf{P}_\Omega\mathbf{P}_i = \mathbf{P}_i) \\
&= (\mathbf{I}_n - \mathbf{P}_\Omega)\mathbf{X}\boldsymbol{\beta} \\
&= \mathbf{0} \quad (\text{as } \mathbf{P}_\Omega\mathbf{X} = \mathbf{X}).
\end{aligned}$$

Hence $\boldsymbol{\theta} \in \Omega$ and $\boldsymbol{\theta} \in$ LHS. Using the alternative definition for the orthogonality of hypotheses, we now have the following equivalent statements for all $i, j, i \neq j$.

$$\begin{aligned}
\omega_i^p \perp \omega_j^p &\iff \omega_i^p \subset \cap_{j:j\neq i}\omega_j \\
&\iff \mathcal{C}[(\mathbf{I}_n - \mathbf{P}_i)\mathbf{X}_i] \subset \mathcal{C}[\mathbf{X}_0, \mathbf{X}_i] \\
&\iff \mathcal{C}[\mathbf{P}_i\mathbf{X}_i] \subset \mathcal{C}[\mathbf{X}_0, \mathbf{X}_i] \\
&\iff \mathcal{C}[\mathbf{P}_i(\mathbf{I}_n - \mathbf{P}_0 + \mathbf{P}_0)\mathbf{X}_i] \subset \mathcal{C}[\mathbf{X}_0, \mathbf{X}_i] \\
&\iff \mathcal{C}[\mathbf{P}_i(\mathbf{I}_n - \mathbf{P}_0)\mathbf{X}_i] \subset \mathcal{C}[\mathbf{X}_0, \mathbf{X}_i] \quad \text{(by (i) and (ii) as } \mathbf{P}_i\mathbf{P}_0 = \mathbf{P}_0) \\
&\iff \mathcal{C}[(\mathbf{P}_i - \mathbf{P}_0\mathbf{P}_i)\mathbf{X}_i] \subset \mathcal{C}[\mathbf{X}_0, \mathbf{X}_i] \quad \text{(as } \mathbf{P}_i\mathbf{P}_0 = \mathbf{P}_0\mathbf{P}_i) \\
&\iff \mathcal{C}[(\mathbf{I}_n - \mathbf{P}_0)\mathbf{P}_i\mathbf{X}_i] \subset \mathcal{C}[\mathbf{X}_i] \quad \text{(by (ii))} \\
&\iff \mathcal{C}[\mathbf{P}_i(\mathbf{I}_n - \mathbf{P}_0)\mathbf{X}_i] \subset \mathcal{C}[\mathbf{X}_i] \\
&\iff \mathbf{P}_i(\mathbf{I}_n - \mathbf{P}_0)\mathbf{X}_i = \mathbf{0} \quad \text{(as } \mathcal{C}[\mathbf{X}_i^*] \cap \mathcal{C}[\mathbf{X}_i] = \mathbf{0}) \\
&\iff \mathbf{X}_j'(\mathbf{I}_n - \mathbf{P}_0)\mathbf{X}_i = \mathbf{0}.
\end{aligned}$$

This completes the proof.

Since $\mathbf{I}_n - \mathbf{P}_0$ is idempotent, the above conditions are equivalent to

$$(\mathbf{I}_n - \mathbf{P}_0)\mathbf{X}_i \perp (\mathbf{I}_n - \mathbf{P}_0)\mathbf{X}_j.$$

A number of special cases of the above theorem follow. For example, if $\mathbf{X}_0 = \mathbf{0}$, then the conditions reduce to $\mathbf{X}_i'\mathbf{X}_j$ or, if $\mathbf{X}_i = \mathbf{x}_i$, to $\mathbf{X}$ having orthogonal columns, as we found in Example 6.3 above. If $\mathbf{X}_0 = \mathbf{1}_n$ and $\mathbf{X}_i = \mathbf{x}_i$, then $\mathbf{P}_0 = \mathbf{1}_n\mathbf{1}_n'/n$ and the conditions reduce to $\sum_{r=1}^n (x_{ir} - \overline{x}_i)(x_{jr} - \overline{x}_j) = 0$ (all $i, j, i \neq j$), where $\mathbf{x}_i$ has elements x_{ir} and $\overline{x}_i$ is the mean of the elements of $\mathbf{x}_i$.

6.4 Orthogonality in Complete Two-Factor Layouts

Consider a two-factor analysis of variance with factor A and B at I and J levels respectively, and suppose that n_{ij} observations $y_{ij1}, y_{ij2}, \ldots, y_{ijn_{ij}}$ are made on the combination ϕ_{ij} of the ith level of A with the jth level of B. This gives us the model $y_{ijk} = \theta_{ijk} + \varepsilon_{ijk}$ for $k = 1, 2, \ldots, n_{ij}$, $i = 1, 2, \ldots, I$, and $j = 1, 2, \ldots, J$, where $\theta_{ijk} = \phi_{ij}$: the random "errors" ε_{ijk} are all independently distributed as $N[0, \sigma^2]$. We now split up the i, jth cell mean ϕ_{ij} into an overall mean μ, an effect α_i due to the ith level of A, an effect β_j due to the jth level of B, and an interaction term γ_{ij}, so that $\phi_{ij} = \mu + \alpha_i + \beta_j + \gamma_{ij}$. If we write the element sets $\{y_{ijk}\}$, $\{\theta_{ijk}\}$, and $\{\varepsilon_{ijk}\}$ as single n-dimensional vectors (where $n = n_{\cdot\cdot} = \sum_i \sum_j n_{ij}$), we can express the above model in the form $\mathbf{y} = \boldsymbol{\theta} + \boldsymbol{\varepsilon}$ with $G : \boldsymbol{\theta} = \mathbf{X}\boldsymbol{\delta}$, where

$$\boldsymbol{\delta} = (\mu, \alpha_1, \ldots, \alpha_I, \beta_1, \ldots, \beta_J, \gamma_{11}, \gamma_{12}, \ldots, \gamma_{1J}, \gamma_{21}, \gamma_{22}, \ldots, \gamma_{I1}, \ldots, \gamma_{IJ})'.$$

Since we have replaced the IJ uniquely defined parameters ϕ_{ij} by $1+I+J+IJ$ new parameters, $\boldsymbol{\delta}$ will not be identifiable and we must introduce some identifiability constraints. The form of these constraints will depend on what "weights" we choose for defining these parameters. For example, the observations for certain i, j cells may be more important than the others, and therefore we would want to give more weight to these observations. Thus we may define our parameters as follows.

The means for the ith level of A and the jth level of B are defined to be $a_i = \sum_s v_s\phi_{is}$ and $b_j = \sum_r u_r\phi_{rj}$, where all the $u_i \geq 0$, $v_j \geq 0$ and $\sum_r u_r = \sum_s v_s$=1. The general mean μ is defined to be

$$\mu = \sum_i u_i a_i = \sum_j v_j b_j = \sum_i \sum_j u_i v_j \phi_{ij}.$$

The main effect of the ith level of A is defined by

$$\alpha_i = a_i - \mu = \sum_s v_s\phi_{is} - \sum_r \sum_s u_r v_s \phi_{rs}$$

and the main effect of the jth level of B is defined by

$$\beta_j = b_j - \mu = \sum_r u_r\phi_{rj} - \sum_r \sum_s u_r v_s \phi_{rs}.$$

Now $\phi_{ij} = \mu + \alpha_i + \beta_j + \gamma_{ij}$, where

$$\gamma_{ij} = \phi_{ij} - \sum_j v_j\phi_{ij} - \sum_i u_i\phi_{ij} + \sum_i \sum_j u_i v_j \phi_{ij}$$

is called the interaction between the ith level of A and the jth level of B. We have effectively imposed the identifiability constraints

$$\sum_i u_i\alpha_i = \sum_j v_j\beta_j = \sum_i u_i\gamma_{ij} = \sum_j v_j\gamma_{ij} = 0.$$

Finally, let $\nu_{ijk} = \theta_{ijk} - \overline{\theta}_{ij\cdot}$ where $\overline{\theta}_{ij\cdot} = \sum_k \theta_{ijk}/n_{ij}$, so that our model now takes the form

$$\theta_{ijk} = \mu + \alpha_i + \beta_j + \gamma_{ij} + \nu_{ijk},$$

where the μ, α_i, β_j, and γ_{ij} are defined above with $\overline{\theta}_{ij\cdot}$ replacing ϕ_{ij} in the definitions. We now consider testing the hypotheses H_i $(i = 1, 2, 3, 4)$ versus G, where

$$\begin{aligned}
G &: \nu_{ijk} = 0, \\
H_1 &: \nu_{ijk} = 0, \quad \gamma_{ij} = 0 \quad \text{(interactions zero)} \\
H_2 &: \nu_{ijk} = 0, \quad \alpha_i = 0 \quad \text{(main effects of } A \text{ zero)} \\
H_3 &: \nu_{ijk} = 0, \quad \beta_j = 0 \quad \text{(main effects of } B \text{ zero)} \\
H_4 &: \nu_{ijk} = 0, \quad \mu = 0,
\end{aligned}$$

and in the following theorem we derive necessary and sufficient conditions for this system of hypotheses to be orthogonal.

Theorem 6.5 *The hypotheses H_i $(i = 1, 2, 3, 4)$ are orthogonal with respect to G if and only if*

$$n_{ij} = n_{i\cdot}n_{\cdot j}/n, \quad u_i = n_{i\cdot}/n, \text{ and } v_j = n_{\cdot j}/n \text{ for all } i, j,$$

where $n_{i\cdot} = \sum_j n_{ij}$, $n_{\cdot j} = \sum_i n_{ij}$, and $n = n_{\cdot\cdot} = \sum_i \sum_j n_{ij}$.

Proof If $\boldsymbol{\theta}$ is the n-dimensional vector with elements θ_{ijk} we can express the hypotheses in the form

$$G : \mathbf{A}\boldsymbol{\theta} = \mathbf{0} \quad \text{and} \quad H_r : \mathbf{A}\boldsymbol{\theta} = \mathbf{0},\ \mathbf{A}_r\boldsymbol{\theta} = \mathbf{0} \quad (r = 1, 2, 3, 4).$$

For example, we wish to express the conditions

$$\nu_{ijk} = \theta_{ijk} - \sum_{\ell=1}^{n_{ij}} \theta_{ij\ell}/n_{ij} = 0 \tag{6.4}$$

in the form $\mathbf{A}\boldsymbol{\theta} = \mathbf{0}$. The matrix $\mathbf{A}$ would be $n \times n$ and the row corresponding to Eq. (6.4) would have the (r_0, s_0, t_0) element of the form

$$\delta_{ir_0}\delta_{js_0}\delta_{kt_0} - \delta_{ir_0}\delta_{js_0}/n_{r_0s_0}, \tag{6.5}$$

where δ_{ab} is the Kronecker delta. Now

$$\gamma_{ij} = \overline{\theta}_{ij\cdot} - \sum_s v_s\overline{\theta}_{is\cdot} - \sum_r u_r\overline{\theta}_{rj\cdot} + \sum_r \sum_s u_r v_s \overline{\theta}_{rs\cdot},$$

and the row of $\mathbf{A}_1$ corresponding to $\gamma_{i_1j_1} = 0$ has its (r_1, s_1, t_1) element as

$$(\delta_{i_1r_1}\delta_{j_1s_1} - v_{s_1}\delta_{i_1r_1} - u_{r_1}\delta_{j_1s_1} + u_{r_1}v_{s_1})/n_{r_1s_1}. \tag{6.6}$$

Similarly the (r_2, s_2, t_2) element of row $\alpha_{i_2} = 0$ for the matrix $\mathbf{A}_2$ is

$$(\delta_{i_2 r_2} v_{s_2} - u_{r_2} v_{s_2})/n_{r_2 s_2}, \tag{6.7}$$

the (r_3, s_3, t_3) element of row $\beta_{j_3} = 0$ for $\mathbf{A}_3$ is

$$(\delta_{j_3 s_3} u_{r_3} - u_{r_3} v_{s_3})/n_{r_3 s_3}, \tag{6.8}$$

and the (r_4, s_4, t_4) element of $\mu = 0$ is

$$u_{r_4} v_{s_4}/n_{r_4 s_4}. \tag{6.9}$$

By multiplying together (6.5) and (6.6), putting $r_1 = r_0$, $s_1 = s_0$, $t_1 = t_0$ and summing on r_0, s_0, t_0 ($t_0 = 1, 2, \ldots, n_{r_0 s_0}$; $r_0 = 1, 2, \ldots, I$; $s_0 = 1, 2, \ldots, J$) we have $\mathbf{A}_1\mathbf{A}' = \mathbf{0}$. Similarly, $\mathbf{A}_2\mathbf{A}'$, $\mathbf{A}_3\mathbf{A}'$, and $\mathbf{A}_4\mathbf{A}'$ are all zero matrices since (6.5) is the only term above containing t_0, and this summed on t_0 is zero. Thus by Theorem 6.2, the hypotheses are orthogonal if and only if $\mathbf{A}_p\mathbf{A}_q'$ is the zero matrix for all $p, q, p \neq q$, and we now show that these matrix conditions hold if and only if

$$n_{ij} = n_{i\cdot}n_{\cdot j}/n, \quad u_i = n_{i\cdot}/n, \text{ and } v_j = n_{\cdot j}/n \text{ for all } i, j.$$

Sufficiency. If the above conditions on the n_{ij}, u_i, and v_j hold, then (6.6) becomes

$$n(\delta_{i_1 r_1}/n_{r_1\cdot} - 1/n)(\delta_{j_1 s_1}/n_{\cdot s_1} - 1/n). \tag{6.10}$$

Therefore by multiplying (6.7) and (6.10) together, putting $r_2 = r_1$, $s_2 = s_1$, $t_2 = t_1$, and summing on r_1, s_1, t_1, we have $\mathbf{A}_1\mathbf{A}_2' = \mathbf{0}$. In a similar manner it can be shown that $\mathbf{A}_1\mathbf{A}_3', \ldots, \mathbf{A}_3\mathbf{A}_4'$ are all zero matrices. Hence the hypotheses are orthogonal and the conditions are sufficient.

Necessity. Given that $\mathbf{A}_1\mathbf{A}_4' = \mathbf{0}$, we multiply (6.6) and (6.9) together, set $r_4 = r_1$, $s_4 = s_1$, $t_4 = t_1$ and then sum on r_1, s_1, t_1. This gives us an element of $\mathbf{A}_1\mathbf{A}_4'$ so that

$$\frac{u_{i_1} v_{j_1}}{n_{i_1 j_1}} - \sum_{s_1} \left\{ \frac{u_{i_1} v_{s_1}^2}{n_{i_1 s_1}} \right\} - \sum_{r_1} \left\{ \frac{v_{j_1} u_{r_1}^2}{n_{r_1 j_1}} \right\} + \sum_{r_1} \sum_{s_1} \left\{ \frac{u_{r_1}^2 v_{s_1}^2}{n_{r_1 s_1}} \right\} = 0. \tag{6.11}$$

Similarly, from $\mathbf{A}_2\mathbf{A}_4' = \mathbf{0}$ and $\mathbf{A}_3\mathbf{A}_4' = \mathbf{0}$ we obtain

$$u_{i_2} \sum_{s_2} \left\{ \frac{v_{s_2}^2}{n_{i_2 s_2}} \right\} - \sum_{r_2} \sum_{s_2} \left\{ \frac{u_{r_2}^2 v_{s_2}^2}{n_{r_2 s_2}} \right\} = 0 \tag{6.12}$$

and

$$v_{j_3} \sum_{r_3} \left\{ \frac{u_{r_3}^2}{n_{r_3 j_3}} \right\} - \sum_{r_3} \sum_{s_3} \left\{ \frac{u_{r_3}^2 v_{s_3}^2}{n_{r_3 s_3}} \right\} = 0. \tag{6.13}$$

From these last two equations we note that $u_{i_2} > 0$ and $v_{j_3} > 0$ for all i_2 and j_3. Adding (6.13) to (6.11), putting $j_1 = j_3$, and dividing by u_{i_1}, we obtain

$$\frac{v_{j_1}}{n_{i_1 j_1}} - \sum_{s_1} \left\{ \frac{v_{s_1}^2}{n_{i_1 s_1}} \right\} = 0. \tag{6.14}$$

Multiplying this equation by $n_{i_1 j_1}$, summing on j_1, and using $\sum_{j_1} v_{j_1} = 1$, give us

$$1 - n_{i_1 \cdot} \sum_{s_1} \left\{ \frac{v_{s_1}^2}{n_{i_1 s_1}} \right\} = 0.$$

Substituting this back into (6.14) leads to $v_{j_1} = n_{i_1 j_1}/n_{i_1 \cdot}$ for all i_1, that is,

$$v_{j_1} = \frac{\sum_{i_1} n_{i_1 j_1}}{\sum_{i_1} n_{i_1 \cdot}} = \frac{n_{\cdot j_1}}{n}.$$

In a similar manner it can be shown that

$$u_{i_1} = \frac{n_{i_1 j_1}}{n_{\cdot j_1}} = \frac{n_{i_1 \cdot}}{n}.$$

Now by multiplying (6.7) and (6.8) together, putting $r_3 = r_2$, $s_3 = s_2$, $t_3 = t_2$, and summing on r_2, s_2, t_2 give us an element of $\mathbf{A}_2 \mathbf{A}_3'$. Thus if $\mathbf{A}_2 \mathbf{A}_3' = \mathbf{0}$ we have, substituting the expressions given above for u_{r_2} and v_{s_2},

$$\begin{aligned} 0 &= \sum_{r_2} \sum_{s_2} \sum_{t_2} \left\{ \left(\frac{\delta_{i_2 r_2}}{n_{r_2 \cdot}} - \frac{1}{n} \right) \left(\frac{\delta_{j_3 s_2}}{n_{\cdot s_2}} - \frac{1}{n} \right) \right\} \\ &= \frac{n_{i_2 j_3}}{n_{i_2 \cdot} n_{\cdot j_3}} - \frac{1}{n} \end{aligned}$$

for every i_2 and j_3. Therefore a necessary condition for orthogonality is that we have $n_{ij} = n_{i \cdot} n_{\cdot j}/n$ for all i, j and this completes the proof of the theorem

We note that when we have equal numbers of observations per cell, that is $n_{ij} = K$ for all i, j, the conditions for orthogonality are automatically satisfied, provided we use equal weights. In this case the identifiability constraints reduce to the equations $\sum_i \alpha_i = \sum_j \beta_j = \sum_i \gamma_{ij} = \sum_j \gamma_{ij} = 0$.

When the hypotheses are orthogonal, we have from Theorem 6.3 in Sect. 6.3 that the least squares estimates of the μ, $\{\alpha_i\}$, $\{\beta_j\}$, and $\{\gamma_{ij}\}$ for $\boldsymbol{\theta} \in \Omega$ are uncorrelated group-wise and are readily derived, as we shall see below, by using the Gauss-Markov theorem. In the remainder of this section we will assume the conditions for orthogonality hold, namely

$$u_i = n_{i\cdot}/n, \quad v_j = n_{\cdot j}/n, \quad n_{ij} = n_{i\cdot}n_{\cdot j}/n \quad \text{for all } i, j.$$

To find the least squares estimates of the parameters μ, α_i etc. for G we require the least squares estimates of $\mathbf{A}_r\boldsymbol{\theta}$ ($r = 1, 2, 3, 4$, $\boldsymbol{\theta} \in \Omega$). From the Gauss-Markov Theorem 3.6 in Sect. 3.5 these are given by $\mathbf{A}_r\mathbf{P}_\Omega\mathbf{y}$, which is just $\mathbf{A}_r\mathbf{y}$ since $\mathbf{P}_\Omega = \mathbf{I}_n - \mathbf{A}'(\mathbf{AA})^{-1}\mathbf{A}$ and $\mathbf{A}_r\mathbf{A}' = \mathbf{0}$ (Theorem 6.2). Therefore the least squares estimates $\hat{\mu}$, $\hat{\alpha}_i$ etc. can be written down immediately from the definitions of the parameters by replacing $\boldsymbol{\theta}$ by $\mathbf{y}$ and using the above conditions for orthogonality: thus

$$\begin{aligned}
\hat{\mu} &= \sum_i \sum_j n_{i\cdot}n_{\cdot j}\overline{y}_{ij\cdot}/n^2 \\
&= \sum_i \sum_j \sum_k y_{ijk}/n \\
&= \overline{y}_{\cdots} \text{ say,} \\
\hat{\gamma}_{ij} &= \overline{y}_{ij\cdot} - \sum_j n_{\cdot j}\overline{y}_{ij\cdot}/n - \sum_i n_{i\cdot}\overline{y}_{ij\cdot}/n + \overline{y}_{\cdots} \\
\hat{\alpha}_i &= \sum_j n_{\cdot j}\overline{y}_{ij\cdot} - \overline{y}_{\cdots} \text{ and} \\
\hat{\beta}_j &= \sum_i n_{i\cdot}\overline{y}_{ij\cdot} - \overline{y}_{\cdots}.
\end{aligned}$$

Suppose we wish to test $H_1 : \nu_{ijk} = 0, \gamma_{ij} = 0$, or $\boldsymbol{\theta} \in \omega_1$, say, then we require the least squares estimates $\mathbf{A}_r\hat{\boldsymbol{\theta}}_1$ of $\mathbf{A}_r\boldsymbol{\theta}$ ($r = 2, 3, 4$) for $\boldsymbol{\theta} \in \omega_1$, namely $\mathbf{A}_r\mathbf{P}_1\mathbf{y}$, where $\mathbf{P}_1 = \mathbf{P}_{\omega_1}$. Now $\mathbf{I}_n - \mathbf{P}_1$ represents the projection onto $\mathcal{C}[(\mathbf{A}', \mathbf{A}_1')]$ (by Theorem 1.1 in Sect. 1.2) which is orthogonal to $\mathcal{C}[\mathbf{A}_r']$ for $r = 2, 3, 4$. Therefore we have the result $(\mathbf{I}_n - \mathbf{P}_1)\mathbf{A}_r' = \mathbf{0}$ or $\mathbf{A}_r\mathbf{P}_1 = \mathbf{A}_r$ and

$$\mathbf{A}_r\hat{\boldsymbol{\theta}}_1 = \mathbf{A}_r\mathbf{y} = \mathbf{A}_r\mathbf{P}_\Omega\mathbf{y} = \mathbf{A}_r\hat{\boldsymbol{\theta}} \quad \text{for } r = 2, 3, 4.$$

This confirms that the least squares estimates of μ, α_i, and β_j remain unchanged when $\gamma_{ij} = 0$ and do not have to be recalculated. As already noted, this follows from the fact that the groups of estimates are uncorrelated and therefore independent under the assumptions of normality. Hence the least squares estimates of any one group, say the $\{\alpha_i\}$, are independent of the least squares estimates of the parameters in the other groups μ, $\{\beta_j\}$, and $\{\gamma_{ij}\}$.

The numerator sums of squares, $\| \hat{\boldsymbol{\theta}} - \hat{\boldsymbol{\theta}}_1 \|^2$, for the F-test of H_1 is simply $\sum_i \sum_j \sum_k \hat{\gamma}_{ij}^2 = \sum_i \sum_j n_{ij}\hat{\gamma}_{ij}^2$ as the i,j,kth element of $\hat{\boldsymbol{\theta}} - \hat{\boldsymbol{\theta}}_1$ is $(\hat{\mu} + \hat{\alpha}_i + \hat{\beta}_j + \hat{\gamma}_{ij}) - (\hat{\mu} + \hat{\alpha}_i + \hat{\beta}_j)$ or $\hat{\gamma}_{ij}$. Similarly we have

$$\| \hat{\boldsymbol{\theta}}_1 - \hat{\boldsymbol{\theta}}_{12} \|^2 = \| \hat{\boldsymbol{\theta}} - \hat{\boldsymbol{\theta}}_2 \|^2 = \sum_i n_{i\cdot}\hat{\alpha}_i^2$$

$$\| \hat{\boldsymbol{\theta}}_{12} - \hat{\boldsymbol{\theta}}_{123} \|^2 = \| \hat{\boldsymbol{\theta}} - \hat{\boldsymbol{\theta}}_3 \|^2 = \sum_j n_{\cdot j}\hat{\beta}_j^2$$

$$\| \hat{\boldsymbol{\theta}}_{123} - \hat{\boldsymbol{\theta}}_{1234} \| = \| \hat{\boldsymbol{\theta}} - \hat{\boldsymbol{\theta}}_4 \|^2 = n\hat{\mu}^2$$

and

$$\begin{aligned}\hat{\theta}_{ijk} &= \hat{\mu} + \hat{\alpha}_i + \hat{\beta}_j + \hat{\gamma}_{ij} \\ &= \overline{y}_{ij\cdot}.\end{aligned}$$

We note that $\hat{\boldsymbol{\theta}}_{1234} = \mathbf{0}$ and therefore $\| \hat{\boldsymbol{\theta}}_{123} \|^2 = n\hat{\mu}^2$. Thus corresponding to the decomposition

$$\theta_{ijk} = \mu + \alpha_i + \beta_j + \gamma_{ij} + \nu_{ijk}$$

we have a similar decomposition

$$y_{ijk} = \hat{\mu} + \hat{\alpha}_i + \hat{\beta}_j + \hat{\gamma}_{ij} + y_{ijk} - \overline{y}_{ij\cdot}.$$

Squaring both sides and summing on i,j,k we find that the cross-product terms vanish because of orthogonality, giving

$$\begin{aligned}\sum_i \sum_j \sum_k y_{ijk}^2 &= n\hat{\mu}^2 + \sum_i n_{i\cdot}\hat{\alpha}_i^2 + \sum_j n_{\cdot j}\hat{\beta}_j^2 \\ &+ \sum_i \sum_j n_{ij}\hat{\gamma}_{ij}^2 + \sum_i \sum_j \sum_k (y_{ijk} - \overline{y}_{ij\cdot})^2.\end{aligned}$$

In general we usually consider the total variation about the mean namely,

$$\begin{aligned}\mathbf{y}'(\mathbf{I}_n - \mathbf{P}_{123})\mathbf{y} &= \sum_i \sum_j \sum_k y_{ijk}^2 - n\hat{\mu}^2 \\ &= \sum_i \sum_j \sum_k (y_{ijk} - \overline{y}_{\cdots})^2\end{aligned}$$

and construct the following analysis of variance table for the sums of squares (SS). Here the term "row" and "column" refer to the levels of the factors A and

Table 6.1 Two-way ANOVA table

Source	SS	df	MSS
Between rows	$\sum_i n_{i\cdot}\hat{\alpha}_i^2$	$I-1$	$MSS(2)$
Between columns	$\sum_j n_{\cdot j}\hat{\beta}_j^2$	$J-1$	$MSS(3)$
Interactions	$\sum_i \sum_j n_{ij}\hat{\gamma}_{ij}^2$	$(I-1)(J-1)$	$MSS(1)$
Residual	$\sum_i \sum_j \sum_k (y_{ijk} - \overline{y}_{ij\cdot})^2$	$n-IJ$	MSS
Corrected total	$\sum_i \sum_j \sum_k (y_{ijk} - \overline{y}_{\cdots})^2$	$n-1$	
Correction for the mean	$n\overline{y}_{\cdots}^2$	1	
Total	$\sum_i \sum_j \sum_k y_{ijk}^2$	n	

B respectively. The test statistic for testing H_r is simply

$$F = \frac{MSS(r)}{MSS} \quad r = 1, 2, 3,$$

where MSS, as usual, denotes the appropriate *SS* divided by its degrees of freedom (Table 6.1).

In general we are not interested in testing $H_4 : \mu = 0$, but if we accept hypothesis H_{123} we may be interested in finding confidence intervals for μ. These can be calculated from $\sqrt{n}(\overline{y}_{\cdots} - \mu)/\sqrt{MSS}$, which has the t-distribution with $n-IJ$ degrees of freedom.

The column giving the degrees of freedom for each *SS* is obtained by calculating the number of independent constraints in $\mathbf{A}_p\boldsymbol{\theta}$ for $p = 1, 2, 3$. Thus $\boldsymbol{\alpha} = \mathbf{A}_2\boldsymbol{\theta}$ has $(I-1)$ independent constraints as there exists one identifiability condition $\sum_i u_i\alpha_i = 0$. Similarly, as $\sum_i u_i\gamma_{ij} = 0$ for $j = 1, 2, \ldots, J$ and $\sum_j v_j\gamma_{ij} = 0$ for $i = 1, 2, \ldots, I$ with $\sum_i \sum_j u_i v_j \gamma_{ij} = 0$ in common, we see that $\boldsymbol{\gamma} = \mathbf{A}_1\boldsymbol{\theta}$ has $(I+J-1)$ identifiability conditions giving us $IJ - I - J + 1$ or $(I-1)(J-1)$ independent constraints for the γ_{ij}. The degrees of freedom can also be obtained from the traces associated with the *SS*, where the trace is the sum of the coefficients of the terms y_{ijk}^2. The expected value of a quadratic $\mathbf{y}'\mathbf{C}\mathbf{y}$ is σ^2 trace[C] $+ \boldsymbol{\theta}'\mathbf{C}\boldsymbol{\theta}$ so that it can be shown, for example, that

$$E[\sum_i n_{i\cdot}\hat{\alpha}_i^2] = \sigma^2(I-1) + \sum_i n_{i\cdot}\alpha_i^2.$$

If $n_{ij} = K$ for all i, j (which is usually referred to as a balanced design), it can be shown that the diagonal elements of each of the projection matrices $\mathbf{P}_i$ for the hypothesis $H_i : \boldsymbol{\theta} \in \omega_i$ $(i = 1, 2, 3)$ and of $\mathbf{P}_\Omega$ are all equal. This means that the corresponding F tests are quadratically balanced, which implies some robustness to non-normality (by Theorem 5.3 in Sect. 5.2 and the following discussion). If we set

$u_i = 1/I$ and $v_j = 1/J$ so that the identifiability conditions are now $\sum_i \alpha_i = 0$ and $\sum_j \beta_j = 0$ etc., we again have orthogonality as the orthogonality conditions are satisfied. Clearly having equal numbers per cell is the ideal situation.

6.5 Orthogonality in Complete p-Factor Layouts

The ideas developed in the previous section can be extended to complete layouts with more than two factors. We have the following theorem (cf. Seber, 1964).

Theorem 6.6 *A p-factor analysis of variance model with $n_{i_1 i_2 i_3 \cdots i_p}$ observations per cell has orthogonal hypotheses if and only if*

$$n_{i_1 i_2 i_3 \cdots i_p} = \frac{(n_{i_1 \cdots \cdot})(n_{\cdot i_2 \cdots})\cdots(n_{\cdots i_p})}{n_{\cdots\cdot}^{p-1}} \quad (\text{for all } i_1, i_2, \ldots, i_p), \tag{6.15}$$

where a "dot" signifies summing on that subscript. For example, $n_{\cdots\cdot}$ (= n, say) is the sum of all the observations, namely

$$n_{\cdots\cdot} = \sum_{i_1}\sum_{i_2}\cdots\sum_{i_p} n_{i_1 i_2 i_3 \cdots i_p}.$$

Proof As the notation becomes very complicated we shall prove this theorem for just $p = 3$, with a change in notation for clarity, and then briefly indicate the generalizations needed for $p > 3$. We use a different approach from the case $p = 2$ by beginning directly with three-way layout model

$$y_{ijk\ell} = \theta_{ijk\ell} + \varepsilon_{ijk\ell}$$

for $i = 1, 2, \ldots, I; j = 1, 2, \ldots, J; k = 1, 2, \ldots, K; \ell = 1, 2, \ldots, n_{ijk}$ and

$$\theta_{ijk\ell} = \phi_{ijk} = \overline{\theta}_{ijk\cdot}.$$

Let $\nu_{ijk\ell} = \theta_{ijk\ell} - \overline{\theta}_{ijk\cdot}$. We now define the following interactions and main effects:

$$\begin{aligned}
\pi_{ijk} &= \overline{\theta}_{ijk\cdot} - \overline{\theta}_{ij\cdot\cdot} - \overline{\theta}_{i\cdot k\cdot} - \overline{\theta}_{\cdot jk\cdot} + \overline{\theta}_{i\cdots} + \overline{\theta}_{\cdot j\cdot\cdot} + \overline{\theta}_{\cdot\cdot k\cdot} - \overline{\theta}_{\cdots\cdot} \\
\gamma_{ij}^{(12)} &= \overline{\theta}_{ij\cdot\cdot} - \overline{\theta}_{i\cdots} - \overline{\theta}_{\cdot j\cdot\cdot} + \overline{\theta}_{\cdots\cdot} \\
\gamma_{ik}^{(13)} &= \overline{\theta}_{i\cdot k\cdot} - \overline{\theta}_{i\cdots} - \overline{\theta}_{\cdot\cdot k\cdot} + \overline{\theta}_{\cdots\cdot} \\
\gamma_{jk}^{(23)} &= \overline{\theta}_{\cdot jk\cdot} - \overline{\theta}_{\cdot j\cdot\cdot} - \overline{\theta}_{\cdot\cdot k\cdot} + \overline{\theta}_{\cdots\cdot} \\
\alpha_i^{(1)} &= \overline{\theta}_{i\cdots} - \overline{\theta}_{\cdots\cdot}
\end{aligned}$$

$$\alpha_j^{(2)} = \overline{\theta}_{\cdot j \cdot\cdot} - \overline{\theta}_{\cdot\cdot\cdot\cdot}$$
$$\alpha_k^{(3)} = \overline{\theta}_{\cdot\cdot k \cdot} - \overline{\theta}_{\cdot\cdot\cdot\cdot}$$

together with $\mu = \overline{\theta}_{\cdot\cdot\cdot\cdot}$, where $\overline{\theta}_{ijk\cdot} = \sum_\ell \theta_{ijk\ell}/n_{ijk}$, $\overline{\theta}_{ij\cdot\cdot} = \sum_k \sum_\ell \theta_{ijk\ell}/n_{ij\cdot}$, $\overline{\theta}_{i\cdot\cdot\cdot} = \sum_j \sum_k \sum_\ell \theta_{ijk\ell}/n_{i\cdot\cdot}$ etc. We note that this time weights are automatically imposed, for example $\sum_i \alpha_i^{(1)} n_{i\cdot\cdot}/n = 0$. The $\{\pi\}$ are called second order interactions, the $\{\gamma\}$ first order interactions, and the $\{\alpha\}$ are the main effects for each factor. Then

$$\theta_{ijk\ell} = \mu + \alpha_i^{(1)} + \alpha_j^{(2)} + \alpha_k^{(3)} + \gamma_{ij}^{(12)} + \gamma_{ik}^{(13)} + \gamma_{jk}^{(23)} + \pi_{ijk} + \nu_{ijk\ell},$$

and we assume $G : \nu_{ijk\ell} = \theta_{ijk\ell} - \overline{\theta}_{ijk\cdot} = 0$ for all i, j, k, ℓ. Given G, let H_1, H_2, and H_3 be the individual hypotheses of no main effects, H_{12}, H_{13}, and H_{23} the hypotheses of zero first order interactions, H_{123} the hypothesis of zero second order interactions, i.e. $\pi_{ijk} = 0$, and $H_0 : \mu = 0$. We denote the matrices corresponding to the null space representations of these hypotheses by $\mathbf{A}_r$, $\mathbf{A}_{rs}$, $\mathbf{A}_{123}$, and $\mathbf{A}_0 = \mathbf{1}_n'$, respectively; also $\mathbf{A}$ denotes the matrix corresponding to G. Now it is seen that, apart from μ, all the other parameters are contrasts in $\theta_{ijk\ell}$ so that $\mathbf{A}\mathbf{1}_n$, $\mathbf{A}_r\mathbf{1}_n$, $\mathbf{A}_{rs}\mathbf{1}_n$, and $\mathbf{A}_{123}\mathbf{1}_n$ are all zero, which means that each matrix post-multiplied by $\mathbf{A}_0'$ is zero. This is obviously true for a p-factor layout. To apply Theorem 6.2 we now need to show that all the matrix products $\mathbf{A}_r\mathbf{A}_s'$, $\mathbf{A}_r\mathbf{A}_{st}'$ etc. are zero if and only if (6.15) is true.

Sufficiency. Assuming (6.15) to be true for $p = 3$, we have

$$n_{ijk} = \frac{n_{i\cdot\cdot} n_{\cdot j \cdot} n_{\cdot\cdot k}}{n^2} \tag{6.16}$$

and summing on i gives us

$$n_{\cdot jk} = \frac{n_{\cdot j \cdot} n_{\cdot\cdot k}}{n} \tag{6.17}$$

together with two similar expressions obtained by summing on j and k respectively. Since $\nu_{ijk\ell}$ is the only parametric expression containing ℓ, and $\nu_{ijk\cdot} = 0$, we have, by taking matrix products and summing on ℓ first, that all the matrices post-multiplied by $\mathbf{A}'$ are zero. For example, if $\boldsymbol{\theta}$ has elements $\theta_{ijk\ell}$ (stacked according to the order i, j, j, k, ℓ), then $\mathbf{A}\boldsymbol{\theta} = \mathbf{0}$ implies that $\mathbf{A}$ is an $n \times n$ matrix with its row corresponding to $\nu_{i_0 j_0 k_0 \ell_0}$ having its (r_0, s_0, t_0, u_0) element

$$\delta_{i_0 r_0} \delta_{j_0 s_0} \delta_{k_0 t_0} \delta_{\ell_0 u_0} - \delta_{i_0 r_0} \delta_{j_0 s_0} \delta_{k_0 t_0} / n_{r_0 s_0 t_0}, \tag{6.18}$$

where δ_{ab} is the Kronecker delta. Summing (6.18) on ℓ_0 gives us zero. Similarly the row of $\mathbf{A}_{123}$ corresponding to $\pi_{i_1 j_1 k_1}$ has (r_1, s_1, t_1, u_1) element

$$\frac{\delta_{i_1 r_1}\delta_{j_1 s_1}\delta_{k_1 t_1}}{n_{r_1 s_1 t_1}} - \frac{\delta_{i_1 r_1}\delta_{j_1 s_1}}{n_{r_1 s_1 \cdot}} - \frac{\delta_{i_1 r_1}\delta_{k_1 t_1}}{n_{r_1 \cdot t_1}} - \frac{\delta_{j_1 s_1}\delta_{k_1 t_1}}{n_{\cdot s_1 t_1}} + \frac{\delta_{i_1 r_1}}{n_{r_1 \cdot\cdot}} + \frac{\delta_{j_1 s_1}}{n_{\cdot s_1 \cdot}} + \frac{\delta_{k_1 t_1}}{n_{\cdot\cdot t_1}} - \frac{1}{n}. \tag{6.19}$$

Using equations like (6.16) we find that (6.19) reduces to

$$n^2 \left(\frac{\delta_{i_1 r_1}}{n_{r_1 \cdot\cdot}} - \frac{1}{n}\right)\left(\frac{\delta_{j_1 s_1}}{n_{\cdot s_1 \cdot}} - \frac{1}{n}\right)\left(\frac{\delta_{k_1 t_1}}{n_{\cdot\cdot t_1}} - \frac{1}{n}\right). \tag{6.20}$$

With $\mathbf{A}_{12}$, the (r_2, s_2, t_2, u_2) element of the row corresponding to $\gamma_{i_2 j_2}^{(12)}$ is

$$\frac{\delta_{i_2 r_2}\delta_{j_2 s_2}}{n_{r_2 s_2 \cdot}} - \frac{\delta_{i_2 r_2}}{n_{r_2 \cdot\cdot}} - \frac{\delta_{j_2 s_2}}{n_{\cdot s_2 \cdot}} + \frac{1}{n},$$

and, using equations like (6.17), the above equation reduces to

$$n \left(\frac{\delta_{i_2 r_2}}{n_{r_2 \cdot\cdot}} - \frac{1}{n}\right)\left(\frac{\delta_{j_2 s_2}}{n_{\cdot s_2 \cdot}} - \frac{1}{n}\right). \tag{6.21}$$

Finally, $\mathbf{A}_1$ has the (r_3, s_3, t_3, u_3) element of the row corresponding to $\alpha_{i_3}^{(1)}$ given by

$$\frac{\delta_{i_3 r_3}}{n_{r_3 \cdot\cdot}} - \frac{1}{n}. \tag{6.22}$$

We see then that all the elements of the matrices $\mathbf{A}_r$, $\mathbf{A}_{rs}$, and $\mathbf{A}_{rst}$ factorize into one or more of the following types of brackets:

$$\left(\frac{\delta_{ir}}{n_{r\cdot\cdot}} - \frac{1}{n}\right), \quad \left(\frac{\delta_{js}}{n_{\cdot s \cdot}} - \frac{1}{n}\right), \quad \text{and} \quad \left(\frac{\delta_{kt}}{n_{\cdot\cdot t}} - \frac{1}{n}\right). \tag{6.23}$$

When we form all the matrix product pairs $\mathbf{A}_1\mathbf{A}_{12}'$, $\mathbf{A}_{23}\mathbf{A}_{123}'$ etc. we find that, ignoring any power of n, the product of corresponding elements is also a product of terms like those given by (6.23) with at least one of the above types of brackets occurring only once. (This is true for general p-factor models.) For example, considering $\mathbf{A}_{12}\mathbf{A}_{23}'$ we multiply the (r_2, s_2, t_2, u_2) term of a row of $\mathbf{A}_{12}$ (cf. (6.21)) by the (r_3, s_3, t_3, u_3) term of a row of $\mathbf{A}_{23}$ and set $r_3 = r_2$, $s_3 = s_2$, $t_3 = t_2$, and $u_3 = u_2$ to get

$$n^2 \left(\frac{\delta_{i_2 r_2}}{n_{r_2 \cdot\cdot}} - \frac{1}{n}\right)\left(\frac{\delta_{j_2 s_2}}{n_{\cdot s_2 \cdot}} - \frac{1}{n}\right)\left(\frac{\delta_{j_3 s_2}}{n_{\cdot s_2 \cdot}} - \frac{1}{n}\right)\left(\frac{\delta_{k_3 t_2}}{n_{\cdot\cdot t_2}} - \frac{1}{n}\right),$$

where we have a "j" term occurring twice and the "i" and "k" terms occurring only once each. We now sum over r_2, s_2, t_2 and u_2 to get an element of $\mathbf{A}_{12}\mathbf{A}'_{23}$. In fact to get the result that we want we only need to sum the first bracket over r_2 and u_2 as it is the only term containing r_2, namely

$$\begin{aligned}\sum_{r_2=1}^{I}\sum_{u_2=1}^{n_{r_2s_2t_2}}\left(\frac{\delta_{i_2r_2}}{n_{r_2\cdot\cdot}}-\frac{1}{n}\right)&=\sum_{r_2=1}^{I}n_{r_2s_2t_2}\left(\frac{\delta_{i_2r_2}}{n_{r_2\cdot\cdot}}-\frac{1}{n}\right)\\&=\frac{n_{i_2s_2t_2}}{n_{i_2\cdot\cdot}}-\frac{n_{\cdot s_2t_2}}{n}\\&=0,\end{aligned}$$

since combining (6.16) and (6.17) gives us

$$n_{ijk}=\frac{n_{i\cdot\cdot}n_{\cdot jk}}{n}$$

for all i,j,k. Using similar arguments we see that all the matrix products are zero and the hypotheses are orthogonal.

Necessity. We assume that all the matrix products are zero. From $\mathbf{A}_1\mathbf{A}'_2=\mathbf{0}$ it can be shown, after some algebra, that

$$\begin{aligned}0&=\sum_{r_3=1}^{I}\sum_{s_3=1}^{J}\sum_{t_3=1}^{K}\sum_{u_3=1}^{n_{r_3s_3t_3}}\left(\frac{\delta_{i_3r_3}}{n_{r_3\cdot\cdot}}-\frac{1}{n}\right)\left(\frac{\delta_{j_3s_3}}{n_{\cdot s_3\cdot}}-\frac{1}{n}\right)\\&=\sum_{r_3=1}^{I}\sum_{s_3=1}^{J}\sum_{t_3=1}^{K}n_{r_3s_3t_3}\left(\frac{\delta_{i_3r_3}}{n_{r_3\cdot\cdot}}-\frac{1}{n}\right)\left(\frac{\delta_{j_3s_3}}{n_{\cdot s_3\cdot}}-\frac{1}{n}\right)\\&=\frac{n_{i_3j_3\cdot}}{n_{i_3\cdot\cdot}n_{\cdot j_3\cdot}}-\frac{1}{n}.\end{aligned}$$

Hence

$$n_{ij\cdot}=\frac{n_{i\cdot\cdot}n_{\cdot j\cdot}}{n}\tag{6.24}$$

for all i and j, and we have similar expressions for $n_{\cdot jk}$ and $n_{i\cdot k}$.

From $\mathbf{A}_{12}\mathbf{A}'_3=\mathbf{0}$ and using expressions like (6.24) it can be shown that

$$\begin{aligned}0&=\sum_{r_2=1}^{I}\sum_{s_2=1}^{J}\sum_{t_2=1}^{K}\left(\frac{\delta_{i_2r_2}\delta_{j_2s_2}}{n_{r_2s_2\cdot}}-\frac{\delta_{i_2r_2}}{n_{r_2\cdot\cdot}}-\frac{\delta_{j_2s_2}}{n_{\cdot s_2\cdot}}+\frac{1}{n}\right)\left(\frac{\delta_{k_2t_2}}{n_{\cdot\cdot t_2}}-\frac{1}{n}\right)\\&=\frac{n_{i_2j_2k_2}}{n_{i_2j_2\cdot}n_{\cdot\cdot k_2}}-\frac{1}{n}.\end{aligned}$$

Hence

$$n_{ijk} = \frac{n_{ij\cdot}n_{\cdot\cdot k}}{n},$$

which combined with (6.24) gives us

$$n_{ijk} = \frac{n_{i\cdot\cdot}n_{\cdot j\cdot}n_{\cdot\cdot k}}{n},$$

and the result is proved.

The above proof can be extended to any higher-way layout and we demonstrate this using a 4-factor experiment. The matrices involved would now be of the form $\mathbf{A}_0 = \mathbf{1}_n$ as before along with $\mathbf{A}_i$, $\mathbf{A}_{ij}$, $\mathbf{A}_{ijk}$, and $\mathbf{A}_{1234}$ representing main effects, first order interactions, second order interactions, and third order interactions, respectively. To prove sufficiency we need to show that the appropriate products of all pairs of matrices, for example $\mathbf{A}_{12}\mathbf{A}_{1234}$ are zero. In the latter case we would get "i" and '"j" factors occurring twice each (cf. (6.23)) as subscripts (1,2) occur in both matrices, and the other two factors occur only once each. We then sum on the number of observations and on a factor occurring only once to get zero. To prove necessity for $p = 4$, we assume $\mathbf{A}_1\mathbf{A}_2'$, $\mathbf{A}_{12}\mathbf{A}_3'$, and $\mathbf{A}_{123}\mathbf{A}_4$ are all zero matrices to prove the following (using a more general notation):

$$\begin{aligned} n_{i_1 i_2\cdot\cdot} &= (n_{i_1\cdot\cdot\cdot})(n_{\cdot i_2\cdot\cdot})/n_{\cdot\cdot\cdot\cdot} \\ n_{i_1 i_2 i_3\cdot} &= (n_{i_1 i_2\cdot\cdot})(n_{\cdot\cdot i_3\cdot})/n_{\cdot\cdot\cdot\cdot} \\ n_{i_1 i_2 i_3 i_4} &= (n_{i_1 i_2 i_3\cdot})(n_{\cdot\cdot\cdot i_4})/n_{\cdot\cdot\cdot\cdot}, \end{aligned}$$

which combined give

$$n_{i_1 i_2 i_3 i_4} = (n_{i_1\cdot\cdot\cdot})(n_{\cdot i_2\cdot\cdot})(n_{\cdot\cdot i_3\cdot})(n_{\cdot\cdot\cdot i_4})/n^3,$$

our required result. The conditions for orthogonality are automatically satisfied if we have equal numbers of observations per cell, i.e. $n_{i_1 i_2 i_3 i_4} = n_0$ for all i_1,i_2, i_3, and i_4. For further background reading, particularly with regard to the formulation of interactions, see Seber and Lee (2003, section 8.6).

6.6 Orthogonality in Randomized Block Designs

Consider a randomized block design with I treatments and J blocks. Let y_{ij} be the observation for the ith treatment on the jth block, and assume the model $E[y_{ij}] = \theta_{ij} = \mu + \alpha_i + \beta_j$ for $i = 1, 2, \ldots, I$ and $j = 1, 2, \ldots, J$. This is the same model as for the two-way layout of Sect. 6.4, except that the treatment $\times$ block interactions are assumed to be negligible and there is only one observation per cell. As before we use weights $\{u_i\}$ and $\{v_j\}$ giving the identifiability conditions $\sum_i u_i\alpha_i = \sum_j v_j\beta_j = 0$. We now consider the hypotheses

$$H_1 : \theta_{ij} = \mu + \beta_j, \quad H_2 : \theta_{ij} = \mu + \alpha_i, \quad H_3 : \theta_{ij} = \alpha_i + \beta_j$$

and find what weights we must use for H_1, H_2, and H_3 to be orthogonal with respect to $G : \theta_{ij} = \mu + \alpha_i + \beta_j$.

Theorem 6.7 *The hypotheses H_i $(i = 1, 2, 3)$ are orthogonal with respect to G if and only if $u_i = 1/I$ and $v_j = 1/J$ for all i and j.*

Proof We shall use a more direct proof than that in Theorem 6.5.

(a) *Necessity*. From the alternative form of the definition of orthogonality we have that the hypotheses are orthogonal if and only if $\omega_i^{\perp} \cap \Omega \subset \omega_j$ for $j \neq i$. Thus a necessary condition for orthogonality is that $\omega_3^{\perp} \cap \Omega \subset \omega_1 \cap \omega_2$. The vector space $\omega_3^{\perp} \cap \Omega$ is defined by the set of $\theta_{ij} = \mu + \alpha_i + \beta_j$ such that $\boldsymbol{\theta} \perp \omega_3$, i.e.,

$$\sum_i \sum_j (\alpha_i^* + \beta_j^*)(\mu + \alpha_i + \beta_j) = 0 \tag{6.25}$$

for every α_i^* and β_j^* satisfying the constraints $\sum_i u_i\alpha_i^* = \sum_j v_j\beta_j^* = 0$. If this set of θ_{ij} also belongs to $\omega_1 \cap \omega_2$, the θ_{ij} must be constant with respect to i and j so that $\alpha_i = \alpha$, $\beta_j = \beta$, say. Equation (6.25) now becomes

$$\sum_i \sum_j (\alpha_i^* + \beta_j^*)(\mu + \alpha + \beta) = 0.$$

Since we are concerned with nontrivial vectors, $(\mu + \alpha + \beta) \neq 0$, and by putting the $\{\beta_j *\}$ equal to zero we see that $\sum_i \alpha_i^* = 0$. In the same way $\sum_j \beta_j^* = 0$ so that we have shown that the identifiability conditions must take the form $\sum_i \alpha_i = 0$ and $\sum_j \beta_j = 0$.

(b) *Sufficiency*. If $\sum_i \alpha_i = \sum_j \beta_j = 0$, then the vector space $\omega_1^{\perp} \cap \Omega$ is defined by the set of θ_{ij} such that

$$\begin{aligned} 0 &= \sum_i \sum_j (\mu^* + \beta_j^*)(\mu + \alpha_i + \beta_j) \\ &= \mu^*(IJ\mu) + I \sum_j \beta_j^* \beta_j, \end{aligned}$$

for every μ^* and β_j^*. Hence $\mu = 0$ and $\beta_j = 0$, giving $\theta_{ij} = \alpha_i$. This implies that $\omega_1^{\perp} \cap \Omega \subset \omega_2$. The other requirements follow in a similar manner. The above method of proof was suggested by Dr. S. D. Silvey.

We now turn our attention to the least squares estimation of the unknown parameters, and by way of variation we give a slightly different approach from that used in the complete two-way layout.

Let $\sum_i \alpha_i = 0$ and $\sum_j \beta_j = 0$, then we find that

$$\mu = \overline{\theta}_{\cdot\cdot}, \quad \alpha_i = \overline{\theta}_{i\cdot} - \overline{\theta}_{\cdot\cdot}, \quad \text{and} \quad \beta_j = \overline{\theta}_{\cdot j} - \overline{\theta}_{\cdot\cdot}. \tag{6.26}$$

Hence

$$\begin{aligned} y_{ij} - \theta_{ij} &= (\overline{y}_{\cdot\cdot} - \mu) + (\overline{y}_{i\cdot} - \overline{y}_{\cdot\cdot} - \alpha_i) + (\overline{y}_{\cdot j} - \overline{y}_{\cdot\cdot} - \beta_j) \\ &\quad + (y_{ij} - \overline{y}_{i\cdot} - \overline{y}_{\cdot j} + \overline{y}_{\cdot\cdot}). \end{aligned}$$

Squaring both sides, summing on i and j, and using the identifiability constraints, we find that the cross-product terms vanish (because of orthogonality) giving

$$\begin{aligned} \sum_i \sum_j (y_{ij} - \theta_{ij})^2 &= IJ(\overline{y}_{\cdot\cdot} - \mu)^2 + J\sum_i (\overline{y}_{i\cdot} - \overline{y}_{\cdot\cdot} - \alpha_i)^2 \\ &\quad + I\sum_j (\overline{y}_{\cdot j} - \overline{y}_{\cdot\cdot} - \beta_j)^2 + \sum_i \sum_j (y_{ij} - \overline{y}_{i\cdot} - \overline{y}_{\cdot j} + \overline{y}_{\cdot\cdot})^2. \end{aligned}$$

Thus minimizing $\sum_i \sum_j (y_{ij} - \theta_{ij})^2$ with respect to the μ, $\{\alpha_i\}$, and $\{\beta_j\}$ gives us the least squares estimates

$$\hat{\mu} = \overline{y}_{\cdot\cdot}, \quad \hat{\alpha}_i = \overline{y}_{i\cdot} - \overline{y}_{\cdot\cdot}, \quad \text{and} \quad \hat{\beta}_j = \overline{y}_{\cdot j} - \overline{y}_{\cdot\cdot}, \tag{6.27}$$

which are of the same form as (6.26) but with $\boldsymbol{\theta}$ replaced by $\mathbf{y}$. Also the above estimates are unchanged if we put some of the parameters equal to zero. This means we do not have to recalculate the estimates for testing the hypotheses H_1 and H_2, a feature we have seen of orthogonality. The analysis of variance table follows with $MSS = SS/(df)$ (Table 6.2).

Table 6.2 ANOVA table for randomized blocks

Source	SS	df	MSS
Between treatments	$J\sum_i(\overline{y}_{i\cdot} - \overline{y}_{\cdot\cdot})^2$	$I-1$	$MSS(1)$
Between Blocks	$I\sum_j(\overline{y}_{\cdot j} - \overline{y}_{\cdot\cdot})^2$	$J-1$	$MSS(2)$
Residual	$\sum_i \sum_j (y_{ij} - \overline{y}_{i\cdot} - \overline{y}_{\cdot j} + \overline{y}_{\cdot\cdot})^2$	$(I-1)(J-1)$	MSS
Corrected total	$\sum_i \sum_j (y_{ij} - \overline{y}_{\cdot\cdot})^2$	$IJ-1$	
Correction for the mean	$IJ\overline{y}_{\cdot\cdot}^2$	1	
Total	$\sum_i \sum_j y_{ij}^2$	IJ	

The test statistic for testing H_r ($r = 1, 2$) is simply

$$F = \frac{MSS(r)}{MSS}.$$

Before concluding this section, let us consider the model

$$y_{ijk*\dots*} = \mu + \alpha_i + \beta_j + \tau_k + \cdots + \varepsilon_{ijk*\dots*}$$

with identifiability conditions $\sum_i u_i\alpha_i = \sum_j v_j\beta_j = \sum_k w_k\tau_k = \cdots = 0$. The hypotheses of interest are H_1: all the α_i zero, H_2: all the β_j zero, etc., and we can add the hypothesis $H_0 : \mu = 0$. As the proof of Theorem 6.5 can be generalized to deal with this as in the next section, we have that the hypotheses H_0, H_1, H_2,... are orthogonal with respect to G if and only if the identifiability conditions take the form $\sum_i \alpha_i = \sum_j \beta_j = \sum_k \tau_k = \cdots = 0$. The Latin and hyper-Latin square designs and factorial designs with no interactions are special cases of this general model. We look at the Latin square next.

6.7 Orthogonality in Latin Square Designs

An $n \times n$ Latin square design is a design method for comparing three factors A, B, and C at n levels for each factor. An example of a 5×5 Latin square is

$$\begin{matrix} 1 & 2 & 3 & 4 & 5 \\ 2 & 3 & 4 & 5 & 1 \\ 3 & 4 & 5 & 1 & 2 \\ 4 & 5 & 1 & 2 & 3 \\ 5 & 1 & 2 & 3 & 4. \end{matrix}$$

Note that each number occurs once in each row and once in each column. Here factor A has five levels given by the row number, factor B has five levels given by the column number, and factor C has five levels given by the number. For example the entry "4" in the (2,3) position represents factor A at level 2, factor B at level 3, and factor C at level 4. We can obtain other Latin squares by permuting rows, columns, and numbers. The one chosen has "5" down one of the diagonals, which may lead to bias. In setting up such an experiment one usually chooses a Latin square at random from an appropriate set. This randomization goes some way to help achieve any underlying normal distribution assumptions and reduce interaction effects. We won't be going into such details as our focus is on orthogonality.

The model we assume for the Latin square is

$$y_{ijk} = \theta_{ijk} + \varepsilon_{ijk}, \quad (i, j, k) \in S,$$

where y_{ijk} is the observation on the treatment combination of factor A at level i, factor B at level j, and factor C at level k. The triples (i,j,k) take on just n^2 values determined by the Latin square chosen. This set of n^2 values of the triple (i,j,k) is denote by the set S. For the above model, the set of n^2 random variables $\{\varepsilon_{ijk}\}$ are assumed to be independently and identically distributed as $N[0,\sigma^2]$. Our model $G : \boldsymbol{\theta} \in \Omega$ is the set of all $\boldsymbol{\theta}$ such that

$$\theta_{ijk} = \mu + \alpha_i + \beta_j + \gamma_k \quad (i,j,k) \in S,$$

where $\sum_i u_i\alpha_i = \sum_j v_j\beta_j = \sum_k w_k\gamma_k = 0$ are identifiability constraints. The hypotheses of interest are

$$\begin{aligned} H_1 &: \theta_{ijk} = \mu + \beta_j + \gamma_k \quad (\alpha_i = 0 \quad \text{for all } i) \\ H_2 &: \theta_{ijk} = \mu + \alpha_i + \gamma_k \quad (\beta_j = 0 \quad \text{for all } j) \\ H_3 &: \theta_{ijk} = \mu + \alpha_i + \beta_j \quad (\gamma_k = 0 \quad \text{for all } k) \end{aligned}$$

and we add

$$H_4 : \theta_{ijk} = \alpha_i + \beta_j + \gamma_k \quad (\mu = 0).$$

We now establish necessary and sufficient conditions for orthogonality of the hypotheses.

Theorem 6.8 *The hypotheses H_i $(i = 1,2,3,4)$ are orthogonal with respect to G if and only if $u_i = v_j = w_k = 1/n$ for all $(i,j,k) \in S$.*

Proof (Necessity) To prove necessity we assume that the hypotheses are orthogonal, that is $\omega_i^p \subset \omega_j$ for all $i, j, i \neq j$, i.e., $\omega_4^p \subset \omega_1 \cap \omega_2 \cap \omega_3$. Now $\omega_4^\perp \cap \Omega$ is defined by the set of $\theta_{ijk} = \mu + \alpha_i + \beta_j + \gamma_k$ such that $\boldsymbol{\theta} \perp \omega_4$, that is,

$$\sum_{(ijk)\in S} (\alpha_i^* + \beta_j^* + \gamma_k^*)(\mu + \alpha_i + \beta_j + \gamma_k) = 0. \tag{6.28}$$

for all α_i^*, β_j^*, and γ_k^* satisfying $\sum_i u_i\alpha_i^* = \sum_j v_j\beta_j^* = \sum_k w_k\gamma_k^* = 0$. If this set of θ_{ijk} also belongs to $\omega_1 \cap \omega_2 \cap \omega_3$, then θ_{ijk} must be constant with respect to i, j, and k so that $\alpha_i = \alpha$, $\beta_j = \beta$, and $\gamma_k = \gamma$, say. Then (6.28) now becomes

$$\sum_{(i,j,k)\in S} (\alpha_i^* + \beta_j^* + \gamma_k^*)(\mu + \alpha + \beta + \gamma) = 0.$$

For nontrivial vectors we have $(\mu + \alpha + \beta + \gamma) \neq 0$, so that putting the $\{\beta_j^*\}$ and $\{\gamma_k^*\}$ all equal to zero, we have that $\sum_i \alpha_i^* = 0$. Using a similar argument we find that $\sum_j \beta_j^* = \sum_k \gamma_k^* = 0$. Hence our identifiability constraints take the form $\sum_i \alpha_i = \sum_j \beta_j = \sum_k \gamma_k = 0$, and the constraints are necessary.

(*Sufficiency.*) If $\sum_i \alpha_i = \sum_j \beta_j = \sum_k \gamma_k = 0$ then $\omega_1^\perp \cap \Omega$ is the set of all $\boldsymbol{\theta}$ such that $\{\theta_{ijk} : \theta_{ijk} = \mu + \alpha_i + \beta_j + \gamma_k\}$ and

$$\begin{aligned} 0 &= \sum_{(i,j,k)\in S} (\mu^* + \beta_j^* + \gamma_k^*)(\mu + \alpha_i + \beta_j + \gamma_k) \\ &= n^2\mu^*\mu + n\sum_j \beta_j^*\beta_j + n\sum_k \gamma_k^*\gamma_k \end{aligned}$$

for all μ^*, β_j^*, and γ_k^*. Hence $\mu = 0$, $\beta_j = 0$ for all j, and $\gamma_k = 0$ for all k giving us $\theta_{ijk} = \alpha_i$ so that $\omega_1^\perp \cap \Omega \subset \omega_2 \cap \omega_3$. Using a similar argument, by cycling the subscripts, we see that $\omega_2^\perp \cap \Omega \subset \omega_1 \cap \omega_3$ and $\omega_3^\perp \cap \Omega \subset \omega_1 \cap \omega_2$. Now $\omega_4^\perp \cap \Omega$ is the set of all $\boldsymbol{\theta}$ such that $\{\theta_{ijk} : \mu + \alpha_i + \beta_j + \gamma_k\}$ and

$$\begin{aligned} 0 &= \sum_{(i,j,k)\in S} (\alpha_i^* + \beta_j^* + \gamma_k^*)(\mu + \alpha_i + \beta_j + \gamma_k) \\ &= \sum_i \alpha_i^*\alpha_i + \sum_j \beta_j^*\beta_j + \sum_k \gamma_k^*\gamma_k \end{aligned}$$

for all α_i^*, β_j^*, and γ_k^*. Hence $\alpha_i = 0$ for all i, $\beta_j = 0$ for all j, and $\gamma_k = 0$ for all k so that $\theta_{ij} = \mu$ and $\omega_4^\perp \cap \Omega \subset \omega_1 \cap \omega_2 \cap \omega_3$. Bearing in mind that $\omega_i^p \subset \omega_j$ if and only if $\omega_j^p \subset \omega_i$ we see that $\omega_i^p \subset \omega_j$ for all $i, j, i \neq j$ and the hypotheses are orthogonal.

We now find the least squares estimates. For $\boldsymbol{\theta} \in \Omega$, we minimize

$$\sum_{(i,j,k)\in S} (y_{ijk} - \alpha_i - \beta_j - \gamma_k)^2 + \lambda_1(\sum_i \alpha_i - 1) + \lambda_2(\sum_j \beta_j - 1) + \lambda_3(\sum_k \gamma_k - 1) \quad (6.29)$$

subject to the identifiability constraints, where the λ_i are the Lagrange multipliers. Differentiating (6.29) with respect to μ and dividing by -2, we get

$$\begin{aligned} 0 &= \sum_{(i,j,k)\in S} (y_{ijk} - \hat{\mu} - \hat{\alpha}_i - \hat{\beta}_j - \hat{\gamma}_k) \\ &= \sum_{(i,j,k)\in S} y_{ijk} - n^2\hat{\mu} - n\sum_i \hat{\alpha}_i - n\sum_j \hat{\beta}_j - n\sum_k \hat{\gamma}_k \\ &= \sum_{(i,j,k)\in S} y_{ijk} - n^2\hat{\mu} \end{aligned}$$

or $\hat{\mu} = \overline{y}_{\cdots}$, the mean of the n^2 observations. Differentiating (6.29) with respect to α_i gives us

$$0 = \sum_{(j,k)\in S_i} -2(y_{ijk} - \hat{\mu} - \hat{\alpha}_i - \hat{\beta}_j - \hat{\gamma}_k) + \lambda_1,$$

where S_i is the set of n pairs (j, k) for which $(i, j, k) \in S$ and i has a fixed value. If we also sum the above equation over i we see that $\lambda_1 = 0$, which is what we expect from the general theory of identifiability constraints. From the above equation we get

$$\begin{aligned} 0 &= \sum_{(j,k)\in S_i} y_{ijk} - n\hat{\mu} - n\hat{\alpha}_i - \sum_j \hat{\beta}_j - \sum_k \hat{\gamma}_k \\ &= \sum_{(j,k)\in S_i} y_{ijk} - n\hat{\mu} - n\hat{\alpha}_i, \end{aligned}$$

or

$$\hat{\alpha}_i = \overline{y}_{i\cdot\cdot} - \overline{y}_{\cdots},$$

where $\overline{y}_{i\cdot\cdot}$ is the mean of the n observations for which factor A is at level i. By symmetry we have that $\hat{\beta}_j = \overline{y}_{\cdot j\cdot} - \overline{y}_{\cdots}$ and $\hat{\gamma}_k = \overline{y}_{\cdot\cdot k} - \overline{y}_{\cdots}$. Now the residual sum of squares Q is

$$\sum_{(i,j,k)\in S} (y_{ijk} - \hat{\mu} - \hat{\alpha}_i - \hat{\beta}_j - \hat{\gamma}_k)^2 = \sum_{(i,j,k)\in S} (y_{ijk} - \overline{y}_{i\cdot\cdot} - \overline{y}_{\cdot j\cdot} - \overline{y}_{\cdot\cdot k} + 2\overline{y}_{\cdots})^2.$$

To test $H_1 : \alpha_i = 0$ for all i we minimize $\sum_{(i,j,k)\in S}(y_{ijk} - \mu - \beta_j - \gamma_k)^2$ with respect to μ, β_j, and γ_k, and we get the same least squares estimates as before (because of orthogonality), namely $\hat{\mu}$, $\{\hat{\beta}_j\}$ and $\{\hat{\gamma}_k\}$. Now

$$\begin{aligned} y_{ijk} - \theta_{ijk} = (\overline{y}_{\cdots} - \mu) + &\ (\overline{y}_{i\cdot\cdot} - \overline{y}_{\cdots} - \alpha_i) + (\overline{y}_{\cdot j\cdot} - \overline{y}_{\cdots} - \beta_j) + (\overline{y}_{\cdot\cdot k} - \overline{y}_{\cdots} - \gamma_k) \\ + &\ (y_{ijk} - \overline{y}_{i\cdot\cdot} - \overline{y}_{\cdot j\cdot} - \overline{y}_{\cdot\cdot k} + 2\overline{y}_{\cdots}) \end{aligned}$$

Table 6.3 ANOVA table for Latin Square

Source	SS	df	MSS
Factor A	$n\sum_i(\overline{y}_{i\cdot\cdot}-\overline{y}_{\cdot\cdot\cdot})^2$	$n-1$	$MSS(1)$
Factor B	$n\sum_j(\overline{y}_{\cdot j\cdot}-\overline{y}_{\cdot\cdot\cdot})^2$	$n-1$	$MSS(2)$
Factor C	$n\sum_k(\overline{y}_{\cdot\cdot k}-\overline{y}_{\cdot\cdot\cdot})^2$	$n-1$	$MSS(3)$
Residual	$\sum_{(i,j,k)\in S}(y_{ijk}-\overline{y}_{i\cdot\cdot}-\overline{y}_{\cdot j\cdot}-\overline{y}_{\cdot\cdot k}+2\overline{y}_{\cdot\cdot\cdot})^2$	n^2-3n+2	MSS
Corrected total	$\sum_{(i,j,k)\in S}(y_{ijk}-\overline{y}_{\cdot\cdot\cdot})^2$	n^2-1	
Correction for the mean	$n^2\overline{y}^2_{\cdot\cdot\cdot}$	1	
Total	$\sum_{(i,j,k)\in S}y^2_{ijk}$	n^2	

Squaring both sides, summing on i, j and k, and using the identifiability constraints, we find that the cross-product terms vanish (because of the orthogonality) giving

$$\begin{aligned}\sum_{(i,j,k)\in S}(y_{ijk}-\theta_{ijk})^2 &= n^2(\overline{y}_{\cdot\cdot\cdot}-\mu)^2+n\sum_i(\overline{y}_{i\cdot\cdot}-\overline{y}_{\cdot\cdot\cdot}-\alpha_i)^2\\ &\quad +n\sum_j(\overline{y}_{\cdot j\cdot}-\overline{y}_{\cdot\cdot\cdot}-\beta_j)^2+\sum_k(\overline{y}_{\cdot\cdot k}-\overline{y}_{\cdot\cdot\cdot}-\gamma_k)^2\\ &\quad +\sum_{(i,j,k)\in S}(y_{ijk}-\overline{y}_{i\cdot\cdot}-\overline{y}_{\cdot j\cdot}-\overline{y}_{\cdot\cdot k}+2\overline{y}_{\cdot\cdot\cdot})^2.\end{aligned}$$

We can now obtain the least squares estimates and the residual sum of squares Q_i for each H_i by inspection. For example, for H_1, the hypothesis sum of squares $Q_1-Q=n\sum_i(\overline{y}_{i\cdot\cdot}-\overline{y}_{\cdot\cdot\cdot})^2$, which leads to $MSS(1)$ in the Table 6.3 above. Alternatively we can use the result

$$Q_1-Q=\|(\hat{\theta}-\hat{\theta}_{H_1})\|^2=\sum_{(i,j,k)\in S}\{(\hat{\mu}+\hat{\alpha}_i+\hat{\beta}_j+\hat{\gamma}_k)-(\hat{\mu}+\hat{\beta}_j+\hat{\gamma}_k)\}^2=n\sum_i\hat{\alpha}_i^2.$$

The analysis of variance table is given above.

6.8 Non-orthogonal Hypotheses

We see from Sect. 6.4 that we don't have orthogonality of the hypotheses with a two-way layout when there are unequal numbers of observations per mean and certain conditions are not satisfied. In this case hypothesis testing is not so straightforward as different parameterizations are used by different computer packages. We also have the problem that least squares estimates have to be recalculated when some of the parameters are put equal to zero. These are important practical issues discussed in Seber and Lee (2003, section 8.3) for example, but are not part of the main theme of this monograph, which is concerned with broad principles.

References

Darroch, J. N., & Silvey, S. D. (1963). On testing more than one hypothesis. *Annals of Mathematical Statistics, 34*, 555–567.

Davis, P. (1975). *Interpolation and approximation.* New York: Wiley.

Seber, G. A. F. (1964). Orthogonality in analysis of variance. *Annals of Mathematical Statistics, 35*(2), 705–710.

Seber, G. A. F., & Lee, A. J. (2003). *Linear regression analysis* (2nd ed.) New York: Wiley.

Chapter 7
Enlarging the Model

7.1 Introduction

Sometimes after a linear model has been fitted it is realized that more explanatory (x) variables need to be added, as in the following examples.

Example 7.1 In an industrial experiment in which the response (y) is the yield and the explanatory variables are temperature, pressure, etc., we may wish to determine what values of the x-variables are needed to produce a certain yield. However, it may be realized that another variable, say concentration, needs to be incorporated in the regression model. This can be readily done by simply using a standard regression computational package. In this case the added variable is quantitative and is readily added into the original model.

Example 7.2 Consider an experiment that involves finding what variables determine a person's performance on a given task. Suppose quantitative variables such as height, weight, and age are used as well as the qualitative variable gender. In this case gender can be incorporated into the initial regression model using an indicator variable which takes just two values, one for female and zero for male. After fitting a model it is decided that another qualitative variable with r possible unordered categories needs to be added. This can be done, for example by adding $r-1$ indicator variables.

Example 7.3 A more common application when one might add to a model arises in the topic of analysis of covariance where we combine qualitative information as in an analysis of variance model with quantitative information as in regression models. For example, suppose we compare the effect of four teaching methods on the performance of students in a test. Students were selected randomly to form four equal-sized groups giving us a one-way analysis of variance model to test for differences in the four group means. It was then decided that another quantitative variable Intelligence Quotient (IQ) needed to be introduced as it was possible that

G.A.F. Seber, *The Linear Model and Hypothesis*, Springer Series in Statistics,
DOI 10.1007/978-3-319-21930-1_7

the IQs were not randomly spread among the groups. Mathematically we started with model $y_{ij} = \mu_i + \varepsilon_{ij}$, where y_{ij} is the score of the ith student in the jth group ($i = 1, 2, 3, 4; j = 1, 2, \ldots, J$) and the ε_{ij} are independently and identically distributed as $N[0, \sigma^2]$. The hypothesis of interest is $H : \mu_i = \mu$ ($i = 1, 2, 3, 4$). If z_{ij} is the IQ of the same student, then a possible new model might be

$$y_{ij} = \mu_i + \gamma_i z_{ij} + \varepsilon_{ij}$$

when IQ is taken into account. This change amounts to fitting a straight line to the data from each group. Any test of H would amount to comparing the means, but allowing for any IQ effect. The variable z is usually referred to as a *concomitant* variable. Several other hypotheses now present themselves such as $\gamma_i = \gamma$ ($i = 1, 2, 3, 4$) in which the slopes of the lines are the same, and perhaps followed by the hypothesis that $\gamma = 0$. We might even go a step further and consider the model

$$y_{ij} = \mu_i + \gamma_i z_{ij} + \delta_i z_{ij}^2 + \varepsilon_{ij},$$

which gives us a quadratic model for each group. Such models can be readily fitted using a standard regression package. However there are some algebraic methods that can be used to assist with model fitting and show the usefulness of projection methods that we now consider. We set up a general model in the next section.

7.2 Least Squares Estimation

Given $\mathbf{y} = \boldsymbol{\theta} + \boldsymbol{\varepsilon}$, suppose our linear model $G : \boldsymbol{\theta} \in \Omega$, with $\dim[\Omega] = p$, is modified to $\tilde{G} : \boldsymbol{\theta} \in \tilde{\Omega} = \Omega \oplus \mathcal{C}[\mathbf{Z}]$, where $\boldsymbol{\varepsilon}$ is $N_n[\mathbf{0}, \sigma^2 \mathbf{I}_n]$, $\Omega \cap \mathcal{C}[\mathbf{Z}] = \mathbf{0}$, and $\mathbf{Z}$ is $n \times r$ of rank r. Instead of calculating the least squares estimates for $\boldsymbol{\theta}$ and $\boldsymbol{\gamma}$ in the new model $\tilde{G}$, it is often more helpful to obtain least squares estimates for G first and then modify them to give the estimates for $\tilde{G}$. Suppose $\tilde{\boldsymbol{\theta}}$ and $\tilde{\boldsymbol{\gamma}}$ are the least squares estimates for $\tilde{G}$, then

$$\tilde{\boldsymbol{\theta}} + \mathbf{Z}\tilde{\boldsymbol{\gamma}} = \mathbf{P}_{\tilde{\Omega}} \mathbf{y}. \tag{7.1}$$

Since $\Omega \subset \tilde{\Omega}$, $\tilde{\Omega}^{\perp} \subset \Omega^{\perp}$, and from Theorem 4.2 in Sect. 4.2, $\mathbf{P}_{\tilde{\Omega}} - \mathbf{P}_{\Omega}$ is the projection onto $\tilde{\Omega} \cap \Omega^{\perp}$. From the last line of Theorem 4.3 with $\mathbf{A}_1' = \mathbf{Z}$

$$\tilde{\Omega} \cap \Omega^{\perp} = (\Omega \oplus \mathcal{C}[\mathbf{Z}]) \cap \Omega^{\perp} = \mathcal{C}[\mathbf{RZ}],$$

where $\mathbf{R} = \mathbf{I}_n - \mathbf{P}_{\Omega} = \mathbf{P}_{\Omega^{\perp}}$. Since $\mathcal{C}[\mathbf{Z}] \cap \Omega = \mathbf{0}$ and $\mathbf{Z}$ has full rank, it follows from Theorem 4.4 and (4.7) with Ω replaced by $\Omega^{\perp}$ and $\mathbf{A}_1'$ replaced by $\mathbf{Z}$ that

$$\mathbf{P}_{\tilde{\Omega}} - \mathbf{P}_{\Omega} = \mathbf{RZ}(\mathbf{Z}'\mathbf{RZ})^{-1}\mathbf{Z}'\mathbf{R}. \tag{7.2}$$

Premultiplying (7.1) by $\mathbf{Z}'\mathbf{R}$, using $\mathbf{R}\tilde{\theta} = \mathbf{0}$ (since $\tilde{\theta} \in \Omega$), applying $\mathbf{R}\mathbf{P}_\Omega = \mathbf{0}$, and noting that $\mathbf{R}^2 = \mathbf{R}$ gives us

$$\begin{aligned}\mathbf{Z}'\mathbf{R}\mathbf{Z}\tilde{\gamma} &= \mathbf{Z}'\mathbf{R}\mathbf{P}_{\tilde{\Omega}}\mathbf{y}\\ &= \mathbf{Z}'\mathbf{R}[\mathbf{P}_\Omega + \mathbf{R}\mathbf{Z}(\mathbf{Z}'\mathbf{R}\mathbf{Z})^{-1}\mathbf{Z}'\mathbf{R}]\mathbf{y}\\ &= \mathbf{Z}'\mathbf{R}\mathbf{y}.\end{aligned}$$

Hence

$$\tilde{\gamma} = (\mathbf{Z}'\mathbf{R}\mathbf{Z})^{-1}\mathbf{Z}'\mathbf{R}\mathbf{y} \tag{7.3}$$

and, using (7.1), (7.2), and (7.3),

$$\begin{aligned}\tilde{\theta} &= \mathbf{P}_{\tilde{\Omega}}\mathbf{y} - \mathbf{Z}\tilde{\gamma}\\ &= [\mathbf{P}_\Omega + \mathbf{R}\mathbf{Z}(\mathbf{Z}'\mathbf{R}\mathbf{Z})^{-1}\mathbf{Z}'\mathbf{R} - \mathbf{Z}(\mathbf{Z}'\mathbf{R}\mathbf{Z})^{-1}\mathbf{Z}'\mathbf{R}]\mathbf{y}\\ &= [\mathbf{P}_\Omega + (\mathbf{I}_n - \mathbf{P}_\Omega)\mathbf{Z}(\mathbf{Z}'\mathbf{R}\mathbf{Z})^{-1}\mathbf{Z}'\mathbf{R} - \mathbf{Z}(\mathbf{Z}'\mathbf{R}\mathbf{Z})^{-1}\mathbf{Z}'\mathbf{R}]\mathbf{y}\\ &= \mathbf{P}_\Omega(\mathbf{y} - \mathbf{Z}\tilde{\gamma}) \qquad (7.4)\\ &= \hat{\theta} - \mathbf{P}_\Omega\mathbf{Z}\tilde{\gamma}.\end{aligned}$$

The above result suggests the following two-stage procedure. First, we assume $\gamma = \mathbf{0}$ and obtain $\hat{\theta} = \mathbf{P}_\Omega\mathbf{y}$ and the residual sum of squares $\mathbf{y}'\mathbf{R}\mathbf{y}$. Second, minimize $(\mathbf{y} - \mathbf{Z}\gamma)'\mathbf{R}(\mathbf{y} - \mathbf{Z}\gamma)$ with respect to γ by differentiating it to get (cf. A.20)

$$-\mathbf{Z}'\mathbf{R}\mathbf{y} + \mathbf{Z}'\mathbf{R}\mathbf{Z}\tilde{\gamma} = \mathbf{0} \tag{7.5}$$

or

$$\tilde{\gamma} = (\mathbf{Z}'\mathbf{R}\mathbf{Z})^{-1}\mathbf{Z}'\mathbf{R}\mathbf{y},$$

which is (7.3). Third, the estimate $\tilde{\theta}$ is now obtained by replacing $\mathbf{y}$ by $\mathbf{y} - \mathbf{Z}\tilde{\gamma}$ in $\hat{\theta} = \mathbf{P}_\Omega\mathbf{y}$, as in (7.4). The correct residual sum of squares for the enlarged model is then simply the actual minimum of $(\mathbf{y} - \mathbf{Z}\gamma)'\mathbf{R}(\mathbf{y} - \mathbf{Z}\gamma)$ as

$$\begin{aligned}(\mathbf{y} - \mathbf{Z}\tilde{\gamma})'\mathbf{R}(\mathbf{y} - \mathbf{Z}\tilde{\gamma}) &= \mathbf{y}'\mathbf{R}\mathbf{y} - 2\tilde{\gamma}'\mathbf{Z}'\mathbf{R}\mathbf{y} + \tilde{\gamma}'\mathbf{Z}'\mathbf{R}\mathbf{Z}\tilde{\gamma}\\ &= \mathbf{y}'\mathbf{R}\mathbf{y} - \tilde{\gamma}'\mathbf{Z}\mathbf{R}\mathbf{y} \quad \text{(by (7.5))}\\ &= \mathbf{y}'[\mathbf{R} - \mathbf{R}\mathbf{Z}(\mathbf{Z}'\mathbf{R}\mathbf{Z})^{-1}\mathbf{Z}'\mathbf{R}]\mathbf{y} \quad \text{(by (7.3))}\\ &= \mathbf{y}'[\mathbf{R} - (\mathbf{P}_{\tilde{\Omega}} - \mathbf{P}_\Omega)]\mathbf{y} \quad \text{(by (7.2))}\\ &= \mathbf{y}(\mathbf{I}_n - \mathbf{P}_{\tilde{\Omega}})\mathbf{y}.\end{aligned}$$

We note from Theorem 1.5(iii) in Sect. 1.6 that

$$\begin{aligned}\text{Var}[\tilde{\gamma}] &= (\mathbf{Z}'\mathbf{R}\mathbf{Z})^{-1}\mathbf{Z}'\mathbf{R}\text{Var}[\mathbf{y}]\mathbf{R}\mathbf{Z}(\mathbf{Z}'\mathbf{R}\mathbf{Z})^{-1} \\ &= \sigma^2(\mathbf{Z}'\mathbf{R}\mathbf{Z})^{-1} \\ &= \sigma^2\mathbf{M}, \end{aligned} \tag{7.6}$$

say. Using $\mathbf{P}_\Omega\mathbf{R} = \mathbf{0}$,

$$\text{Cov}[\hat{\theta}, \tilde{\gamma}] = \sigma^2\mathbf{P}_\Omega\mathbf{R}\mathbf{Z}(\mathbf{Z}'\mathbf{R}\mathbf{Z})^{-1} = \mathbf{0} \tag{7.7}$$

so that from (7.4)

$$\begin{aligned}\text{Var}[\tilde{\theta}] &= \text{Var}[\hat{\theta}] + \text{Var}[\mathbf{P}_\Omega\mathbf{Z}\tilde{\gamma}] \\ &= \sigma^2\{\mathbf{P}_\Omega + \mathbf{P}_\Omega\mathbf{Z}\mathbf{M}\mathbf{Z}'\mathbf{P}_\Omega\}. \end{aligned} \tag{7.8}$$

We note that the residuals for the enlarged model are, from (7.2),

$$\begin{aligned}(\mathbf{I}_n - \mathbf{P}_{\tilde{\Omega}})\mathbf{y} &= \mathbf{R}\mathbf{y} - \mathbf{R}\mathbf{Z}(\mathbf{Z}'\mathbf{R}\mathbf{Z})^{-1}\mathbf{Z}'\mathbf{R}\mathbf{y} \\ &= \mathbf{R}[\mathbf{I}_n - \mathbf{Z}(\mathbf{Z}'\mathbf{R}\mathbf{Z})^{-1}\mathbf{Z}']\mathbf{R}\mathbf{y} \\ &= \mathbf{R}\mathbf{S}\mathbf{R}\mathbf{y}, \quad \text{say}. \end{aligned} \tag{7.9}$$

The above equation forms the basis of an algorithm due to Wilkinson (1970) for fitting analysis of variance models by regression methods. The steps are

Step 1: Compute the residuals $\mathbf{R}\mathbf{y}$.
Step 2: Use the operator that Wilkinson called a sweep to produce a vector of apparent residuals $\mathbf{R}\mathbf{y} - \mathbf{Z}\tilde{\gamma}$ ($= \mathbf{S}\mathbf{R}\mathbf{y}$).
Step 3: Applying the operator $\mathbf{R}$ once again, reanalyze the apparent residuals to produce the correct residuals $\mathbf{R}\mathbf{S}\mathbf{R}\mathbf{y}$.

7.3 Hypothesis Testing

One of the first hypotheses of interest is $H_\gamma : \gamma = \mathbf{0}$ and the F-statistic for testing this is given by (4.1), namely

$$F = \frac{n - \dim[\tilde{\Omega}]}{\text{rank}[\mathbf{Z}]} \cdot \frac{\mathbf{y}'(\mathbf{P}_{\tilde{\Omega}} - \mathbf{P}_\Omega)\mathbf{y}}{\mathbf{y}'(\mathbf{I} - \mathbf{P}_{\tilde{\Omega}})\mathbf{y}},$$

where $\dim[\tilde{\Omega}] = p + r$ and $\text{rank}[\mathbf{Z}] = r$. If the test is not significant and we accept the hypothesis, then we are back to our usual model G. However, if the test is significant, we would then test some other hypothesis of the form $\tilde{H} : E[\mathbf{y}] \in \tilde{\omega} = \omega \oplus \mathcal{C}[\mathbf{Z}]$ and repeat the above procedure using ω instead Ω. The F-statistic would then be

$$F = \frac{n - \dim[\tilde{\Omega}]}{\dim[\Omega] - \dim[\omega]} \cdot \frac{\mathbf{y}'(\mathbf{P}_{\tilde{\Omega}} - \mathbf{P}_{\tilde{\omega}})\mathbf{y}}{\mathbf{y}'(\mathbf{I} - \mathbf{P}_{\tilde{\Omega}})\mathbf{y}}.$$

7.4 Regression Extensions

Suppose our original model for Ω is a regression model with $\boldsymbol{\theta} = \mathbf{X}\boldsymbol{\beta}$, where $\mathbf{X}$ is $n \times p$ of rank p and the columns of $\mathbf{X}$ are linearly independent of the columns of $\mathbf{Z}$. Now $\tilde{\boldsymbol{\theta}} = \mathbf{X}\tilde{\boldsymbol{\beta}}$ so that from (7.4) with $\mathbf{X}'\mathbf{P}_\Omega = \mathbf{X}'$

$$\begin{aligned}
\tilde{\boldsymbol{\beta}} &= (\mathbf{X}'\mathbf{X})^{-1}\mathbf{X}'\tilde{\boldsymbol{\theta}} \\
&= (\mathbf{X}'\mathbf{X})^{-1}\mathbf{X}'\mathbf{P}_\Omega(\mathbf{y} - \mathbf{Z}\tilde{\boldsymbol{\gamma}}) \\
&= (\mathbf{X}'\mathbf{X})^{-1}\mathbf{X}'(\mathbf{y} - \mathbf{Z}\tilde{\boldsymbol{\gamma}}) \\
&= \hat{\boldsymbol{\beta}} - (\mathbf{X}'\mathbf{X})^{-1}\mathbf{X}'\mathbf{Z}\tilde{\boldsymbol{\gamma}} \\
&= \hat{\boldsymbol{\beta}} - \mathbf{L}\tilde{\boldsymbol{\gamma}}, \quad \text{say.}
\end{aligned} \tag{7.10}$$

Of particular interest in model fitting is the case of fitting one extra explanatory variable. Suppose our original model G is denoted by

$$E[\mathbf{y}] = (\mathbf{x}^{(0)}, \mathbf{x}^{(1)}, \ldots, \mathbf{x}^{(p-1)})\boldsymbol{\beta},$$

where $\mathbf{x}^{(j)}$ is the $(j+1)$th column of $\mathbf{X}$, and we wish to add an extra variable with column $\mathbf{x}^{(p)}$ and parameter β_p so that $\mathbf{Z}\boldsymbol{\gamma} = \mathbf{x}^{(p)}\beta_p$. From the previous section we find that the least squares estimates for this enlarged model are readily calculated since $\mathbf{Z}'\mathbf{R}\mathbf{Z}$ $(= \mathbf{x}^{(p)\prime}\mathbf{R}\mathbf{x}^{(p)})$ is only a 1×1 matrix, that is a scalar. Hence

$$\tilde{\beta}_p = \tilde{\gamma} = (\mathbf{Z}'\mathbf{R}\mathbf{Z})^{-1}\mathbf{Z}'\mathbf{R}\mathbf{y} = \frac{\mathbf{x}^{(p)\prime}\mathbf{R}\mathbf{y}}{\mathbf{x}^{(p)\prime}\mathbf{R}\mathbf{x}^{(p)}},$$

and from (7.10)

$$\tilde{\boldsymbol{\beta}} = (\tilde{\beta}_0, \tilde{\beta}_1, \ldots, \tilde{\beta}_{p-1})' = \hat{\boldsymbol{\beta}} - (\mathbf{X}'\mathbf{X})^{-1}\mathbf{X}'\mathbf{x}^{(p)}\tilde{\beta}_p = \hat{\boldsymbol{\beta}} - \mathbf{L}\tilde{\beta}_p.$$

From (7.7) we have

$$\mathrm{Cov}[\hat{\boldsymbol{\beta}}, \mathbf{L}\tilde{\beta}_p] = \mathrm{Cov}[(\mathbf{X}'\mathbf{X})^{-1}\mathbf{X}'\hat{\boldsymbol{\theta}}, \mathbf{L}\tilde{\beta}_p] = (\mathbf{X}'\mathbf{X})^{-1}\mathbf{X}'\mathrm{Cov}[\hat{\boldsymbol{\theta}}, \tilde{\beta}_p]\mathbf{L}' = \mathbf{0}.$$

Hence using Example 3.1 (iii) of Sect. 3.3 and (7.6), we have from the above equation,

$$\begin{aligned}\text{Var}[\tilde{\beta}] &= \text{Var}[\hat{\beta}] + \text{Var}[\mathbf{L}\tilde{\beta}_p] \\ &= \sigma^2[(\mathbf{X}'\mathbf{X})^{-1} + \mathbf{LML}'] \\ &= \sigma^2[(\mathbf{X}'\mathbf{X})^{-1} + \mathbf{ll}'m],\end{aligned}$$

where $\mathbf{l} = (\mathbf{X}'\mathbf{X})^{-1}\mathbf{X}'\mathbf{x}^{(p)}$ (a vector) and, from (7.6), $m = \mathbf{x}^{p\prime}\mathbf{R}\mathbf{x}^{(p)}$ (a scalar). Now

$$\begin{aligned}\text{Cov}[\tilde{\beta}, \tilde{\beta}_p] &= \text{Cov}[\hat{\beta} - \mathbf{L}\tilde{\beta}_p, \tilde{\beta}_p] \\ &= \text{Cov}[\hat{\beta}, \tilde{\beta}_p] - \mathbf{l}\text{var}[\tilde{\beta}_p] \\ &= -\sigma^2\mathbf{l}m.\end{aligned}$$

If $\tilde{\delta} = (\tilde{\beta}', \tilde{\beta}_p)'$, then

$$\text{Var}[\tilde{\delta}] = \sigma^2 \begin{bmatrix} (\mathbf{X}'\mathbf{X})^{-1} + \mathbf{ll}' & -\mathbf{l}m \\ -\mathbf{l}'m & m \end{bmatrix}.$$

Since the "corrections" involved in updating the original regression model are readily made, the above method can be used in stepwise methods for regression models. In particular, Wilkinson's algorithm from (7.9) can be used. Methods for adding and deleting cases and variables are given in Seber and Lee (2003, Sect. 11.6).

7.5 Analysis of Covariance Extensions

Example 7.4 (One way ANCOVA) We revisit Example 7.3 in Sect. 7.1 where we have the balanced model

$$E[y_{ij}] = \mu_i + \gamma_i z_{ij}, \quad (i = 1, 2, \ldots, I; j = 1, 2, \ldots, J)$$

with $n = IJ$. Applying the theory of Sects. 7.2 and 7.3 and using the results from Example 5.1 in Sect. 5.2, the least squares estimate of μ_i for $G : y_{ij} = \mu_i + \varepsilon_{ij}$ is $\hat{\mu}_i = \bar{y}_{i\cdot}$ with residual sum of squares $R_{yy} = \sum_i \sum_j (y_{ij} - \bar{y}_{i\cdot})^2$. Replacing y_{ij} by $y_{ij} - \gamma_i x_{ij}$ in R_{yy} gives us

$$R_{yy} - 2\sum_i \sum_j \gamma_i (y_{ij} - \bar{y}_{i\cdot})(z_{ij} - \bar{z}_{i\cdot}) + \sum_i \sum_j \gamma_i^2 (z_{ij} - \bar{z}_{i\cdot})^2.$$

Differentiating this expression with respect to γ_i give us the least squares estimate of γ_i for the extended model, namely

$$\tilde{\gamma}_i = \frac{\sum_j (y_{ij} - \overline{y}_{i\cdot})(z_{ij} - \overline{z}_{i\cdot})}{\sum_j (z_{ij} - \overline{z}_{i\cdot})^2} = \frac{R_{yzi}}{R_{zzi}},$$

say. The residual sum of squares for the extended model is then

$$\begin{aligned}\mathbf{y}'(\mathbf{I}_n - \mathbf{P}_{\tilde{\Omega}})\mathbf{y} &= R_{yy} - 2\sum_i \tilde{\gamma}_i R_{yzi} + \sum_i \tilde{\gamma}_i^2 R_{zzi} \\ &= R_{yy} - \sum_i \frac{R_{yzi}^2}{R_{zzi}}.\end{aligned}$$

To test $\tilde{H} : \gamma_i = \gamma$ for all $i = 1, 2, \ldots, I$ we find the least squares estimate of γ by minimizing $\sum_i \sum_i (y_{ij} - \overline{y}_{i\cdot} - \gamma(z_{ij} - \overline{z}_{i\cdot}))^2$ to get

$$\tilde{\gamma}_H = \frac{\sum_i \sum_j (y_{ij} - \overline{y}_{i\cdot})(z_{ij} - \overline{z}_{i\cdot})}{\sum_i \sum_j (z_{ij} - \overline{z}_{i\cdot})^2} = \frac{R_{yz}}{R_{zz}},$$

say, and we find that

$$\mathbf{y}'(\mathbf{I}_n - \mathbf{P}_{\tilde{\omega}})\mathbf{y} = R_{yy} - \frac{R_{yz}^2}{R_{zz}}.$$

Example 7.5 (Randomized block design) If we wish to extend an experimental design such as the randomized block design to

$$E[y_{ij}] = \theta_{ij} + \gamma z_{ij} = \mu + \alpha_i + \beta_j + \gamma z_{ij},$$

then we find that identifiability conditions such as $\sum_i \alpha_i = \sum_j \beta_j = 0$ need to be incorporated into the model. This can be readily done in general as follows. Suppose we use a regression formulation $\boldsymbol{\theta} = \mathbf{X}\boldsymbol{\beta}$ for our experimental design, with identifiability conditions of $\mathbf{H}\boldsymbol{\beta} = \mathbf{0}$. Enlarging this model to $E[\mathbf{y}] = \boldsymbol{\phi} = \mathbf{X}\boldsymbol{\beta} + \mathbf{Z}\boldsymbol{\gamma}$ we find that the conditions $\mathbf{H}\boldsymbol{\beta} = \mathbf{0}$ are still necessary and sufficient for identifiability in the enlarged model, as we might expect. In fact, as we have

$$\begin{pmatrix} \boldsymbol{\phi} \\ \mathbf{0} \end{pmatrix} = \begin{pmatrix} \mathbf{X} & \mathbf{Z} \\ \mathbf{H} & \mathbf{0} \end{pmatrix} \begin{pmatrix} \boldsymbol{\beta} \\ \boldsymbol{\gamma} \end{pmatrix},$$

the rows of $(\mathbf{X}, \mathbf{Z})$ are linearly independent of the rows of $(\mathbf{H}, \mathbf{0})$ as the rows of $\mathbf{X}$ are linearly independent of the rows of $\mathbf{H}$. From Theorem 3.5 in Sect. 3.4 we have

$$\mathbf{P}_{\Omega} = \mathbf{X}(\mathbf{X}'\mathbf{X} + \mathbf{H}'\mathbf{H})^{-1}\mathbf{X}' = \mathbf{X}(\mathbf{G}'\mathbf{G})^{-1}\mathbf{X}' \quad \text{and} \quad \hat{\boldsymbol{\beta}} = (\mathbf{G}'\mathbf{G})^{-1}\mathbf{X}'\mathbf{y}.$$

Since $\tilde{\boldsymbol{\theta}} = \mathbf{X}\tilde{\boldsymbol{\beta}}$ and $\mathbf{H}\tilde{\boldsymbol{\beta}} = \mathbf{0}$, we get

$$\mathbf{X}'\tilde{\boldsymbol{\theta}} = \mathbf{X}'\mathbf{X}\tilde{\boldsymbol{\beta}} = (\mathbf{X}'\mathbf{X} + \mathbf{H}'\mathbf{H})\tilde{\boldsymbol{\beta}}$$

and

$$\tilde{\boldsymbol{\beta}} = (\mathbf{X}'\mathbf{X} + \mathbf{H}'\mathbf{H})^{-1}\mathbf{X}'\tilde{\boldsymbol{\theta}} = (\mathbf{G}'\mathbf{G})^{-1}\mathbf{X}'\tilde{\boldsymbol{\theta}}.$$

Hence from (7.4)

$$\begin{aligned}\tilde{\boldsymbol{\beta}} &= (\mathbf{G}'\mathbf{G})^{-1}\mathbf{X}'\mathbf{P}_{\Omega}(\mathbf{y} - \mathbf{Z}\tilde{\boldsymbol{\gamma}}) \\ &= (\mathbf{G}'\mathbf{G})^{-1}\mathbf{X}'(\mathbf{y} - \mathbf{Z}\tilde{\boldsymbol{\gamma}}),\end{aligned}$$

which is $\hat{\boldsymbol{\beta}}$ with $\mathbf{y}$ replaced by $\mathbf{y} - \mathbf{Z}\tilde{\boldsymbol{\gamma}}$. Since the Lagrange multipliers associated with the identifiability conditions are zero (by Theorem 3.5), this means that the general three-stage method of the previous section will apply to this example. We demonstrate this by finding the squares estimates for our randomized block extended model. One method of doing this has already been given in Sect. 6.6. We now use the normal equations instead by differentiating $\sum_i \sum_j (y_{ij} - \mu - \alpha_i - \beta_j)^2$ with respect to each parameter and ignoring the Lagrange multipliers associated with the identifiable conditions. The answer is

$$\hat{\mu} = \overline{y}_{\cdot\cdot}, \quad \widehat{\alpha_i} = \overline{y}_{i\cdot} - \overline{y}_{\cdot\cdot}, \quad \text{and} \quad \widehat{\beta_j} = \overline{y}_{\cdot j} - \overline{y}_{\cdot\cdot},$$

with residual sum of squares

$$R_{yy} = \sum_i \sum_j (y_{ij} - \hat{\mu} - \hat{\alpha}_i - \hat{\beta}_j)^2 = \sum_i \sum_j (y_{ij} - \overline{y}_{i\cdot} - \overline{y}_{\cdot j} + \overline{y}_{\cdot\cdot})^2.$$

We now replace y_{ij} by $y_{ij} - \gamma z_{ij}$ in R_{yy} and differentiate with respect to γ to get $\tilde{\gamma} = R_{yz}/R_{zz}$, where

$$\begin{aligned}R_{yz} &= \sum_i \sum_j (y_{ij} - \overline{y}_{i\cdot} - \overline{y}_{\cdot j} + \overline{y}_{\cdot\cdot})(z_{ij} - \overline{z}_{i\cdot} - \overline{z}_{\cdot j} + \overline{z}_{\cdot\cdot}) \\ &= \sum_i \sum_j y_{ij}(z_{ij} - \overline{z}_{i\cdot} - \overline{z}_{\cdot j} + \overline{z}_{\cdot\cdot}).\end{aligned}$$

The residual sum of squares for the extended model is then the minimum value

$$\sum_i \sum_j (y_{ij} - \overline{y}_{i\cdot} - \overline{y}_{\cdot j} + \overline{y}_{\cdot\cdot} - \tilde{\gamma}(z_{ij} - \overline{z}_{i\cdot} - \overline{z}_{\cdot j} + \overline{z}_{\cdot\cdot}))^2 = R_{yy} - \frac{R_{yz}^2}{R_{zz}}.$$

7.6 Missing Observations

In some experimental situations observations are "lost", for example a test tube is broken, a flood damages part of an agricultural experiment, animals die, and patients withdraw from a medical trial because of some unpredictable event such as having to move from the district or having an accident. When this happens, an experimental design usually becomes unbalanced so that, for example, we lose some robustness that we have shown with balanced designs in previous chapters. In recent years there has been a complete change in methods for handling missing data, beginning with the classification of types of missing data given by Rubin (1976) and Little and Rubin (2002). They described three types of processes leading to missing observations: missing completely at random (MCAR), missing at random (MAR), and missing not at random (MNAR, sometimes referred to as NMAR). Here MCAR assumes that the observed data can be regarded as a random subsample of the hypothetically complete data sample. It means that the probability of a missing observation on a variable y is unrelated to other measured variables and to the values of y itself. The term MAR is a bit of a misnomer as the mechanism is not strictly random but describes systematic "missingness," where the propensity for missing is related to other measured variables but not to the underlying values of y. Finally, data are MNAR if the probability of missing is systematically related to the hypothetical values that are missing. It is often hard to know whether we have MAR or MNAR. In this Section I am only going to consider MCAR of which examples were given at the beginning of the section. In this case some of the traditional methods of analysis are satisfactory, in particular list-wise deletion; also known as complete-case analysis. Many of more complex missing data problems do not fit into the MCAR category so that the complete cases method produces biased estimates when the MCAR assumption does not hold. The appropriate methods are then multiple imputation and maximum likelihood (e.g., Baraldi and Enders 2010; Graham 2012). In using the complete-cases method we have two basic strategies: (1) carry out the statistical analysis with the data that we have and use the incomplete model, or (2) add artificial numbers to replace the missing data so that we now use the properties of a balanced design, but choose the numbers so that the final statistical analysis gives results that are the same as those obtained through (1). The second method essentially means choosing the artificial data so that the final residuals for those missing data points are zero. We now develop the theory based on Kruskal (1960) for this second approach using projection matrices. Later we also use an analysis of covariance method due to Bartlett to produce the same estimates.

We begin with our usual model $\mathbf{y} = \boldsymbol{\theta} + \boldsymbol{\varepsilon}$, where, by relabelling the y_i, we have

$$\mathbf{y} = \begin{pmatrix} \mathbf{y}_{n-m} \\ \mathbf{0} \end{pmatrix} + \begin{pmatrix} \mathbf{0} \\ \mathbf{y}_m \end{pmatrix} = \mathbf{y}^{(1)} + \mathbf{y}^{(2)},$$

where only the $n - m$ values $\mathbf{y}_{n-m}$ are observed: $\boldsymbol{\theta}$ is partitioned in the same way. We can therefore write $\mathbb{R}^n$ as the direct sum of two vector spaces V_1 and V_2, that is

$\mathbb{R}^n = V_1 \bigoplus V_2$, and $\mathbf{y}^{(i)} = \mathbf{P}^{(i)}\mathbf{y}$, where $\mathbf{P}^{(i)}$ represents the orthogonal projection of $\mathbb{R}^n$ on V_i. Here

$$\mathbf{P}^{(1)} = \begin{pmatrix} \mathbf{I}_{n-m} & \mathbf{0} \\ \mathbf{0} & \mathbf{0} \end{pmatrix} \quad \text{and} \quad \mathbf{P}^{(2)} = \begin{pmatrix} \mathbf{0} & \mathbf{0} \\ \mathbf{0} & \mathbf{I}_m \end{pmatrix}.$$

Let $\Omega_i \equiv \mathbf{P}^{(i)}\Omega$ and $\mathrm{E}[\mathbf{y}^{(i)}] = \boldsymbol{\theta}^{(i)} = \mathbf{P}^{(i)}\boldsymbol{\theta}$. Then $\Omega_1 \perp \Omega_2$ and in general we have $\Omega \neq \Omega_1 \bigoplus \Omega_2$, although $\Omega \subset \Omega_1 \bigoplus \Omega_2$.

To find the least squares estimates we first minimize

$$\| \mathbf{y} - \boldsymbol{\theta} \|^2 = \| \mathbf{y}^{(1)} - \boldsymbol{\theta}^{(1)} \|^2 + \| \mathbf{y}^{(2)} - \boldsymbol{\theta}^{(2)} \|^2$$

subject to $\boldsymbol{\theta} \in \Omega$ to get $\hat{\boldsymbol{\theta}} = \mathbf{P}_\Omega \mathbf{y}$ and the usual residual sum of squares, namely $\| (\mathbf{I}_n - \mathbf{P}_\Omega)\mathbf{y} \|^2$, and then minimize this sum of squares with respect to $\mathbf{y}^{(2)}$ to get $\hat{\mathbf{y}}^{(2)}$, which is substituted back into $\hat{\boldsymbol{\theta}}$. Hence from the first step we get

$$\hat{\boldsymbol{\theta}} = \hat{\boldsymbol{\theta}}^{(1)} + \hat{\boldsymbol{\theta}}^{(2)} = \mathbf{P}_\Omega \mathbf{y} = \mathbf{P}_\Omega(\mathbf{y}^{(1)} + \mathbf{y}^{(2)}),$$

and from the second step

$$\hat{\mathbf{y}}^{(2)} = \hat{\boldsymbol{\theta}}^{(2)}.$$

Combining these two equations,

$$\hat{\boldsymbol{\theta}}^{(2)} = \mathbf{P}^{(2)}\hat{\boldsymbol{\theta}} = \mathbf{P}^{(2)}\mathbf{P}_\Omega(\mathbf{y}^{(1)} + \hat{\boldsymbol{\theta}}^{(2)}). \tag{7.11}$$

We now ask, when does (7.11) have a unique solution for $\hat{\boldsymbol{\theta}}^{(2)}$? To answer this, let $\mathbf{P}_\Omega^{(i)}$ represent the projection onto Ω_i. Now

$$\hat{\boldsymbol{\theta}} = \mathbf{P}_\Omega\hat{\boldsymbol{\theta}} = \mathbf{P}_\Omega(\hat{\boldsymbol{\theta}}^{(1)} + \hat{\boldsymbol{\theta}}^{(2)}), \tag{7.12}$$

and $\mathbf{P}_\Omega^{(1)}\mathbf{y}^{(1)} \in \Omega_1 \subset V_1$ (as well as $\mathbf{y}^{(1)}$), which implies that $(\mathbf{I}_n - \mathbf{P}_\Omega^{(1)})\mathbf{y}^{(1)} \in V_1$, and is therefore perpendicular to V_2. This means that $(\mathbf{I}_n - \mathbf{P}_\Omega^{(1)})\mathbf{y}^{(1)}$ is orthogonal to both Ω_1 and Ω_2 and hence to Ω, so that $\mathbf{P}_\Omega(\mathbf{I}_n - \mathbf{P}_\Omega^{(1)})\mathbf{y}^{(1)} = \mathbf{0}$. Therefore

$$\hat{\boldsymbol{\theta}} = \mathbf{P}_\Omega(\mathbf{y}^{(1)} + \hat{\mathbf{y}}^{(2)}) \tag{7.13}$$

$$= \mathbf{P}_\Omega(\mathbf{P}_\Omega^{(1)}\mathbf{y}^{(1)} + \hat{\boldsymbol{\theta}}^{(2)}). \tag{7.14}$$

Subtracting (7.14) from (7.12) gives us

$$\mathbf{P}_\Omega(\hat{\boldsymbol{\theta}}^{(1)} - \mathbf{P}_\Omega^{(1)}\mathbf{y}^{(1)}) = \mathbf{0}.$$

Thus $\hat{\boldsymbol{\theta}}^{(1)} - \mathbf{P}_{\Omega}^{(1)}\mathbf{y}^{(1)}$, which belongs to Ω_1, is orthogonal to Ω, and this can only happen if

$$\hat{\boldsymbol{\theta}}^{(1)} = \mathbf{P}_{\Omega}^{(1)}\mathbf{y}^{(1)}.$$

(To see this suppose $(\mathbf{a}_1', \mathbf{a}_2')' \in \Omega$ so that $(\mathbf{a}_1', \mathbf{0}')' \in \Omega_1$. If this vector is perpendicular to Ω then $\mathbf{a}_1'\mathbf{a}_1 = 0$; that is $\mathbf{a}_1 = \mathbf{0}$.) We see then that the above two-step least-squares procedure corresponds to first minimizing $(\mathbf{y}^{(1)} - \boldsymbol{\theta}^{(1)})'(\mathbf{y}^{(1)} - \boldsymbol{\theta}^{(1)})$, subject to $\boldsymbol{\theta}^{(1)} \in \Omega_1$, and then putting $\hat{\mathbf{y}}^{(2)} = \hat{\boldsymbol{\theta}}^{(2)}$, where $\hat{\boldsymbol{\theta}}^{(2)}$ is chosen such that $\hat{\boldsymbol{\theta}} = \hat{\boldsymbol{\theta}}^{(1)} + \hat{\boldsymbol{\theta}}^{(2)}$ belongs to Ω. The residual sum of squares is then

$$Q = \mathbf{y}^{(1)\prime}(\mathbf{I}_n - \mathbf{P}_{\Omega}^{(1)})\mathbf{y}^{(1)} = (\mathbf{y}^{(1)\prime} - \hat{\boldsymbol{\theta}}^{(1)})'(\mathbf{y}^{(1)\prime} - \hat{\boldsymbol{\theta}}^{(1)}) = \mathbf{z}'(\mathbf{I}_n - \mathbf{P}_{\Omega})\mathbf{z},$$

where $\mathbf{z} = \mathbf{y}^{(1)} + \hat{\boldsymbol{\theta}}^{(2)}$. This last result follows from the fact that $(\mathbf{I}_n - \mathbf{P}_{\Omega})$ is idempotent, and using (7.13) give us

$$\begin{aligned}(\mathbf{I}_n - \mathbf{P}_{\Omega})(\mathbf{y}^{(1)} + \hat{\boldsymbol{\theta}}^{(2)}) &= \mathbf{y}^{(1)} + \hat{\boldsymbol{\theta}}^{(2)} - \hat{\boldsymbol{\theta}} \\ &= \mathbf{y}^{(1)} - \hat{\boldsymbol{\theta}}^{(1)}.\end{aligned}$$

Obviously $\hat{\boldsymbol{\theta}}^{(2)}$ can only be unique if, corresponding to every $\boldsymbol{\theta}^{(1)} \in \Omega_1$, there exists a unique $\boldsymbol{\theta}^{(2)} \in \Omega_2$ such that $\boldsymbol{\theta}^{(1)} + \boldsymbol{\theta}^{(2)} \in \Omega$. Now $\boldsymbol{\theta}^{(2)}$ will be unique if and only if there is no non-zero $\boldsymbol{\phi}^{(2)} \in \Omega_2$ such that $\mathbf{0} + \boldsymbol{\phi}^{(2)} \in \Omega$, for then $\boldsymbol{\theta}^{(2)}$ and $\boldsymbol{\theta}^{(2)} + \boldsymbol{\phi}^{(2)}$ both correspond to $\boldsymbol{\theta}^{(1)}$. Thus the condition for uniqueness is that

$$\dim[\Omega] = \dim[\Omega_1],$$

and as an exercise we verify that the above condition implies that (7.11) has a unique solution for $\boldsymbol{\theta}^{(2)}$.

Suppose two solutions $\mathbf{u}$ and $\mathbf{v}$ exist, then from (7.11)

$$\mathbf{u} - \mathbf{v} = \mathbf{P}^{(2)}\mathbf{P}_{\Omega}(\mathbf{u} - \mathbf{v}). \tag{7.15}$$

Now if $\mathbf{P}_W$ represents the projection on any vector space W, then $\mathbf{z} = \mathbf{P}_W\mathbf{z} + (\mathbf{I}_n - \mathbf{P}_W)\mathbf{z}$ for every $\mathbf{z}$ and therefore $\| \mathbf{z} \| \geq \| \mathbf{P}_W\mathbf{z} \|$ with equality if and only if $\mathbf{z} \in W$. Applying this twice to Eq. (7.15) give us that $(\mathbf{u} - \mathbf{v}) \in \Omega$ and $\mathbf{P}_{\Omega}(\mathbf{u} - \mathbf{v}) \in V_2$. Hence $\mathbf{P}_{\Omega}(\mathbf{u} - \mathbf{v})$ is in both V_2 and Ω; it is therefore zero as $\dim[\Omega] = \dim[\Omega_1]$, and $(\mathbf{u} - \mathbf{v}) \perp \Omega$. Thus $\mathbf{u} = \mathbf{v}$, establishing the uniqueness of $\boldsymbol{\theta}^{(2)}$.

To test the hypothesis $H : \boldsymbol{\theta} \in \omega$, a $p - q$ subspace of Ω, we simply go through the same procedure as described above with the estimate $\hat{\theta}_H^{(2)}$ given by (7.11) with Ω replaced by ω, namely

$$\hat{\theta}_H^{(2)} = \mathbf{P}^{(2)}\mathbf{P}_{\omega}(\mathbf{y}^{(1)} + \dot{\theta}_H^{(2)}). \tag{7.16}$$

Once again this equation will have unique solution if $\dim[\omega] = \dim[\omega_1]$, where $\omega_1 \equiv \mathbf{P}^{(1)}\omega$. The residual sum of squares for the hypothesis is

$$Q_H = \mathbf{y}^{(1)\prime}(\mathbf{I}_n - \mathbf{P}_\omega^{(1)})\mathbf{y}^{(1)} = \mathbf{z}_H'(\mathbf{I}_n - \mathbf{P}_\omega)\mathbf{z}_H,$$

where $\mathbf{z}_H = \mathbf{y}^{(1)} + \hat{\boldsymbol{\theta}}_H^{(2)}$. The corresponding degrees of freedom for the residual sums of squares Q and Q_H are $n - p - m$ and $n - p + q - m$, respectively, m degrees of freedom being lost due to the estimation of $\mathbf{y}_m$. The F-statistic is

$$F = \frac{(n-p-m)}{q}\frac{(Q_H - Q)}{Q} = \frac{(n-p-m)}{q}\frac{\mathbf{y}^{(1)\prime}(\mathbf{P}_\Omega^{(1)} - \mathbf{P}_\omega^{(1)})\mathbf{y}^{(1)}}{\mathbf{y}^{(1)\prime}(\mathbf{I}_n - \mathbf{P}_\Omega^{(1)})\mathbf{y}^{(1)}}.$$

Example 7.5 We revisit the randomized block design, namely

$$E[y_{ij}] = \theta_{ij} = \mu + \alpha_i + \beta_j, \quad (i = 1, 2, \ldots, I; j = 1, 2, \ldots, J),$$

where $\sum_i \alpha_i = 0$ and $\sum_j \beta_j = 0$. From (6.27), the least squares estimate of $\hat{\theta}_{ij}$ is given by

$$\hat{\theta}_{ij} = \hat{\mu} + \hat{\alpha}_i + \hat{\beta}_j = \overline{y}_{i\cdot} + \overline{y}_{\cdot j} - \overline{y}_{\cdot\cdot}.$$

We assume that the observation y_{IJ} is missing under the MCAR scenario, and denote its estimate by u. From the above theory, u is also the least squares estimate of θ_{IJ}. With $\boldsymbol{\theta}^{(2)} = (0, 0, \ldots, u)'$, (7.11) becomes

$$u = \hat{\theta}_{IJ} = \frac{y_{I*} + u}{J} + \frac{y_{*J} + u}{I} - \frac{y_{**} + u}{IJ},$$

where the "star" notation denotes summation on the observed variables; for example $y_{I*} = \sum_{j=1}^{J-1} y_{Ij}$ and $y_{**} = IJ\overline{y}_{\cdot\cdot} - y_{IJ}$. Solving for u gives us

$$u = [Iy_{I*} + Jy_{*J} - y_{**}]/[(I-1)(J-1)],$$

and since $E[u] = \mu + \alpha_i + \beta_j = \theta_{IJ}$, u is an unbiased estimator of θ_{IJ}. To obtain the residual sum of squares we evaluate $RSS = \sum_i \sum_j (y_{ij} - \overline{y}_{i\cdot} - \overline{y}_{\cdot j} + \overline{y}_{\cdot\cdot})^2$ with y_{IJ} replaced by its estimate u.

Suppose we wish to test $H : \alpha_i = \alpha_2 = \ldots = \alpha_I = 0$. Then, when H is true, the least squares estimate of θ_{ij} is

$$\hat{\theta}_{ij} = \hat{\mu} + \hat{\beta}_j = \overline{y}_{\cdot\cdot} + \overline{y}_{\cdot j} - \overline{y}_{\cdot\cdot} = \overline{y}_{\cdot j}.$$

If we denote the new estimate of y_{IJ} under H by u_H, we have by (7.16)

$$u_H = \hat{\theta}_{IJH} = (y_{*J} + u_H)/I \tag{7.17}$$

or $u_H = y_{*J}/(I-1)$. To obtain the residual sum of squares under H, we calculate $RSS_H = \sum_i \sum_j (y_{ij} - \bar{y}_{\cdot j})^2$ with y_{IJ} replaced by u_H. Then

$$F = \frac{[(I-1)(J-1)-1]}{(I-1)} \frac{(RSS_H - RSS)}{RSS}.$$

Bartlett (1937) suggested using an analysis of covariance method for handling missing observations that we now use assuming the MCAR scenario. The method is to assume that the missing observations are zero and then introduce concomitant variables having a value of -1 corresponding to the missing observations, and zero values elsewhere. For example, in the above randomized block example we assume the model

$$y_{ij} = \mu + \alpha_i + \beta_j + \gamma z_{ij},$$

where $y_{IJ} = 0$ and $z_{ij} = -\delta_{iI}\delta_{jJ}$. To find $\hat{\gamma}$, the least squares estimate of γ, we first assume $\gamma = 0$ and obtain $RSS = \sum_i \sum_i (y_{ij} - \bar{y}_{i\cdot} - \bar{y}_{\cdot j} + \bar{y}_{\cdot\cdot})^2$. Then setting $y_{IJ} = 0$ and replacing y_{ij} by $y_{ij} - \gamma z_{ij}$, we minimize RSS with respect to γ (cf. Example 7.5 in Sect. 7.5). We therefore minimize

$$\sum_i \sum_i \left\{ (y_{ij} - \bar{y}_{i\cdot} - \bar{y}_{\cdot j} + \bar{y}_{\cdot\cdot}) - \gamma (\sum_i \sum_i (z_{ij} - \bar{z}_{i\cdot} - \bar{z}_{\cdot j} + \bar{z}_{\cdot\cdot}) \right\}^2,$$

or $R_{yy} - 2\gamma R_{yz} + \gamma^2 R_{zz}$, say, giving $\hat{\gamma} = R_{yz}/R_{zz}$. With $y_{IJ} = 0$ we find that

$$R_{yz} = (Iy_{I*} + Jy_{*J} - y_{**})/IJ \quad \text{and} \quad R_{zz} = (I-1)(J-1)/IJ$$

so that

$$\begin{aligned} \hat{\gamma} &= (Iy_{I*} + Jy_{*J} - y_{**})/(I-1)(J-1) \\ &= u. \end{aligned}$$

Hence the covariance method leads to the same estimate u above. The reason for this follows from the fact that all we are effectively doing is replacing y_{IJ} by γ and minimizing the residual sum of squares with respect to γ. We note that we have the alternative form

$$RSS = R_{yy} - \hat{\gamma}^2 R_{zz}, \quad \text{where } y_{IJ} = 0 \text{ in } R_{yy}.$$

In conclusion we find that the covariance method will lead to the same F-statistic as before. However, the variance of $\hat{\gamma}$ for the covariance method will be greater by σ^2 than the variance of u for the previous method, for although y_{IJ} is put equal to zero it will still have a variance of σ^2.

References

Baraldi, A. N., & Enders, C. K. (2010). An introduction to modern missing data analyses. *Journal of School Psychology, 48*, 5–37.

Bartlett, M. S. (1937). Some examples of statistical methods of research in agriculture and applied biology. *Journal of the Royal Statistical Society Supplement, 4*, 137–170.

Graham, J. W. (2012). *Missing data: Analysis and design*. New York: Springer.

Kruskal, W. (1960). The coordinate-free approach to Gauss-Markov estimation and its application to missing and extra observations. In *Proceedings of the 4th Berkeley symposium in mathematical statistics and probability*, Berkeley (Vol. 1, pp. 435–451).

Little, R. J. A., & Rubin, D. B. (2002). *Statistical analysis with missing data* (2nd ed.). New York: Wiley.

Rubin, D. B. (1976). Inference and missing data. *Biometrika, 63*(3), 581–592.

Seber, G. A. F., & Lee, A. J. (2003). *Linear regression analysis* (2nd ed.). New York: Wiley.

Wilkinson, G. N. (1970). A general recursive procedure for analysis of variance. *Biometrika, 57*, 19–46.

Chapter 8
Nonlinear Regression Models

8.1 Introduction

Nonlinear models arise when $E[\mathbf{y}]$ is a nonlinear function of unknown parameters. Hypotheses about these parameters may be linear or nonlinear. Such models tend to be used when they are suggested by theoretical considerations or used to build non-linear behavior into a model. Even when a linear approximation works well, a nonlinear model may still be used to retain a clear interpretation of the parameters. Once we have established a nonlinear relationship the next problem is how to incorporate the "error" term ε. Sometimes a nonlinear relationship can be transformed into a linear one but in doing so we may end up with an error term that has awkward properties. In this case it is usually better to work with the non-linear model. These kinds of problems are demonstrated by several examples.

A simple example of a non-linear model is

$$y_i = \beta_0 + \beta_1 e^{\beta_2 x} + \varepsilon,$$

which is nonlinear in β_2. If β_0 is zero, we have the choice of two models

$$y_i = \beta_1 e^{\beta_2 x} + \varepsilon \quad \text{or} \quad y_i = \beta_1 e^{\beta_2 x} \varepsilon,$$

depending on whether we think the error is additive or multiplicative. In the latter case we have the possibility of using a log transformation

$$\log y = \log \beta_1 + \beta_2 x + \log \varepsilon.$$

with its usefulness depending on the distribution of $\log \varepsilon$.

As a further example, theoretical chemistry predicts that for a given sample of gas kept at constant temperature, the volume v and pressure p of the gas satisfy the relationship $pv^\gamma = c$, where γ is a constant depending on the gas. Setting $y = p$

G.A.F. Seber, *The Linear Model and Hypothesis*, Springer Series in Statistics,
DOI 10.1007/978-3-319-21930-1_8

and $x = v^{-1}$, we have a linear model $y \approx cx^{\gamma}$, where any error term will be due to experimental error. Once again we have the possibility of a *log* transformation. We can then use the model to estimate the value of the gas constant γ.

We have considered just two simple models. However the subject of nonlinear modeling is a large and complex one and the associated inference theory depends very much on complex assumptions that are discussed in detail by Seber and Wild (1989, chapter 12). In this chapter we shall focus on the role and interplay of asymptotic linear theory.

8.2 Estimation

We use the general model

$$y_i = f(\mathbf{x}_i; \boldsymbol{\theta}) + \varepsilon_i = f_i(\boldsymbol{\theta}) + \varepsilon_i \quad (i = 1, 2, \ldots, n),$$

or

$$\mathbf{y} = \boldsymbol{\phi} + \boldsymbol{\varepsilon} = \mathbf{f}(\boldsymbol{\theta}) + \boldsymbol{\varepsilon},$$

where $\mathbf{f}(\boldsymbol{\theta}) = (f_1(\boldsymbol{\theta}), f_2(\boldsymbol{\theta}), \ldots, f_n(\boldsymbol{\theta}))'$, $\mathbf{x}_i$ is a $k \times 1$ vector of explanatory variables, $\boldsymbol{\theta}$ is a p-dimensional vector, and $\boldsymbol{\theta}_T$, the true value of $\boldsymbol{\theta}$, is known to belong to Θ, a subset of $\mathbb{R}^p$. (We use the notation $\boldsymbol{\theta}_T$ to fit in with this and later chapters on asymptotic theory.) For example, if we have the model

$$y_i = \alpha_1 e^{\beta_1 x_{i1}} + \alpha_2 e^{\beta_2 x_{i2}} + \varepsilon_i,$$

then $\mathbf{x}_i = (x_{i1}, x_{i2})'$, $k = 2$, $\boldsymbol{\theta} = (\alpha_1, \alpha_2, \beta_1, \beta_2)'$, and $p = 4$.

Let $\mathbf{F}(\boldsymbol{\theta}) = \partial \mathbf{f}(\boldsymbol{\theta})/\partial \boldsymbol{\theta}'$ with (i, r)th element $\partial f_i(\boldsymbol{\theta})/\partial \theta_r$. We shall make the following regularity assumptions.

A(1). The ε_i are independently and identically distributed with mean zero and variance σ^2.

A(2). For each i, $f_i(\boldsymbol{\theta}) = f(\mathbf{x}_i; \boldsymbol{\theta})$ is a continuous function of $\boldsymbol{\theta}$ for $\boldsymbol{\theta} \in \Theta$.

A(3). Θ is a closed, bounded (i.e., compact) subset of $\mathbb{R}^p$. (Such an assumption is not too much of a restriction as parameters are usually bounded by the physical constraints of the system being modeled. Also actual computations are discrete so that Θ can be regarded as a set with a finite number of elements (Wu, 1981).)

A(4). Let

$$C_n(\boldsymbol{\theta}, \boldsymbol{\theta}_1) = \sum_{i=1}^{n} f_i(\boldsymbol{\theta}) f_i(\boldsymbol{\theta}_1) \text{ and } D_n(\boldsymbol{\theta}, \boldsymbol{\theta}_1) = \sum_{i=1}^{n} [f_i(\boldsymbol{\theta}) - f_i(\boldsymbol{\theta}_1)]^2.$$

Then as $n \to \infty$, $n^{-1}C_n(\boldsymbol{\theta}, \boldsymbol{\theta}_1)$ converges uniformly for all $\boldsymbol{\theta}$ and $\boldsymbol{\theta}_1$ in Θ to a function $C(\boldsymbol{\theta}, \boldsymbol{\theta}_1)$ (which is continuous if A(2) and A(3) hold). Also we have $D(\boldsymbol{\theta}, \boldsymbol{\theta}_T) = 0$ if and only $\boldsymbol{\theta} = \boldsymbol{\theta}_T$.

A(5). $\boldsymbol{\theta}_T$ is an interior point of Ω. Therefore there exists an open neighborhood of $\boldsymbol{\theta}_T$ in Θ, say Θ_T.

A(6). The first and second derivatives

$$\partial f_i(\boldsymbol{\theta})/\partial \theta_r \quad \text{and } \partial^2 f_i(\boldsymbol{\theta})/\partial \theta_r \partial \theta_s \quad (r, s = 1, 2, \ldots, p),$$

exist and are continuous for all $\boldsymbol{\theta} \in \Theta_T$.

A(7). The matrix

$$n^{-1} \sum_{i=1}^{n} \frac{\partial f_i(\boldsymbol{\theta})}{\partial \boldsymbol{\theta}} \frac{\partial f_i(\boldsymbol{\theta})}{\partial \boldsymbol{\theta}'} = n^{-1}\mathbf{F}'(\boldsymbol{\theta})\mathbf{F}(\boldsymbol{\theta})$$

converges to some matrix $\boldsymbol{\Phi}(\boldsymbol{\theta})$ uniformly in $\boldsymbol{\theta}$ for $\boldsymbol{\theta} \in \Theta_T$ as $n \to \infty$.

A(8). The matrix

$$n^{-1} \sum_{i=1}^{n} \left[\frac{\partial^2 f_i(\boldsymbol{\theta})}{\partial \theta_r \partial \theta_s} \right]^2$$

converges uniformly in $\boldsymbol{\theta}$ for $\boldsymbol{\theta} \in \Theta_T$ $(r, s = 1, 2, \ldots, p)$ as $n \to \infty$.

A(9). $\boldsymbol{\Phi}_T = \boldsymbol{\Phi}(\boldsymbol{\theta}_T)$ is nonsingular.

The least squares estimate $\hat{\boldsymbol{\theta}}$ of $\boldsymbol{\theta}$ is obtained by minimizing

$$Q(\boldsymbol{\theta}) = \sum_{i=1}^{n} \{y_i - f(\mathbf{x}_i; \boldsymbol{\theta})\}^2.$$

In contrast to the linear situation, $Q(\boldsymbol{\theta})$ may have several relative minima in addition to the absolute minimum. Given assumptions A(1) to A(4) above, we find that $\hat{\boldsymbol{\theta}}$ exists and $\hat{\boldsymbol{\theta}}$ and $\hat{\sigma}^2 = Q(\hat{\boldsymbol{\theta}})/n$ are (strongly) consistent estimators of $\boldsymbol{\theta}_T$ and σ^2 respectively. Differentiating with respect to $\boldsymbol{\theta}$, we find from assumption A(5) and n sufficiently large that $\hat{\boldsymbol{\theta}}$ is an interior point of Θ and satisfies the equation

$$\frac{\partial Q(\boldsymbol{\theta})}{\partial \theta_j} = 2 \sum_{i=1}^{n} [y_i - f_i(\boldsymbol{\theta})] \frac{\partial f_i(\boldsymbol{\theta})}{\partial \theta_j} = 0, \quad (j = 1, 2, \ldots, p). \tag{8.1}$$

This gives us

$$\begin{aligned} \mathbf{0} &= \hat{\mathbf{F}}'\{\mathbf{y} - \mathbf{f}(\hat{\boldsymbol{\theta}})\} \\ &= \hat{\mathbf{F}}'\hat{\boldsymbol{\varepsilon}}, \end{aligned}$$

the *normal equations* for the nonlinear model, where $\hat{\mathbf{F}} = \mathbf{F}(\hat{\boldsymbol{\theta}})$ and $\hat{\boldsymbol{\varepsilon}}$ is the vector of residuals. If $\hat{\mathbf{P}}_{\mathbf{F}} = \hat{\mathbf{F}}(\hat{\mathbf{F}}'\hat{\mathbf{F}})^{-1}\hat{\mathbf{F}}'$, the idempotent matrix projecting $\mathbb{R}^p$ orthogonally on to $\mathcal{C}[\hat{\mathbf{F}}]$, then the above equation can be written as

$$\hat{\mathbf{P}}_{\mathbf{F}}\hat{\boldsymbol{\varepsilon}} = \mathbf{0}.$$

We can also prove from the above assumptions that $n^{1/2}(\hat{\boldsymbol{\theta}} - \boldsymbol{\theta}_T)$ is asymptotically $N_p[\mathbf{0}, \sigma^2\boldsymbol{\Phi}_T^{-1}]$. Since the assumptions imply that $n^{-1}\hat{\mathbf{F}}'\hat{\mathbf{F}}$ is a strongly consistent estimator of $\boldsymbol{\Phi}_T$, then for large n

$$n^{1/2}(\hat{\boldsymbol{\theta}} - \boldsymbol{\theta}_T) \text{ is approximately } N_p[\mathbf{0}, \sigma^2(\frac{1}{n}\mathbf{F}_T'\mathbf{F}_T)^{-1}], \tag{8.2}$$

where $\mathbf{F}_T = \mathbf{F}(\boldsymbol{\theta}_T)$.

If in addition to assumption A(1) above we assume that the ε_i are normally distributed, then using exactly the same method used in Sect. 3.9 we find that $\hat{\boldsymbol{\theta}}$ and $\hat{\sigma}^2$ are also the maximum likelihood estimates of $\boldsymbol{\theta}$ and σ^2.

Before considering some linear approximations we require the following result for future use. From (8.1) we have

$$\frac{\partial^2 Q(\boldsymbol{\theta})}{\partial\theta_j\partial\theta_k} = -2\sum_{i=1}^{n}\left\{[y_i - f_i(\boldsymbol{\theta})]\frac{\partial^2 f_i(\boldsymbol{\theta})}{\partial\theta_j\partial\theta_k} - \frac{\partial f_i(\boldsymbol{\theta})}{\partial\theta_j}\cdot\frac{\partial f_i(\boldsymbol{\theta})}{\partial\theta_k}\right\}$$

so that

$$\mathrm{E}\left[\frac{\partial^2 Q(\boldsymbol{\theta})}{\partial\boldsymbol{\theta}\partial\boldsymbol{\theta}'}\right] = 2\mathbf{F}(\boldsymbol{\theta})'\mathbf{F}(\boldsymbol{\theta}). \tag{8.3}$$

8.3 Linear Approximations

From assumption A(5), if $\boldsymbol{\theta} \in \Theta_T$, where Θ_T is a small neighborhood of $\boldsymbol{\theta}_T$, we have the Taylor expansion

$$f_i(\boldsymbol{\theta}) \approx f_i(\boldsymbol{\theta}_T) + \sum_{j=1}^{p}\left.\frac{\partial f_i}{\partial\theta_j}\right|_{\boldsymbol{\theta}_T}(\theta_j - \theta_{Tj}),$$

or

$$\mathbf{f}(\boldsymbol{\theta}) \approx \mathbf{f}(\boldsymbol{\theta}_T) + \mathbf{F}_T(\boldsymbol{\theta} - \boldsymbol{\theta}_T). \tag{8.4}$$

Hence

$$\begin{aligned} Q(\boldsymbol{\theta}) &= \| \mathbf{y} - \mathbf{f}(\boldsymbol{\theta}) \|^2 \\ &\approx \| \mathbf{y} - \mathbf{f}(\boldsymbol{\theta}_T) - \mathbf{F}_T(\boldsymbol{\theta} - \boldsymbol{\theta}_T) \|^2 \\ &= \| \boldsymbol{\varepsilon} - \mathbf{F}_T\boldsymbol{\beta} \|^2, \end{aligned} \tag{8.5}$$

say, where $\boldsymbol{\varepsilon} = \mathbf{y} - \mathbf{f}(\boldsymbol{\theta}_T)$ and $\boldsymbol{\beta} = \boldsymbol{\theta} - \boldsymbol{\theta}_T$. From the properties of the linear regression model, (8.5) is minimized when $\boldsymbol{\beta}$ is given by (cf. Example 3.1 in Sect. 3.3)

$$\hat{\boldsymbol{\beta}} = (\mathbf{F}_T'\mathbf{F}_T)^{-1}\mathbf{F}_T'\boldsymbol{\varepsilon}.$$

For n sufficiently large, $\hat{\boldsymbol{\beta}}$ is almost certain to be in Θ_T so that $\hat{\boldsymbol{\theta}} - \boldsymbol{\theta}_T \approx \hat{\boldsymbol{\beta}}$ and

$$\hat{\boldsymbol{\theta}} - \boldsymbol{\theta}_T \approx (\mathbf{F}_T'\mathbf{F}_T)^{-1}\mathbf{F}_T'\boldsymbol{\varepsilon}. \tag{8.6}$$

Furthermore, from (8.4) with $\boldsymbol{\theta} = \hat{\boldsymbol{\theta}}$,

$$\begin{aligned} \mathbf{f}(\hat{\boldsymbol{\theta}}) - \mathbf{f}(\boldsymbol{\theta}_T) &\approx \mathbf{F}_T(\hat{\boldsymbol{\theta}} - \boldsymbol{\theta}_T) \\ &\approx \mathbf{F}_T(\mathbf{F}_T'\mathbf{F}_T)^{-1}\mathbf{F}_T'\boldsymbol{\varepsilon} \\ &= \mathbf{P}_{\mathbf{F}}\boldsymbol{\varepsilon}, \end{aligned} \tag{8.7}$$

and

$$\begin{aligned} \mathbf{y} - \mathbf{f}(\hat{\boldsymbol{\theta}}) &\approx \mathbf{y} - \mathbf{f}(\boldsymbol{\theta}_T) - \mathbf{F}_T(\hat{\boldsymbol{\theta}} - \boldsymbol{\theta}_T) \\ &\approx \boldsymbol{\varepsilon} - \mathbf{P}_F\boldsymbol{\varepsilon} \\ &= (\mathbf{I}_n - \mathbf{P}_{\mathbf{F}})\boldsymbol{\varepsilon}. \end{aligned} \tag{8.8}$$

Hence from (8.8) and (8.7) we have

$$\begin{aligned} (n-p)s^2 &= Q(\hat{\boldsymbol{\theta}}) \\ &= \| \mathbf{y} - \mathbf{f}(\hat{\boldsymbol{\theta}}) \|^2 \\ &\approx \| (\mathbf{I}_n - \mathbf{P}_{\mathbf{F}})\boldsymbol{\varepsilon} \|^2 \\ &= \boldsymbol{\varepsilon}'(\mathbf{I}_n - \mathbf{P}_{\mathbf{F}})\boldsymbol{\varepsilon}, \end{aligned} \tag{8.9}$$

and

$$\begin{aligned} \| \mathbf{f}(\hat{\boldsymbol{\theta}}) - \mathbf{f}(\boldsymbol{\theta}_T) \|^2 &\approx \| \mathbf{P}_{\mathbf{F}}\boldsymbol{\varepsilon} \|^2 \\ &= \boldsymbol{\varepsilon}'\mathbf{P}_{\mathbf{F}}\boldsymbol{\varepsilon}. \end{aligned} \tag{8.10}$$

Therefore, using (8.9) and (8.10) we get

$$\begin{aligned} Q(\boldsymbol{\theta}_T) - Q(\hat{\boldsymbol{\theta}}) &\approx \boldsymbol{\varepsilon}'\boldsymbol{\varepsilon} - \boldsymbol{\varepsilon}'(\mathbf{I}_n - \mathbf{P}_{\mathbf{F}})\boldsymbol{\varepsilon} \\ &= \boldsymbol{\varepsilon}'\mathbf{P}_{\mathbf{F}}\boldsymbol{\varepsilon} \\ &\approx (\hat{\boldsymbol{\theta}} - \boldsymbol{\theta}_T)\mathbf{F}_T'\mathbf{F}_T(\hat{\boldsymbol{\theta}} - \boldsymbol{\theta}_T). \end{aligned} \tag{8.11}$$

Within the order of the linear approximation used, we can replace $\mathbf{F}_T$ by $\hat{\mathbf{F}}$ in the above expressions when necessary. Also (8.6) and (8.9) hold to $\mathbf{o}_p(n^{-1/2})$ and $o_p(1)$, respectively (e.g., Gallant, 1987, 258–260). We now have the following theorem.

Theorem 8.1 *Given $\boldsymbol{\varepsilon} \sim N_n[\mathbf{0}, \sigma^2\mathbf{I}_n]$ and regularity conditions A(1) to A(9) above, we have approximately for large n:*

(i) $\hat{\boldsymbol{\theta}} - \boldsymbol{\theta}_T \sim N_p[\boldsymbol{\theta}, \sigma^2\mathbf{C}_T^{-1}]$, *where* $\mathbf{C}_T = \mathbf{F}_T'\mathbf{F}_T$.
(ii) $(n-p)s^2/\sigma^2 \approx \boldsymbol{\varepsilon}'(\mathbf{I}_n - \mathbf{P}_{\mathbf{F}})\boldsymbol{\varepsilon}/\sigma^2 \sim \chi^2_{n-p}$.
(iii) $\hat{\boldsymbol{\theta}}$ *is statistically independent of* s^2.
(iv)

$$\begin{aligned} \frac{[Q(\boldsymbol{\theta}_T) - Q(\hat{\boldsymbol{\theta}})]/p}{Q(\hat{\boldsymbol{\theta}})/(n-p)} &\approx \frac{\boldsymbol{\varepsilon}'\mathbf{P}_{\mathbf{F}}\boldsymbol{\varepsilon}}{\boldsymbol{\varepsilon}'(\mathbf{I}_n - \mathbf{P}_{\mathbf{F}})\boldsymbol{\varepsilon}} \cdot \frac{n-p}{p} \\ &\sim F_{p,n-p}. \end{aligned} \tag{8.12}$$

(v)

$$\frac{(\hat{\boldsymbol{\theta}} - \boldsymbol{\theta}_T)'\mathbf{F}_T'\mathbf{F}_T(\hat{\boldsymbol{\theta}} - \boldsymbol{\theta}_T)}{ps^2} \sim F_{p,n-p}. \tag{8.13}$$

Proof Parts (i) to (iii) follow from the exact linear theory (see Example 3.1 in Sect. 3.3) with $\mathbf{X} = \mathbf{F}_T$. Part (iv) follows from Theorem 4.1 by noting that $\mathbf{I}_n = (\mathbf{I}_n - \mathbf{P}_\Omega) + \mathbf{P}_\Omega$. Part (v) follows from (iv) and (8.11). [The normality of $\boldsymbol{\varepsilon}$ is not needed for the proof of (i).]

We can use the above theorem to test a hypothesis such as $H : \boldsymbol{\theta}_T = \mathbf{c}$. The so-called Wald test uses (8.13), namely

$$(\hat{\boldsymbol{\theta}} - \mathbf{c})'\hat{\mathbf{F}}'\hat{\mathbf{F}}(\hat{\boldsymbol{\theta}} - \mathbf{c})/(ps^2),$$

which is asymptotically distributed as $F_{p,n-p}$ when H is true. We can also use (8.13) to obtain approximate simultaneous confidence intervals. An approximate likelihood ratio test for H is given by (8.12).

8.4 Concentrated Likelihood Methods

We note that σ^2 is a "nuisance" parameter as far as inference about $\boldsymbol{\theta}$ is concerned. There is however a method that uses a useful technique referred to as the method of *concentrated likelihood*, which is a step-wise method of maximum likelihood that side-steps involvement with $v = \sigma^2$.

Suppose we have a general log-likelihood function $L(\boldsymbol{\theta}, v)$ to be maximized with respect to $\boldsymbol{\theta}$ and v, given the data $\mathbf{y}$. We assume that L is uniquely maximized with respect to $\boldsymbol{\theta}$ and v for every $\mathbf{y}$. The first step of the maximization is to find $\gamma_{\max}(\boldsymbol{\theta}, \mathbf{y})$, the unique value of v that maximizes L with respect to v, with $\boldsymbol{\theta}$ being regarded as a constant. The second step consists of finding $\hat{\boldsymbol{\theta}} = \hat{\boldsymbol{\theta}}(\mathbf{y})$, the value of $\boldsymbol{\theta}$ that maximizes $M(\boldsymbol{\theta} \mid \mathbf{y}) \equiv L(\boldsymbol{\theta}, \gamma_{\max}(\boldsymbol{\theta}, \mathbf{y}) \mid \mathbf{y})$. Then

$$\begin{aligned} L[\hat{\boldsymbol{\theta}}(\mathbf{y}), \gamma_{\max}(\hat{\boldsymbol{\theta}}, \mathbf{y}) \mid \mathbf{y}] &= M(\hat{\boldsymbol{\theta}} \mid \mathbf{y}) \\ &\geq M(\boldsymbol{\theta} \mid \mathbf{y}) \\ &= L[\boldsymbol{\theta}, \gamma_{\max}(\boldsymbol{\theta}, \mathbf{y}) \mid \mathbf{y}] \\ &\geq L(\boldsymbol{\theta}, v \mid \mathbf{y}), \end{aligned}$$

and $\hat{\boldsymbol{\theta}}$ and $\hat{v} = \gamma_{\max}(\hat{\boldsymbol{\theta}}, \mathbf{y})$ are the maximum likelihood estimates of $\boldsymbol{\theta}$ and v. The function $M(\boldsymbol{\theta})$ is called the concentrated log-likelihood function because it is concentrated on $\boldsymbol{\theta}$. The usefulness of $M(\boldsymbol{\theta})$ is highlighted by the following theorem.

Theorem 8.2 *Let $L(\boldsymbol{\theta}, v)$ be the log-likelihood defined above. We assume that L is twice differentiable. Define $\boldsymbol{\delta} = (\boldsymbol{\theta}', v)'$, and let $\hat{\boldsymbol{\delta}} = (\hat{\boldsymbol{\theta}}', \hat{v})'$ solve*

$$\frac{\partial L(\boldsymbol{\theta}, v)}{\partial \boldsymbol{\theta}} \Big|_{\hat{\boldsymbol{\delta}}} = \mathbf{0} \quad \textit{and} \quad \frac{\partial L(\boldsymbol{\theta}, v)}{\partial v} \Big|_{\hat{\boldsymbol{\delta}}} = \mathbf{0}. \tag{8.14}$$

Define

$$\begin{aligned} \mathbf{I}(\delta) &= -\frac{\partial^2 L}{\partial \boldsymbol{\delta} \partial \boldsymbol{\delta}'} \\ &= \begin{pmatrix} -\frac{\partial^2 L}{\partial \boldsymbol{\theta} \partial \boldsymbol{\theta}'} & -\frac{\partial^2 L}{\partial \boldsymbol{\theta} \partial v} \\ -\frac{\partial^2 L}{\partial v \partial \boldsymbol{\theta}'} & -\frac{\partial^2 L}{\partial v^2} \end{pmatrix} \\ &= \begin{pmatrix} \mathbf{I}_{\theta\theta} & \mathbf{I}_{\theta v} \\ \mathbf{I}_{v\theta} & I_{vv} \end{pmatrix}, \end{aligned} \tag{8.15}$$

say, and assume that it is positive definite at $\boldsymbol{\delta} = \hat{\boldsymbol{\delta}}$. Also define

$$\mathbf{I}^{-1}(\delta) = \begin{pmatrix} \mathbf{J}_{\theta\theta} & \mathbf{J}_{\theta v} \\ \mathbf{J}_{v\theta} & J_{vv} \end{pmatrix}. \tag{8.16}$$

Suppose for any fixed $\boldsymbol{\theta}$ that $v = \gamma(\boldsymbol{\theta})$ solves

$$\frac{\partial L(\boldsymbol{\theta}, v)}{\partial v} = \mathbf{0}, \tag{8.17}$$

and let $M(\boldsymbol{\theta}) = L[\boldsymbol{\theta}, \gamma(\boldsymbol{\theta})]$. Then:

(i)

$$\left\{\frac{\partial M(\boldsymbol{\theta})}{\partial \boldsymbol{\theta}}\right\}_{\hat{\boldsymbol{\theta}}} = \mathbf{0}.$$

(ii)

$$\left\{-\frac{\partial^2 M(\boldsymbol{\theta})}{\partial \boldsymbol{\theta} \partial \boldsymbol{\theta}'}\right\}_{\hat{\boldsymbol{\theta}}} = \mathbf{J}_{\theta\theta}^{-1}\big|_{\hat{\boldsymbol{\delta}}} = (\mathbf{I}_{\theta\theta} - \mathbf{I}_{\theta v} I_{vv}^{-1} \mathbf{I}_{v\theta})_{\hat{\boldsymbol{\delta}}}. \tag{8.18}$$

Proof Since $\mathbf{I}(\hat{\boldsymbol{\delta}})$ is assumed to be positive definite, it follows from A.9(viii) that $\mathbf{I}(\boldsymbol{\delta})$ is positive definite in a neighborhood $\mathcal{N}$ of $\hat{\boldsymbol{\delta}}$. Also, setting appropriate elements of $\mathbf{x}$ in $\mathbf{x}'\mathbf{I}(\boldsymbol{\delta})\mathbf{x}$ equal to zero we see that in $\mathcal{N}$ the principal submatrix $\mathbf{I}_{\theta\theta}$ is positive definite and therefore nonsingular, and $I_{vv} > 0$. Applying the implicit-function theorem to (8.17) we find in $\mathcal{N}$ that $\gamma(\boldsymbol{\theta})$ is uniquely defined, $\hat{v} = \gamma(\hat{\boldsymbol{\theta}})$, and $\gamma(\boldsymbol{\theta})$ has continuous first order derivatives. We assume that the following expressions are valid for $\boldsymbol{\delta} \in \mathcal{N}$.

(i) For $v = \gamma(\boldsymbol{\theta})$,

$$\begin{aligned}\frac{\partial M(\boldsymbol{\theta})}{\partial \theta_j} &= \sum_i \frac{\partial L(\boldsymbol{\theta}, v)}{\partial \theta_i} \cdot \frac{\partial \theta_i}{\partial \theta_j} + \frac{\partial L(\boldsymbol{\theta}, v)}{\partial v} \cdot \frac{\partial v}{\partial \theta_j} \\ &= \left.\frac{\partial L(\boldsymbol{\theta}, v)}{\partial \theta_j}\right|_{v=\gamma(\boldsymbol{\theta})},\end{aligned}$$

since the second term is zero by (8.17). Hence using $\hat{v} = \gamma(\hat{\boldsymbol{\theta}})$,

$$\begin{aligned}\left.\frac{\partial M(\boldsymbol{\theta})}{\partial \boldsymbol{\theta}}\right|_{\hat{\boldsymbol{\theta}}} &= \left.\frac{\partial L(\boldsymbol{\theta}, \gamma(\boldsymbol{\theta}))}{\partial \boldsymbol{\theta}}\right|_{\hat{\boldsymbol{\theta}}} \\ &= \left.\frac{\partial L(\boldsymbol{\theta}, v)}{\partial \boldsymbol{\theta}}\right|_{\hat{\boldsymbol{\theta}}, \hat{v}} \\ &= \mathbf{0},\end{aligned}$$

by (8.14). Thus (i) is proved.

(ii) Now in $\mathcal{N}$, (8.17) is an identity in $\boldsymbol{\theta}$, so that differentiating with respect to θ_j,

$$0 = \frac{\partial}{\partial \theta_j} \left\{ \left. \frac{\partial L(\boldsymbol{\theta}, v)}{\partial v} \right|_{v=\gamma(\boldsymbol{\theta})} \right\}$$

$$= \left\{ \frac{\partial^2 L(\boldsymbol{\theta}, v)}{\partial \theta_j \partial v} + \frac{\partial^2 L(\boldsymbol{\theta}, v)}{\partial v^2} \cdot \frac{\partial v}{\partial \theta_j} \right\}_{v=\gamma(\boldsymbol{\theta})},$$

that is

$$\mathbf{0} = \left\{ \frac{\partial^2 L(\boldsymbol{\theta}, v)}{\partial \boldsymbol{\theta} \partial v} + \frac{\partial^2 L(\boldsymbol{\theta}, v)}{\partial v^2} \frac{\partial v}{\partial \boldsymbol{\theta}} \right\}_{v=\gamma(\boldsymbol{\theta})}. \tag{8.19}$$

Now $-\partial^2 L(\boldsymbol{\theta}, v)/\partial v^2$ evaluated at $v = \gamma(\boldsymbol{\theta})$ cannot be identified with I_{vv} as the former is a function of $\boldsymbol{\theta}$ ($\gamma(\boldsymbol{\theta})$ being a particular function of $\boldsymbol{\theta}$), whereas the latter is a function of both $\boldsymbol{\theta}$ and v with v unconstrained. However, when $\boldsymbol{\theta} = \hat{\boldsymbol{\theta}}$ we have $\gamma(\hat{\boldsymbol{\theta}}) = \hat{v}$ and the two matrices then have the same value. The same argument applies to the first term in (8.19). Hence from (8.19) we have

$$\left. \frac{\partial \gamma(\boldsymbol{\theta})}{\partial \boldsymbol{\theta}} \right|_{\hat{\boldsymbol{\theta}}} = -(\mathbf{I}_{\boldsymbol{\theta} v} I_{vv}^{-1})_{\hat{\boldsymbol{\theta}}, \hat{v}}. \tag{8.20}$$

Using a similar argument leads to

$$-\frac{\partial^2 M(\boldsymbol{\theta})}{\partial \boldsymbol{\theta} \partial \boldsymbol{\theta}'} = \left\{ -\frac{\partial^2 L(\boldsymbol{\theta}, v)}{\partial \boldsymbol{\theta} \partial \boldsymbol{\theta}'} - \frac{\partial \gamma}{\partial \boldsymbol{\theta}} \frac{\partial^2 L(\boldsymbol{\theta}, v)}{\partial v \partial \boldsymbol{\theta}'} \right\}_{v=\gamma(\boldsymbol{\theta})}.$$

Setting $\boldsymbol{\theta} = \hat{\boldsymbol{\theta}}$ and using (8.20) we have

$$\left(-\frac{\partial^2 M(\boldsymbol{\theta})}{\partial \boldsymbol{\theta} \partial \boldsymbol{\theta}'} \right)_{\hat{\boldsymbol{\theta}}} = (\mathbf{I}_{\boldsymbol{\theta}\boldsymbol{\theta}} - \mathbf{I}_{\boldsymbol{\theta} v} I_{vv}^{-1} \mathbf{I}_{v \boldsymbol{\theta}})_{\hat{\boldsymbol{\delta}}}.$$

Then applying $\mathbf{F}^{-1}$ given by A.17 to $\mathbf{I}^{-1}(\hat{\boldsymbol{\delta}})$ of (8.16) gives us

$$\mathbf{J}_{\boldsymbol{\theta}\boldsymbol{\theta}}|_{\hat{\boldsymbol{\delta}}} = (\mathbf{I}_{\boldsymbol{\theta}\boldsymbol{\theta}} - \mathbf{I}_{\boldsymbol{\theta} v} I_{vv}^{-1} \mathbf{I}_{v \boldsymbol{\theta}})_{\hat{\boldsymbol{\delta}}}^{-1},$$

and (ii) is proved.

From the above theorem we have the following steps to finding the maximum likelihood estimates of $\boldsymbol{\theta}$ and v and their asymptotic variance-covariance matrices.

(1) Differentiate the log likelihood function $L(\boldsymbol{\theta}, v)$ with respect to v and solve the resulting equation for $v = \gamma(\boldsymbol{\theta})$ as a function of $\boldsymbol{\theta}$.
(2) Replace v by $\gamma(\boldsymbol{\theta})$ in $L(\boldsymbol{\theta}, v)$ to get $M(\boldsymbol{\theta})$.

(3) Treat $M(\boldsymbol{\theta})$ as though it were the true log-likelihood function for $\boldsymbol{\theta}$, namely differentiate $M(\boldsymbol{\theta})$ with respect to $\boldsymbol{\theta}$, solve for $\hat{\boldsymbol{\theta}}$, and find the estimated information matrix (8.18). Under general regularity conditions, the latter matrix is an estimate of the asymptotic variance-covariance matrix of $\hat{\boldsymbol{\theta}}$.
(4) $\hat{v}$ is given by $\gamma(\hat{\boldsymbol{\theta}})$.

Example 8.1 The above process is now demonstrated for the normal distribution. We have

$$f(\mathbf{y}) = (2\pi v)^{-n/2} \exp\left(-\frac{1}{2v}\sum_{i=1}^{n}[y_i - f(\mathbf{x}_i;\boldsymbol{\theta})]^2\right),$$

so that taking logarithms and ignoring constants

$$\begin{aligned} L(\boldsymbol{\theta}, v) &= -\frac{n}{2}\log v - \frac{1}{2v}\sum_{i=1}^{n}[y_i - f(\mathbf{x}_i;\boldsymbol{\theta})]^2 \\ &= -\frac{n}{2}\log v - \frac{1}{2v}Q(\boldsymbol{\theta}). \end{aligned}$$

For fixed $\boldsymbol{\theta}$, and differentiating with respect to v, the above expression is maximized when $v = Q(\boldsymbol{\theta})/n$ so that the concentrated log-likelihood function is

$$\begin{aligned} M(\boldsymbol{\theta}) &= L\left\{\boldsymbol{\theta}, \frac{Q(\boldsymbol{\theta})}{n}\right\} \\ &= -\frac{n}{2}\log Q(\boldsymbol{\theta}) + \frac{n}{2}(\log n - 1). \end{aligned}$$

This expression is maximized when $Q(\boldsymbol{\theta})$ is minimized, that is at the least-squares estimator $\hat{\boldsymbol{\theta}}$. To get $\hat{v}$ we replace $\boldsymbol{\theta}$ by $\hat{\boldsymbol{\theta}}$ in $\gamma(\boldsymbol{\theta})$ so that $\hat{v} = Q(\hat{\boldsymbol{\theta}})/n$ is the maximum likelihood estimator of v. Now $[\partial Q(\boldsymbol{\theta})/\partial\boldsymbol{\theta}]_{\hat{\boldsymbol{\theta}}} = \mathbf{0}$, so that

$$\begin{aligned} \left\{-\frac{\partial^2 M(\boldsymbol{\theta})}{\partial\boldsymbol{\theta}\partial\boldsymbol{\theta}'}\right\}_{\hat{\boldsymbol{\theta}}} &= \left\{-\frac{n}{2[Q(\boldsymbol{\theta})]^2}\frac{\partial Q(\boldsymbol{\theta})}{\partial\boldsymbol{\theta}}\cdot\frac{\partial Q(\boldsymbol{\theta})}{\partial\boldsymbol{\theta}'} + \frac{n}{2Q(\boldsymbol{\theta})}\frac{\partial^2 Q(\boldsymbol{\theta})}{\partial\boldsymbol{\theta}\partial\boldsymbol{\theta}'}\right\}_{\hat{\boldsymbol{\theta}}} \\ &= \frac{1}{2\hat{v}}\left\{\frac{\partial^2 Q(\boldsymbol{\theta})}{\partial\boldsymbol{\theta}\partial\boldsymbol{\theta}'}\right\}_{\hat{\boldsymbol{\theta}}} \\ &\approx \frac{1}{2\hat{v}}2\hat{\mathbf{F}}'\hat{\mathbf{F}} \text{ by (8.3)} \\ &= \frac{\hat{\mathbf{F}}'\hat{\mathbf{F}}}{\hat{v}} \end{aligned}$$

so that $\text{Var}[\hat{\boldsymbol{\theta}}]$ is estimated by $\hat{v}(\hat{\mathbf{F}}'\hat{\mathbf{F}})^{-1}$. This is our usual estimator but with s^2 replaced by $\hat{v}$ (cf. (8.13)).

8.5 Large Sample Tests

Three large sample tests are available for testing a nonlinear hypothesis $H : \mathbf{a}(\boldsymbol{\theta}) = [a_1(\boldsymbol{\theta}), a_2(\boldsymbol{\theta}), \ldots, a_q(\boldsymbol{\theta})]' = \mathbf{0}$, the Likelihood ratio (LR) test, Wald's (1943) (W) test, and the Lagrange Multiplier (LM) or Score test (Rao, 1947; Silvey, 1959). If $M(\boldsymbol{\theta})$ is the concentrated log-likelihood function, then the three test statistics are defined as follows:

$$\begin{aligned} \mathrm{LR} &= 2(M(\hat{\boldsymbol{\theta}}) - M(\hat{\boldsymbol{\theta}}_H)), \\ \mathrm{W} &= \mathbf{a}'(\hat{\boldsymbol{\theta}})[\mathbf{A}\mathbf{M}^{-1}\mathbf{A}']_{\hat{\boldsymbol{\theta}}}^{-1}\mathbf{a}(\hat{\boldsymbol{\theta}}) \quad \text{and} \\ \mathrm{LM} &= \left(\frac{\partial M}{\partial \boldsymbol{\theta}'}\mathbf{M}^{-1}\frac{\partial M}{\partial \boldsymbol{\theta}}\right)_{\hat{\boldsymbol{\theta}}_H}, \end{aligned}$$

where

$$\mathbf{A} = \left(\frac{\partial a_i(\boldsymbol{\theta})}{\partial \theta_j}\right) \quad \text{and} \quad \mathbf{M} = -\frac{\partial^2 M(\boldsymbol{\theta})}{\partial \boldsymbol{\theta} \partial \boldsymbol{\theta}'}.$$

Under fairly general conditions (cf. Amemiya 1983: 351; Engle 1984), the above three statistics are asymptotically equivalent and asymptotically distributed as χ_q^2 when H is true. When normal errors are assumed, another method is available since the expected information matrix for $\boldsymbol{\theta}$ and σ^2 is block diagonal for the two parameters. We can then effectively treat σ^2 as though it were a constant, use the log-likelihood L instead of M, derive the three test statistics, and then replace σ^2 by an appropriate estimator. However, if σ^2 is actually a function of $\boldsymbol{\theta}$ in the nonlinear model, then we can use the likelihood function $L(\boldsymbol{\theta})$ directly.

The asymptotic equivalence of the above three statistics can be proved by showing that the nonlinear model can be approximated for large samples by a linear normal model with a linear hypothesis as described by Theorem 4.5 in Sect. 4.3. There we showed that for this model all three test statistics are identical, and it transpires that those three statistics are asymptotically equivalent to the above three large-sample test statistics. The theory showing the asymptotic equivalence to linear theory is essentially spelt out in detail in Chap. 10 with L appropriately replaced by M if necessary, so we won't reproduce it here. When L is used, we usually replace $\mathbf{M}$ by $\mathbf{B} = -n^{-1}\mathrm{E}[\partial^2 L(\boldsymbol{\theta})/\partial\boldsymbol{\theta}\partial\boldsymbol{\theta}']$, the expected information matrix. We can't use $\mathrm{E}[\mathbf{M}]$ because of the result (8.18).

References

Amemiya, T. (1983). Non-linear regression models. In Z. Griliches & M. D. Intriligator (Eds.), *Handbook of econometrics* (Vol. 1, pp. 333–389). Amsterdam: North-Holland.

Engle, R. F. (1984). Wald, likelihood ratio and Lagrange multiplier rests in econometrics. In Z. Griliches & M. D. Intriligator (Eds.), *Handbook of Econometrics* (Vol. 2, pp. 333–389). Amsterdam: North-Holland.

Gallant, A. R. (1987). *Nonlinear statistical models*. New York: Wiley.

Rao, C. R. (1947). Large sample tests of statistical hypotheses concerning several parameters with applications to problems of estimation. *Proceedings of the Cambridge Philosophical Society, 44*, 50–57.

Seber, G. A. F., & Wild, C. J. (1989). *Nonlinear regression*. New York: Wiley. Also reproduced in paperback by Wiley in (2004).

Silvey, S. D. (1959). The Lagrangian multiplier test. *Annals of Mathematical Statistics, 30*, 389–407.

Wald, A. (1943). Tests of statistical hypotheses concerning several parameters when the number of observations is large. *Transactions of the American Mathematical Society, 54*, 426–482.

Wu, C. F. (1981). Asymptotic theory of nonlinear least squares estimation. *Annals of Statistics, 9*, 501–513.

Chapter 9
Multivariate Models

9.1 Notation

Up till now we have been considering various univariate linear models of the form $y_i = \theta_i + \varepsilon_i$ $(i = 1, 2, \ldots, n)$, where $E[\varepsilon_i] = 0$ and the ε_i are independently and identically distributed. We assumed G that $\boldsymbol{\theta} \in \Omega$, where Ω is a p-dimensional vector space in $\mathbb{R}^n$. A natural extension to this is to replace the response variable y_i by a $1 \times d$ row vector of response variables $\mathbf{y}_i'$, and replace the vector $\mathbf{y} = (y_i)$ by the data matrix

$$\mathbf{Y} = \begin{pmatrix} \mathbf{y}_1' \\ \mathbf{y}_2' \\ \vdots \\ \mathbf{y}_n' \end{pmatrix} = (\mathbf{y}^{(1)}, \mathbf{y}^{(2)}, \ldots, \mathbf{y}^{(d)}),$$

say. Here $\mathbf{y}^{(j)}$ $(j = 1, 2, \ldots, d)$ represents n independent observations on the jth variable of $\mathbf{y}$. Writing $\mathbf{y}^{(j)} = \boldsymbol{\theta}^{(j)} + \mathbf{u}^{(j)}$ with $\mathrm{E}[\mathbf{u}^{(j)}] = \mathbf{0}$, we now have d univariate models, which will generally not be independent, and we can combine them into one equation giving us

$$\mathbf{Y} = \boldsymbol{\Theta} + \mathbf{U},$$

where $\boldsymbol{\Theta} = (\boldsymbol{\theta}^{(1)}, \boldsymbol{\theta}^{(2)}, \ldots, \boldsymbol{\theta}^{(d)})$, $\mathbf{U} = (\mathbf{u}^{(1)}, \mathbf{u}^{(2)}, \ldots, \mathbf{u}^{(d)})$, and $\mathrm{E}[\mathbf{U}] = \mathbf{0}$. Of particular interest are vector extensions of experimental designs where each observation is replaced by a vector observation. For example, we can extend the randomized block design

$$\theta_{ij} = \mu + \alpha_i + \tau_j \quad (i = 1, 2, \ldots, I; j = 1, 2, \ldots, J),$$

G.A.F. Seber, *The Linear Model and Hypothesis*, Springer Series in Statistics,
DOI 10.1007/978-3-319-21930-1_9

where $\boldsymbol{\theta} = (\theta_{11}, \theta_{12}, \ldots, \theta_{1J}, \theta_{21}, \theta_{22} \ldots, \theta_{2J}, \ldots, \theta_{I1}, \theta_{I2}, \ldots, \theta_{IJ})'$ to

$$\theta_{ij} = \mu + \alpha_i + \tau_j.$$

In the univariate case we can use a regression model representation $\boldsymbol{\theta} = \mathbf{X}\boldsymbol{\beta}$, where $\mathbf{X}$ is an $n \times p$ design matrix. Since the form of $\mathbf{X}$ depends on the structure of the design, it will be the same for each of the response variables in the multivariate model so that $\boldsymbol{\theta}^{(j)} = \mathbf{X}\boldsymbol{\beta}^{(j)}$, $\boldsymbol{\Theta} = \mathbf{X}(\boldsymbol{\beta}^{(1)}, \boldsymbol{\beta}^{(2)}, \ldots, \boldsymbol{\beta}^{(d)}) = \mathbf{XB}$, say, and

$$\mathbf{Y}_{n \times d} = \mathbf{X}_{n \times p}\mathbf{B}_{p \times d} + \mathbf{U}_{n \times d}. \tag{9.1}$$

If we let $\Omega = \mathcal{C}[\mathbf{X}]$ then our general model G now becomes $\boldsymbol{\theta}^{(j)} \in \Omega$ for each $j = 1, 2, \ldots, d$, that is the columns of $\boldsymbol{\Theta}$ are in Ω. We can now generalize the univariate least squares theory if we use so-called partial (Löwner) ordering for symmetric matrices, namely, we say that $\mathbf{C} \geq \mathbf{D}$ when $\mathbf{C} - \mathbf{D}$ is nonnegative definite (Seber 2008, 219–220). Thus if $\mathbf{C}(\boldsymbol{\Theta})$ is a symmetric matrix-valued function, we say that $\mathbf{C}$ is minimized at $\boldsymbol{\Theta} = \hat{\boldsymbol{\Theta}}$ if $\mathbf{C}(\boldsymbol{\Theta}) \geq \mathbf{C}(\hat{\boldsymbol{\Theta}})$.

By analogy with univariate least squares estimation, we can minimize the matrix $\mathbf{U}'\mathbf{U} = (\mathbf{Y} - \boldsymbol{\Theta})'(\mathbf{Y} - \boldsymbol{\Theta})$ subject to the columns of $\boldsymbol{\Theta}$ belonging to Ω. Now it seems reasonable to apply the univariate method to each column of $\boldsymbol{\Theta}$ and consider $\hat{\boldsymbol{\theta}}^{(j)} = \mathbf{P}_\Omega \mathbf{y}^{(j)}$ or $\hat{\boldsymbol{\Theta}} = \mathbf{P}_\Omega \mathbf{Y}$. Then $\mathbf{P}_\Omega \boldsymbol{\Theta} = \boldsymbol{\Theta}$, and since $\mathbf{P}_\Omega(\mathbf{I}_n - \mathbf{P}_\Omega) = \mathbf{0}$, we have

$$(\mathbf{Y} - \hat{\boldsymbol{\Theta}})'(\hat{\boldsymbol{\Theta}} - \boldsymbol{\Theta}) = \mathbf{Y}'(\mathbf{I}_n - \mathbf{P}_\Omega)\mathbf{P}_\Omega(\mathbf{Y} - \boldsymbol{\Theta}) = \mathbf{0}. \tag{9.2}$$

Hence for all $\boldsymbol{\Theta}$ with columns in Ω

$$\begin{aligned}
\mathbf{C}(\boldsymbol{\Theta}) &= (\mathbf{Y} - \boldsymbol{\Theta})'(\mathbf{Y} - \boldsymbol{\Theta}) \\
&= (\mathbf{Y} - \hat{\boldsymbol{\Theta}} + \hat{\boldsymbol{\Theta}} - \boldsymbol{\Theta})'(\mathbf{Y} - \hat{\boldsymbol{\Theta}} + \hat{\boldsymbol{\Theta}} - \boldsymbol{\Theta}) \\
&= (\mathbf{Y} - \hat{\boldsymbol{\Theta}})'(\mathbf{Y} - \hat{\boldsymbol{\Theta}}) + (\hat{\boldsymbol{\Theta}} - \boldsymbol{\Theta})'(\hat{\boldsymbol{\Theta}} - \boldsymbol{\Theta}) \quad \text{(by Eq. 9.2)} \\
&\geq (\mathbf{Y} - \hat{\boldsymbol{\Theta}})'(\mathbf{Y} - \hat{\boldsymbol{\Theta}}) \qquad (9.3) \\
&= \mathbf{C}(\hat{\boldsymbol{\Theta}}),
\end{aligned}$$

since $(\hat{\boldsymbol{\Theta}} - \boldsymbol{\Theta})'(\hat{\boldsymbol{\Theta}} - \boldsymbol{\Theta}) \geq \mathbf{0}$, and $\hat{\boldsymbol{\Theta}}$ gives the required minimum. Equality occurs in (9.3) only when $(\hat{\boldsymbol{\Theta}} - \boldsymbol{\Theta})'(\hat{\boldsymbol{\Theta}} - \boldsymbol{\Theta}) = \mathbf{0}$ or, by A.9(v), when $\hat{\boldsymbol{\Theta}} = \boldsymbol{\Theta}$. Since $\mathbf{P}_\Omega$ is unique, $\hat{\boldsymbol{\Theta}}$ is unique and it is called the least squares estimator of $\boldsymbol{\Theta}$. The minimum value of $(\mathbf{Y} - \boldsymbol{\Theta})'(\mathbf{Y} - \boldsymbol{\Theta})$ is

$$\begin{aligned}
\mathbf{Q} &= (\mathbf{Y} - \hat{\boldsymbol{\Theta}})'(\mathbf{Y} - \hat{\boldsymbol{\Theta}}) \\
&= \mathbf{Y}'(\mathbf{I}_n - \mathbf{P}_\Omega)^2\mathbf{Y} \\
&= \mathbf{Y}'(\mathbf{I}_n - \mathbf{P}_\Omega)\mathbf{Y}, \qquad (9.4)
\end{aligned}$$

the matrix analogue of $Q = \mathbf{y}'(\mathbf{I}_n - \mathbf{P}_\Omega)\mathbf{y}$, the residual sum of squares for the univariate model.

We now apply the theory to the case $\boldsymbol{\Theta} = \mathbf{XB}$ with $\Omega = \mathcal{C}[\mathbf{X}]$. Referring to the univariate Example 3.1 in Sect. 3.3, we have from Eq. (3.2) and $\mathbf{X}'\mathbf{P}_\Omega = \mathbf{X}'$ that

$$\mathbf{X}'(\mathbf{Y} - \hat{\boldsymbol{\Theta}}) = \mathbf{X}'(\mathbf{I}_n - \mathbf{P}_\Omega)\mathbf{Y} = \mathbf{0}.$$

Hence if $\hat{\mathbf{B}}$ satisfies $\hat{\boldsymbol{\Theta}} = \mathbf{X}\hat{\mathbf{B}}$, it satisfies the equations

$$\mathbf{X}'\mathbf{X}\hat{\mathbf{B}} = \mathbf{X}'\mathbf{Y}, \tag{9.5}$$

the multivariate analogue of the normal equations. The converse is also true. If $\hat{\mathbf{B}}$ satisfies (9.5) then $\mathbf{X}'(\mathbf{Y} - \mathbf{X}\hat{\mathbf{B}}) = \mathbf{0}$, $\mathbf{X}'(\mathbf{y}^{(j)} - \mathbf{X}\hat{\boldsymbol{\beta}}^{(j)}) = \mathbf{0}$ and $\mathbf{y}^{(j)} - \mathbf{X}\hat{\boldsymbol{\beta}}^{(j)} \perp \Omega$ for every j. Now

$$\mathbf{y}^{(j)} = \mathbf{X}\hat{\boldsymbol{\beta}}^{(j)} + \mathbf{y}^{(j)} - \mathbf{X}\hat{\boldsymbol{\beta}}^{(j)} = \mathbf{a} + \mathbf{b},$$

where $\mathbf{a} \in \Omega$ and $\mathbf{b} \in \Omega^\perp$. Since this orthogonal decomposition is unique, we have $\hat{\boldsymbol{\theta}}^{(j)} = \mathbf{a} = \mathbf{X}\hat{\boldsymbol{\beta}}^{(j)}$ and $\hat{\mathbf{B}}$ satisfies $\hat{\boldsymbol{\Theta}} = \mathbf{X}\hat{\mathbf{B}}$.

Extracting the jth column from (9.5) we have $\mathbf{X}'\mathbf{X}\hat{\boldsymbol{\beta}}^{(j)} = \mathbf{X}'\mathbf{y}^{(j)}$, so that as far as least squares estimation is concerned, we can treat each of the d response variables separately, even though the $\mathbf{y}^{(j)}$ are correlated. Therefore any technique for finding $\hat{\boldsymbol{\beta}}$ in the corresponding univariate model can be used to find each $\hat{\boldsymbol{\beta}}^{(j)}$. This means that univariate computational techniques can be readily extended to the multivariate case.

We began this section with a randomized block example in which $\mathbf{X}$ does not have full rank so we need to address this situation. Once again univariate methods carry over naturally. We introduce identifiability restrictions $\mathbf{H}\boldsymbol{\beta}^{(j)} = \mathbf{0}$ $(j = 1, 2, \ldots, d)$ or $\mathbf{HB} = \mathbf{0}$, where the rank of $\mathbf{G} = (\mathbf{X}', \mathbf{H}')'$ is p and the rows of $\mathbf{H}$ are linearly independent of the rows of $\mathbf{X}$ (see Sect. 3.4).

In the case of multivariate regression we would generally not have the same $\mathbf{X}$ matrix for each response variable so that a more appropriate model would then be

$$\mathbf{y}^{(j)} = \mathbf{X}_j\boldsymbol{\beta}^{(j)} + \mathbf{u}^{(j)}.$$

We shall not consider this situation (cf. Seber 1984, Section 8.9 for some details).

Instead of the column representation of the multivariate model

$$\mathbf{y}^{(j)} = \mathbf{X}\boldsymbol{\beta}^{(j)} + \mathbf{u}^{(j)} \quad (j = 1, 2, \ldots, d) \tag{9.6}$$

it is sometimes more convenient to use the ith row representation

$$\mathbf{y}_i = \mathbf{B}'\mathbf{x}_i + \mathbf{u}_i \quad (i = 1, 2, \ldots, n), \tag{9.7}$$

where $\mathbf{x}_i$ is the ith row of $\mathbf{X}$.

9.2 Estimation

So far we have only assumed that $\mathrm{E}[\mathbf{U}] = \mathbf{0}$. Then

$$\mathrm{E}[\hat{\boldsymbol{\Theta}}] = \mathbf{P}_{\Omega}\mathrm{E}[\mathbf{Y}] = \mathbf{P}_{\Omega}\boldsymbol{\Theta} = \boldsymbol{\Theta}, \tag{9.8}$$

and $\hat{\boldsymbol{\Theta}}$ is an unbiased estimator of $\boldsymbol{\Theta}$. If $\mathbf{X}$ has less than full rank and we introduce identifiability restrictions $\mathbf{HB} = \mathbf{0}$ then, by analogy with the univariate case, we have $\hat{\mathbf{B}} = (\mathbf{G}'\mathbf{G})^{-1}\mathbf{X}'\mathbf{Y}$, where $\mathbf{G}'\mathbf{G} = \mathbf{X}'\mathbf{X} + \mathbf{H}'\mathbf{H}$, and

$$\begin{aligned}
\mathrm{E}[\hat{\mathbf{B}}] &= (\mathbf{G}'\mathbf{G})^{-1}\mathbf{X}'\mathrm{E}[\mathbf{Y}] \\
&= (\mathbf{G}'\mathbf{G})^{-1}\mathbf{X}'\mathbf{X}\mathbf{B} \\
&= (\mathbf{G}'\mathbf{G})^{-1}(\mathbf{X}'\mathbf{X} + \mathbf{H}'\mathbf{H})\mathbf{B} \\
&= \mathbf{B}.
\end{aligned}$$

To consider variance properties we generalize the univariate assumption that the ε_i are uncorrelated with common variance σ^2, that is $E[\varepsilon_h\varepsilon_i] = \delta_{hi}\sigma^2$ where the Kronecker delta $\delta_{hi} = 1$ when $h = i$ and 0 otherwise. The multivariate version is that the $\mathbf{u}_i$ are uncorrelated with common variance-covariance matrix $\boldsymbol{\Sigma} = (\sigma_{jk})$, namely

$$\begin{aligned}
\mathrm{Cov}[\mathbf{y}_h, \mathbf{y}_i] &= \mathrm{Cov}[\mathbf{u}_h, \mathbf{u}_i] \\
&= \mathrm{E}[\mathbf{u}_h\mathbf{u}_i'] \\
&= \delta_{hi}\boldsymbol{\Sigma} \quad (h, i = 1, 2, \ldots, n).
\end{aligned} \tag{9.9}$$

Referring to (9.6) we have

$$\mathrm{Cov}[\mathbf{y}^{(j)}, \mathbf{y}^{(k)}] = \mathrm{Cov}[\mathbf{u}^{(j)}, \mathbf{u}^{(k)}] = \sigma_{jk}\mathbf{I}_d, \tag{9.10}$$

and, since $\hat{\beta}^{(j)} = (\mathbf{G}'\mathbf{G})^{-1}\mathbf{X}'\mathbf{y}^{(j)}$,

$$\begin{aligned}
\mathrm{Cov}[\hat{\beta}^{(j)}, \hat{\beta}^{(k)}] &= (\mathbf{G}'\mathbf{G})^{-1}\mathbf{X}'\mathrm{Cov}[\mathbf{y}^{(j)}, \mathbf{y}^{(k)}]\mathbf{X}(\mathbf{G}'\mathbf{G})^{-1} \\
&= \sigma_{jk}(\mathbf{G}'\mathbf{G})^{-1}\mathbf{X}'\mathbf{X}(\mathbf{G}'\mathbf{G})^{-1}.
\end{aligned} \tag{9.11}$$

Here $(\mathbf{G}'\mathbf{G})^{-1}$ is a generalized inverse of $\mathbf{X}'\mathbf{X}$ by A.15(iii). If $\mathbf{X}$ has full rank then $\mathbf{G}$ is replaced by $\mathbf{X}$ and (9.11) reduces to $\sigma_{jk}(\mathbf{X}'\mathbf{X})^{-1}$. The univariate version of the so-called Gauss-Markov theorem can be generalized to the multivariate case as in the following theorem.

Theorem 9.1 *Let* $\mathbf{Y} = \boldsymbol{\Theta} + \mathbf{U}$, *where the rows of* $\mathbf{U}$ *are uncorrelated with mean* $\mathbf{0}$ *and have a common variance-covariance matrix* $\boldsymbol{\Sigma}$; $\boldsymbol{\Theta} = (\boldsymbol{\theta}^{(1)}, \boldsymbol{\theta}^{(2)}, \ldots, \boldsymbol{\theta}^{(d)})$. *Let* $\phi = \sum_{j=1}^{d} \mathbf{b}_j'\boldsymbol{\theta}^{(j)}$ *and let* $\hat{\boldsymbol{\Theta}}$ *be the least squares estimate of* $\boldsymbol{\Theta}$ *subject to the columns of* $\boldsymbol{\Theta}$ *belonging to* Ω. *Then* $\hat{\phi} = \sum_{j=1}^{d} \mathbf{b}_j'\hat{\boldsymbol{\theta}}^{(j)}$ *is the BLUE of* ϕ, *that is, the linear unbiased estimate of* ϕ *with minimum variance.*

Proof From (9.8) $\hat{\phi}$ is an unbiased estimator of ϕ. Since $\hat{\boldsymbol{\Theta}} = \mathbf{P}_\Omega \mathbf{Y}$, $\hat{\boldsymbol{\theta}}^{(j)} = \mathbf{P}_\Omega \mathbf{y}^{(j)}$ and

$$\hat{\phi} = \sum_{j=1}^{d} \mathbf{b}_j'\mathbf{P}_\Omega \mathbf{y}^{(j)} = \sum_{j=1}^{d} (\mathbf{P}_\Omega \mathbf{b}_j)'\mathbf{y}^{(j)}$$

is linear in the elements of $\mathbf{Y}$. Let $\phi^* = \sum_j \mathbf{c}_j'\mathbf{y}^{(j)}$ be any other linear unbiased estimator of ϕ. Then, taking expected values,

$$\sum_j \mathbf{c}_j'\boldsymbol{\theta}^{(j)} = \phi = \sum_j \mathbf{b}_j'\boldsymbol{\theta}^{(j)} \quad \text{for all } \boldsymbol{\theta}^{(j)} \in \Omega$$

so that $(\mathbf{b}_j - \mathbf{c}_j)'\boldsymbol{\theta}^{(j)} = 0$ for all $\boldsymbol{\theta}^{(j)}$ $(j = 1, 2, \ldots, d)$. Hence $(\mathbf{b}_j - \mathbf{c}_j)$ is perpendicular to Ω, and its projection onto Ω is zero; that is, $\mathbf{P}_\Omega(\mathbf{b}_j - \mathbf{c}_j) = \mathbf{0}$, or $\mathbf{P}_\Omega \mathbf{b}_j = \mathbf{P}_\Omega \mathbf{c}_j$ for $(j = 1, 2, \ldots, d)$. We now compare the variances of the two estimators $\hat{\phi}$ and ϕ^*.

$$\begin{aligned}
\text{var}[\hat{\phi}] &= \text{var}\left[\sum_j \mathbf{b}_j'\mathbf{P}_\Omega \mathbf{y}^{(j)}\right] \\
&= \text{cov}\left[\sum_j \mathbf{c}_j'\mathbf{P}_\Omega \mathbf{y}^{(j)}, \sum_k \mathbf{c}_k'\mathbf{P}_\Omega \mathbf{y}^{(k)}\right] \\
&= \sum_j \sum_k \mathbf{c}_j'\mathbf{P}_\Omega \text{Cov}[\mathbf{y}^{(j)}, \mathbf{y}^{(k)}]\mathbf{P}_\Omega \mathbf{c}_k \\
&= \sum_j \sum_k \mathbf{c}_j'\mathbf{P}_\Omega \mathbf{c}_k \sigma_{jk} \quad \text{[by (9.10)]},
\end{aligned}$$

and similarly

$$\text{var}[\phi^*] = \sum_j \sum_k \mathbf{c}_j'\mathbf{c}_k \sigma_{jk}.$$

Setting $\mathbf{C} = (\mathbf{c}_1, \mathbf{c}_2, \ldots, \mathbf{c}_n)$ and $\boldsymbol{\Sigma} = \mathbf{R}\mathbf{R}'$, where $\mathbf{R}$ is nonsingular (see A.9(iii)), we have, since $\mathbf{I}_n - \mathbf{P}_\Omega$ is symmetric and idempotent,

$$\begin{aligned}
\text{var}[\phi^*] - \text{var}[\hat{\phi}] &= \sum_j \sum_k \mathbf{c}_j'(\mathbf{I}_n - \mathbf{P}_\Omega)\mathbf{c}_k \sigma_{jk} \\
&= \text{trace}[\mathbf{C}'(\mathbf{I}_n - \mathbf{P}_\Omega)\mathbf{C}\boldsymbol{\Sigma}] \\
&= \text{trace}[\mathbf{R}'\mathbf{C}'(\mathbf{I}_n - \mathbf{P}_\Omega)'(\mathbf{I}_n - \mathbf{P}_\Omega)\mathbf{C}\mathbf{R}] \quad \text{by (A.1)} \\
&= \text{trace}[\mathbf{D}'\mathbf{D}] \quad \text{say}, \\
&\geq 0,
\end{aligned}$$

since $\mathbf{D}'\mathbf{D}$ is positive semidefinite, and its trace is the sum of its (nonnegative) eigenvalues (by A.9(ix)). Equality occurs only if $\mathbf{D}'\mathbf{D} = \mathbf{0}$ or $\mathbf{D} = \mathbf{0}$ (by A.9(v)), that is if $(\mathbf{I}_n - \mathbf{P}_\Omega)\mathbf{C} = \mathbf{0}$ or if $\mathbf{c}_j = \mathbf{P}_\Omega \mathbf{c}_j = \mathbf{P}_\Omega \mathbf{b}_j$. Thus $\text{var}[\phi^*] \geq \text{var}[\hat{\phi}]$ with equality if and only if $\phi^* = \hat{\phi}$, and $\hat{\phi}$ is the unique estimate with minimum variance. This completes the proof.

The advantage of the above approach is that Ω is not specified. We now turn out attention to the estimation of $\boldsymbol{\Sigma}$. If Ω has dimension p then, by analogy with the univariate case, a natural contender would be (9.4), namely $\mathbf{Q}/(n-p) = \mathbf{Y}'(\mathbf{I}_n - \mathbf{P}_\Omega)\mathbf{Y}/(n-p)$. Since $\mathbf{P}_\Omega$ is symmetric and idempotent, we have from Theorem 1.4 in Sect. 1.5 that $\text{trace}[\mathbf{P}_\Omega] = \text{rank}[\mathbf{P}_\Omega] = p$ so that

$$\text{trace}[\mathbf{I}_n - \mathbf{P}_\Omega] = n - \text{trace}[\mathbf{P}_\Omega] = n - p.$$

Since $\mathbf{P}_\Omega \boldsymbol{\Theta} = \boldsymbol{\Theta}$,

$$\begin{aligned}
\mathbf{Q} &= \mathbf{Y}'(\mathbf{I}_n - \mathbf{P}_\Omega)\mathbf{Y} \\
&= (\mathbf{Y} - \boldsymbol{\Theta})'(\mathbf{I}_n - \mathbf{P}_\Omega)(\mathbf{Y} - \boldsymbol{\Theta}) \\
&= \mathbf{U}'(\mathbf{I}_n - \mathbf{P}_\Omega)\mathbf{U} \\
&= \sum_h \sum_i (\mathbf{I}_n - \mathbf{P}_\Omega)_{hi} \mathbf{u}_h \mathbf{u}_i'.
\end{aligned}$$

Hence, by (9.9),

$$\begin{aligned}
\text{E}[\mathbf{Q}] &= \sum_h \sum_i (\mathbf{I}_n - \mathbf{P}_\Omega)_{hi} \delta_{hi} \boldsymbol{\Sigma} \\
&= \{\text{trace}[(\mathbf{I}_n - \mathbf{P}_\Omega)\mathbf{I}_n]\}\boldsymbol{\Sigma} \\
&= (n-p)\boldsymbol{\Sigma}, \qquad (9.12)
\end{aligned}$$

so that $\mathbf{Q}/(n-p)$ is an unbiased estimator of $\boldsymbol{\Sigma}$.

9.3 Hypothesis Testing

In order to use the geometrical approach in hypothesis testing as in univariate models, some multivariate distribution theory is needed that requires a type of multivariate generalization of the chi-square distribution, namely the Wishart distribution. A number of equivalent definitions are available and the simplest definition is as follows. If $\mathbf{u}_1, \mathbf{u}_2, \ldots, \mathbf{u}_m$ are independently and identically distributed as $N_d[\mathbf{0}, \boldsymbol{\Sigma}]$, where $\boldsymbol{\Sigma}$ is positive definite, then

$$\mathbf{W} = \mathbf{U}'\mathbf{U} = \sum_{i=1}^{m} \mathbf{u}_i\mathbf{u}_i'$$

is said to have the (nonsingular) Wishart distribution with m degrees of freedom. We shall write $\mathbf{W} \sim W_d[m, \boldsymbol{\Sigma}]$, and the definition can be extended if $\boldsymbol{\Sigma}$ is non-negative definite. If $m \geq d$ (which we shall assume), then it can be shown that $\mathbf{W}$ is positive definite and has distinct positive eigenvalues, all with probability 1. Given the above definitions, we list some properties.

Theorem 9.2

(i) *If $\mathbf{C}$ is a $q \times d$ matrix of rank q, then*

$$\mathbf{CWC}' \sim W_q[m, \mathbf{C\Sigma C}'].$$

The distribution is nonsingular if $m \geq q$.

(ii) *For every $\boldsymbol{\ell}$, $\boldsymbol{\ell}'\mathbf{W}\boldsymbol{\ell}/\boldsymbol{\ell}'\boldsymbol{\Sigma}\boldsymbol{\ell} \sim \chi_n^2$.*

(iii) *If $\mathbf{A}$ is an $n \times n$ matrix of rank r, then $\mathbf{U}'\mathbf{AU} \sim W_d[r, \boldsymbol{\Sigma}]$ if and only if $\mathbf{A}^2 = \mathbf{A}$, that is $\mathbf{A}$ is a projection matrix.*

(iv) *Let $\mathbf{W}_i = \mathbf{U}'\mathbf{A}_i\mathbf{U} \sim W_d[m_i, \boldsymbol{\Sigma}]$ for $i = 1, 2$. Then $\mathbf{W}_1$ and $\mathbf{W}_2$ are statistically independent if and only if $\mathbf{AB} = \mathbf{0}$. If $\mathbf{W}_1$ and $\mathbf{W}_2$ are statistically independent, then $\mathbf{W}_1 + \mathbf{W}_2 \sim \mathbf{W}_d(m_1 + m_2, \boldsymbol{\Sigma}]$*

(v) *If $\mathbf{A}$ is an $n \times n$ non-negative definite matrix of rank r and $r \geq d$, then, with probability one, $\mathbf{U}'\mathbf{AU}$ is positive definite with distinct eigenvalues (Okamoto 1973; Eaton and Perlman 1973). Setting $\mathbf{A} = \mathbf{I}_d$ we see that this result applies to any Wishart matrix when $m \geq d$ as a Wishart matrix can be expressed in the $d \times d$ form $\mathbf{U}'\mathbf{U}$ of rank d.*

Proof Proofs are given in Seber (1984, Section 2.3; A2.8, A5.13).

Given $\mathbf{Y} = \boldsymbol{\Theta} + \mathbf{U}$, where the columns of $\boldsymbol{\Theta}$ are in Ω, a p-dimensional subspace of $\mathbb{R}^n$, we wish to test whether the columns are in ω, a $(p-q)$-dimensional subspace of Ω. If $\mathbf{Q}_H = \mathbf{Y}'(\mathbf{I}_n - \mathbf{P}_\omega)\mathbf{Y}$, then by analogy with the univariate model, our interest focusses on $\mathbf{Q}_H - \mathbf{Q} = \mathbf{Y}'(\mathbf{P}_\Omega - \mathbf{P}_\omega)\mathbf{Y}$ and $\mathbf{Q} = \mathbf{Y}'(\mathbf{I}_n - \mathbf{P}_\Omega)\mathbf{Y}$. We know from univariate theory that $(\mathbf{P}_\Omega - \mathbf{P}_\omega)(\mathbf{I}_n - \mathbf{P}_\Omega) = \mathbf{0}$ so that by (iv) in the above theorem, $\mathbf{Q}_H - \mathbf{Q}$ and $\mathbf{Q}$ are statistically independent. Also, since we showed above that

$$\mathbf{Y}'(\mathbf{I}_n - \mathbf{P}_\Omega)\mathbf{Y} = \mathbf{U}'(\mathbf{I}_n - \mathbf{P}_\Omega)\mathbf{U},$$

where $\mathbf{I}_n - \mathbf{P}_\Omega$ is idempotent of rank $n-p$, it follows from (iii) above that $\mathbf{Q} \sim W_d[n-p, \boldsymbol{\Sigma}]$. When H is true, $\mathbf{P}_\omega \boldsymbol{\Theta} = \boldsymbol{\Theta}$ and

$$\begin{aligned}\mathbf{Y}'(\mathbf{P}_\Omega - \mathbf{P}_\omega)\mathbf{Y} &= (\mathbf{Y} - \boldsymbol{\Theta})'(\mathbf{P}_\Omega - \mathbf{P}_\omega)(\mathbf{Y} - \boldsymbol{\Theta}) \\ &= \mathbf{U}'(\mathbf{P}_\Omega - \mathbf{P}_\omega)\mathbf{U}. \end{aligned} \tag{9.13}$$

Since $\mathbf{P}_\Omega - \mathbf{P}_\omega$ is idempotent of rank q, $\mathbf{Q}_H - \mathbf{Q} \sim W_d[q, \boldsymbol{\Sigma}]$ when H is true. Thus we have independent Wishart distributions when H is true. To test H, a natural statistic to use is the likelihood-ratio test statistic. To do this we first need to obtain the maximum likelihood estimates under G and H.

Theorem 9.3 *Given our general linear model* $\mathbf{Y} = \boldsymbol{\Theta} + \mathbf{U}$ *with the rows* $\mathbf{u}_i'$ *of* $\mathbf{U}$ *independently and identically distributed as* $N_d[\mathbf{0}, \boldsymbol{\Sigma}]$, *then* $\hat{\boldsymbol{\Theta}} = \mathbf{P}_\Omega \mathbf{Y}$ *and* $\hat{\boldsymbol{\Sigma}} = (\mathbf{Y} - \hat{\boldsymbol{\Theta}})'(\mathbf{Y} - \hat{\boldsymbol{\Theta}})/n$ *are the maximum likelihood estimates of* $\boldsymbol{\Theta}$ *and* $\boldsymbol{\Sigma}$. *Also the maximum value of the likelihood function is*

$$f(\mathbf{Y}; \hat{\boldsymbol{\Theta}}, \hat{\boldsymbol{\Sigma}}) = (2\pi)^{-nd/2} |\hat{\boldsymbol{\Sigma}}|^{-n/2} e^{-nd/2}.$$

Proof The likelihood function of $\mathbf{Y} = (\mathbf{y}_1, \mathbf{y}_2, \ldots, \mathbf{y}_n)'$ is the product of the density functions of the $\mathbf{y}_i$, namely

$$f(\mathbf{Y}; \boldsymbol{\Theta}, \boldsymbol{\Sigma}) = (2\pi)^{-nd/2} |\boldsymbol{\Sigma}|^{-n/2} \exp\left\{ -\frac{1}{2} \sum_{i=1}^{n} (\mathbf{y}_i - \boldsymbol{\theta}_i)' \boldsymbol{\Sigma}^{-1} (\mathbf{y}_i - \boldsymbol{\theta}_i) \right\},$$

where $\boldsymbol{\theta}_i$ is the ith row of $\boldsymbol{\Theta}$. Since a constant equals its trace and trace[$\mathbf{CD}$] = trace[$\mathbf{DC}$] (by A.1), the last term of the above expression is

$$\begin{aligned}&\text{trace}[\boldsymbol{\Sigma}^{-1} \textstyle\sum_{i=1}^n (\mathbf{y}_i - \boldsymbol{\theta}_i)(\mathbf{y}_i - \boldsymbol{\theta}_i)'] \\ &\quad= \text{trace}[\boldsymbol{\Sigma}^{-1}(\mathbf{Y} - \boldsymbol{\Theta})'(\mathbf{Y} - \boldsymbol{\Theta})] \\ &\quad= \text{trace}[\boldsymbol{\Sigma}^{-1}(\mathbf{Y} - \hat{\boldsymbol{\Theta}})'(\mathbf{Y} - \hat{\boldsymbol{\Theta}}) + \boldsymbol{\Sigma}^{-1}(\hat{\boldsymbol{\Theta}} - \boldsymbol{\Theta})'(\hat{\boldsymbol{\Theta}} - \boldsymbol{\Theta})] \text{ by (9.2)} \\ &\quad= \text{trace}[\boldsymbol{\Sigma}^{-1}\mathbf{Q}] + \text{trace}[(\hat{\boldsymbol{\Theta}} - \boldsymbol{\Theta})\boldsymbol{\Sigma}^{-1}(\hat{\boldsymbol{\Theta}} - \boldsymbol{\Theta})'].\end{aligned}$$

Since $\boldsymbol{\Sigma}$ is positive definite, then so is $\boldsymbol{\Sigma}^{-1}$ and by A.9(iv)

$$b = \text{trace}[(\hat{\boldsymbol{\Theta}} - \boldsymbol{\Theta})\boldsymbol{\Sigma}^{-1}(\hat{\boldsymbol{\Theta}} - \boldsymbol{\Theta})'] \geq \mathbf{0}.$$

Now the log-likelihood function takes the form

$$L(\boldsymbol{\Theta}, \boldsymbol{\Sigma}) = c - \frac{n}{2} \log|\boldsymbol{\Sigma}| - \frac{1}{2} \text{trace}[\boldsymbol{\Sigma}^{-1}\mathbf{Q}] - \frac{1}{2} b,$$

which is maximized for any positive definite $\boldsymbol{\Sigma}$ when $b = 0$, that is when $\boldsymbol{\Theta} = \hat{\boldsymbol{\Theta}}$. Hence $L(\hat{\boldsymbol{\Theta}}, \boldsymbol{\Sigma}) \geq L(\boldsymbol{\Theta}, \boldsymbol{\Sigma})$ for all positive definite $\boldsymbol{\Sigma}$. Now, as $\mathbf{Q}$ is positive definite with probability 1,

$$L(\hat{\mathbf{\Theta}}, \mathbf{\Sigma}) = -\frac{n}{2}\left\{\log|\mathbf{\Sigma}| + \text{trace}[\mathbf{\Sigma}^{-1}\mathbf{Q}/n]\right\}$$

has a maximum at $\mathbf{\Sigma} = \hat{\mathbf{\Sigma}} = \mathbf{Q}/n$ (by A.10). Thus

$$L(\hat{\theta}, \hat{\mathbf{\Sigma}}) \geq L(\hat{\mathbf{\Theta}}, \mathbf{\Sigma}) \geq L(\mathbf{\Theta}, \mathbf{\Sigma}),$$

so that $\hat{\mathbf{\Theta}}$ and $\hat{\mathbf{\Sigma}}$ are the maximum likelihood estimates. When $b = 0$ we have $\text{trace}[\hat{\mathbf{\Sigma}}^{-1}\mathbf{Q}] = n\,\text{trace}[\mathbf{I}_d] = nd$ and

$$f(\hat{\mathbf{\Theta}}, \hat{\mathbf{\Sigma}}) = (2\pi)^{-nd/2}|\hat{\mathbf{\Sigma}}|^{-n/2}e^{-nd/2}. \tag{9.14}$$

To obtain the likelihood ratio test we note that under H the maximum likelihood estimates are $\hat{\mathbf{\Theta}}_H = \mathbf{P}_\omega\mathbf{Y}$ and

$$\hat{\mathbf{\Sigma}}_H = (\mathbf{Y} - \hat{\mathbf{\Theta}}_H)'(\mathbf{Y} - \hat{\mathbf{\Theta}}_H)/n = \mathbf{Y}'(\mathbf{I}_n - \mathbf{P}_\omega)\mathbf{Y}/n = \mathbf{Q}_H/n.$$

Hence, from (9.14), the likelihood ratio is

$$\begin{aligned} \Lambda[H|G] &= \frac{f(\mathbf{Y}; \hat{\mathbf{\Theta}}_H, \hat{\mathbf{\Sigma}}_H)}{f(\mathbf{Y}; \hat{\mathbf{\Theta}}, \hat{\mathbf{\Sigma}})} \\ &= \frac{|\hat{\mathbf{\Sigma}}_H|^{-n/2}}{|\hat{\mathbf{\Sigma}}|^{-n/2}} \end{aligned}$$

and

$$\begin{aligned} \Lambda_W &= \Lambda[H|G]^{2/n} \\ &= \frac{|\hat{\mathbf{\Sigma}}|}{|\hat{\mathbf{\Sigma}}_H|} \\ &= \frac{|\mathbf{Y}'(\mathbf{I}_n - \mathbf{P}_\Omega)\mathbf{Y}|}{|\mathbf{Y}'(\mathbf{I}_n - \mathbf{P}_\omega)\mathbf{Y}|} \\ &= \frac{|\mathbf{Q}|}{|\mathbf{Q}_H|}. \end{aligned} \tag{9.15}$$

To find the distribution of Λ_W, a statistic proposed by Wilks (1932) and a monotonic function of the likelihood ratio test, we can use a multivariate analogue of Cochran's Theorem 4.1 in Sect. 4.1 stated below. This theorem is useful in testing several hypotheses, as we shall see later.

Theorem 9.4 *Let* $\mathbf{u}_1, \mathbf{u}_2, \ldots, \mathbf{u}_n$ *be independently and identically distributed (i.i.d.) as* $N_d[\mathbf{0}, \mathbf{\Sigma}]$ *and let* $\mathbf{A}_i$ $(i = 1, 2, \ldots, m)$ *be a sequence of* $n \times n$ *symmetric matrices* *with ranks* r_i *such that* $\sum_{i=1}^{m} \mathbf{A}_i = \mathbf{I}_n$. *If one (and therefore all, by A.12) of the following conditions hold, namely*

(i) $\sum_{i=1}^{m} r_i = m$, *where* $r_i = \text{rank}[\mathbf{A}_i]$,
(ii) $\mathbf{A}_i\mathbf{A}_j = \mathbf{0}$ *for all* $i, j, i \neq j$,
(iii) $\mathbf{A}_i^2 = \mathbf{A}_i$ *for* $i = 1, 2, \ldots, m$,

then the generalized quadratics $\mathbf{U}'\mathbf{A}_i\mathbf{U}$ *are independently distributed as Wishart distributions* $W_d[r_i, \boldsymbol{\Sigma}]$.

Proof It is convenient to break the proof into two cases.

(*Case 1*: $\boldsymbol{\Sigma} = \mathbf{I}_d$). The method of proof follows the same pattern as for the univariate case. Suppose the $\mathbf{x}_i$ $(i = 1, 2, \ldots, n)$ are independently and identically distributed (i.i.d.) $N_d[\mathbf{0}, \mathbf{I}_d]$ and $\mathbf{X} = (\mathbf{x}_1, \mathbf{x}_2, \ldots, \mathbf{x}_n)' = (\mathbf{x}^{(1)}, \mathbf{x}^{(2)}, \ldots, \mathbf{x}^{(d)}) = (x_{ij})$. Then using the orthogonal matrix $\mathbf{T}$ as in Theorem 4.1 in Sect. 4.1, and making the transformation $\mathbf{T}'\mathbf{x}^{(i)} = \mathbf{z}^{(i)}$, we have $\mathbf{X} = \mathbf{TZ}$ and

$$\begin{aligned}\mathbf{X}'\mathbf{A}_1\mathbf{X} &= \mathbf{Z}'\mathbf{T}'\mathbf{A}_1\mathbf{T}\mathbf{Z} \\ &= \sum_{r=1}^{r_1} \mathbf{z}_r\mathbf{z}_r'.\end{aligned}$$

Similarly

$$\mathbf{X}'\mathbf{A}_2\mathbf{X} = \sum_{r=r_1+1}^{r_1+r_2} \mathbf{z}_r\mathbf{z}_r' \quad \text{etc.}$$

Now the elements of $\mathbf{x}_i$ are i.d.d. as $N[0, 1]$ so that all the x_{ij} are i.i.d. as $N[0, 1]$, as the $\mathbf{x}_i$ are independent. If $\mathbf{Z} = (z_{ij})$, then, since

$$\text{Var}[\mathbf{z}^{(i)}] = \mathbf{T}'\text{Var}[\mathbf{x}^{(i)}]\mathbf{T} = \mathbf{T}'\mathbf{I}_d\mathbf{T} = \mathbf{T}\mathbf{T}' = \mathbf{I}_d,$$

the z_{ij} are i.i.d. $N[0, 1]$ and the $\mathbf{z}_i$ are i.i.d. $N_d[\mathbf{0}, \mathbf{I}_d]$. Hence the $\mathbf{X}'\mathbf{A}_i\mathbf{X}$ are independently distributed as $W_d[r_i, \mathbf{I}_d]$.

(*Case 2*: $\boldsymbol{\Sigma}$ positive definite). Now $\boldsymbol{\Sigma} = \mathbf{V}\mathbf{V}'$ for some nonsingular $\mathbf{V}$ (A.9(iii)), and setting $\mathbf{x}_i = \mathbf{V}^{-1}\mathbf{u}_i$ we have $\text{Var}[\mathbf{x}_i] = \mathbf{V}^{-1}\boldsymbol{\Sigma}\mathbf{V}^{-1\prime} = \mathbf{I}_d$ so that the $\mathbf{x}_i$ are i.i.d. $N_d[\mathbf{0}, \mathbf{I}_d]$. Now $\mathbf{x}_i' = \mathbf{u}_i'\mathbf{V}^{-1\prime}$ or $\mathbf{X} = \mathbf{U}\mathbf{V}^{-1\prime} = \mathbf{U}\mathbf{V}'^{-1}$. Setting $\mathbf{X} = \mathbf{TZ}$ as in case 1, the transformation $\mathbf{U} = \mathbf{X}\mathbf{V}' = \mathbf{TZ}\mathbf{V}'$ gives us

$$\begin{aligned}\mathbf{U}'\mathbf{A}_1\mathbf{U} &= \mathbf{V}\mathbf{Z}'\mathbf{T}'\mathbf{A}_1\mathbf{T}\mathbf{Z}\mathbf{V}' \\ &= \sum_{r=1}^{r_1} \mathbf{V}\mathbf{z}_r(\mathbf{V}\mathbf{z}_r)' \\ &= \sum_{r=1}^{r_1} \mathbf{w}_r\mathbf{w}_r', \quad \text{say,}\end{aligned}$$

which is distributed as $W_d[r_i, \boldsymbol{\Sigma}]$ as the $\mathbf{w}_r$ are i.i.d. $N_d[\mathbf{0}, \mathbf{V}\mathbf{V}']$ that is $N_d[\mathbf{0}, \boldsymbol{\Sigma}]$. Applying the same transformation again we get

$$\mathbf{U}'\mathbf{A}_2\mathbf{U} = \sum_{r=r_1+1}^{r_1+r_2} \mathbf{w}_r\mathbf{w}_r'$$

etc. showing that $\mathbf{U}'\mathbf{A}_i\mathbf{U} \sim W_d[r_i, \boldsymbol{\Sigma}]$ and the $\mathbf{U}'\mathbf{A}_i\mathbf{U}$ are mutually independent. This complete the proof.

We now return to (9.15). When H is true, we have from (9.15) and (9.13)

$$\begin{aligned} \Lambda_W &= \frac{|\mathbf{Q}|}{|\mathbf{Q}_H|} \\ &= \frac{|\mathbf{Q}|}{|\mathbf{Q}_H - \mathbf{Q} + \mathbf{Q}|} \\ &= \frac{|\mathbf{U}'(\mathbf{I}_n - \mathbf{P}_\Omega)\mathbf{U}|}{|\mathbf{U}'(\mathbf{P}_\Omega - \mathbf{P}_\omega)\mathbf{U} + \mathbf{U}'(\mathbf{I}_n - \mathbf{P}_\Omega)\mathbf{U}|}. \end{aligned}$$

Since $\mathbf{I}_n = (\mathbf{I}_n - \mathbf{P}_\Omega) + (\mathbf{P}_\Omega - \mathbf{P}_\omega) + \mathbf{P}_\omega$ is a decomposition into idempotent matrices, Theorem 9.4 applies. Hence, once again, we find that $\mathbf{Q}$ and $\mathbf{Q}_H - \mathbf{Q}$ are independently distributed as $W_d[n-p, \boldsymbol{\Sigma}]$ and $W_d[q, \boldsymbol{\Sigma}]$ respectively and, by Theorem 9.2(iv), $\mathbf{Q}_H$ is $W_d[n-p+q, \boldsymbol{\Sigma}]$, when H is true.

Here Λ_W has a so-called $U_{d,q,n-p}$ distribution when H is true and its properties are discussed in Seber (1984, Sects 2.5.4, 2.5.5). When $(n-p) \geq d$ and $q \geq d$, $\mathbf{Q}$ and $\mathbf{Q}_H$ are both positive definite with probability one (cf. Theorem 9.2(v)) so that $\mathbf{Q}_H$ has a symmetric positive-definite square root $\mathbf{Q}_H^{1/2}$ (see A.9(ii)). Hence,

$$\begin{aligned} \Lambda_W &= |\mathbf{Q}_H^{-1/2}\mathbf{Q}\mathbf{Q}_H^{-1/2}| \\ &= |\mathbf{Q}_H^{-1/2}(\mathbf{Q}_H + \mathbf{Q} - \mathbf{Q}_H)\mathbf{Q}_H^{-1/2}| \\ &= |\mathbf{I}_d - \mathbf{V}| \\ &= |\mathbf{T}'||\mathbf{I}_d - \mathbf{V}||\mathbf{T}| \quad (\mathbf{T} \text{ orthogonal and } \mathbf{T}'\mathbf{V}\mathbf{T} = \mathrm{diag}(\theta_1, \theta_2, \ldots, \theta_d)) \\ &= |\mathbf{I}_d - \mathrm{diag}(\theta_1, \theta_2, \ldots, \theta_d)| \\ &= \prod_{j=1}^{d}(1 - \theta_j), \end{aligned}$$

where $\mathbf{V} = \mathbf{Q}_H^{-1/2}(\mathbf{Q}_H - \mathbf{Q})\mathbf{Q}_H^{-1/2}$ and $(\mathbf{I}_d - \mathbf{V})$ have multivariate Beta distributions, and the θ_j are the ordered eigenvalues of $\mathbf{V}$. Since $\mathbf{V}$ is positive definite with probability one (by A.9(iv)), each $\theta_j > 0$ with probability one. Then θ_j is a root of

$$\begin{aligned}0 &= |\mathbf{V} - \theta_j \mathbf{I}_d| \\ &= |\mathbf{Q}_H^{-1/2}(\mathbf{Q}_H - \mathbf{Q})\mathbf{Q}_H^{-1/2} - \theta_j \mathbf{Q}_H^{-1/2}\mathbf{Q}_H\mathbf{Q}_H^{-1/2}| \\ &= |\mathbf{Q}_H|^{-1}|\mathbf{Q}_H - \mathbf{Q} - \theta_j\mathbf{Q}_H|,\end{aligned}$$

that is, a root of

$$|(\mathbf{Q}_H - \mathbf{Q}) - \theta_j \mathbf{Q}_H| = 0. \tag{9.16}$$

The above matrix, written as $(1 - \theta_j)(\mathbf{Q}_H - \mathbf{Q}) - \theta_j\mathbf{Q}$, is negative definite with probability 1 if $\theta_j \geq 1$ which implies that the above determinant is zero with probability 0 (a contradiction); hence $\theta_j < 1$ with probability one. Since $\mathbf{Q}$ is positive definite with probability one, we can express (9.16) in the form

$$|(\mathbf{Q}_H - \mathbf{Q}) - \phi_j \mathbf{Q}| = 0, \tag{9.17}$$

where $\phi_j = \theta_j/(1 - \theta_j)$ are the eigenvalues of $\mathbf{Q}^{-1/2}(\mathbf{Q}_H - \mathbf{Q})\mathbf{Q}^{-1/2}$. Then conditional on $\mathbf{Q}$, $\mathbf{Q}_H - \mathbf{Q}$ is still $W_d[q, \boldsymbol{\Sigma}]$ (because of independence) and $\mathbf{Q}^{-1/2}(\mathbf{Q}_H - \mathbf{Q})\mathbf{Q}^{-1/2}$ is $W_d[q, \mathbf{Q}^{-1/2}\boldsymbol{\Sigma}\mathbf{Q}^{-1/2}]$ having distinct eigenvalues ϕ_j with probability one. Hence the eigenvalues of $\mathbf{V}$ are also distinct so that we can order them in the form $1 > \theta_1 > \theta_2 > \cdots > \theta_d > 0$.

We note from above that $\Lambda_W = \prod_{j=1}^d (1 - \theta_j) = \prod_{j=1}^d (1 + \phi_j)^{-1}$, a statistic proposed by Wilks (1932). By the likelihood principle (cf. 9.15) we reject H if Λ_W is too small, that is, if $|\mathbf{Q}_H|$ is much greater than $|\mathbf{Q}|$. However there are several other competing test statistics also based on functions of eigenvalues. The key ones as well as Wilks' lambda are as follows:

(1) Wilks' $\Lambda_W = \prod_{j=1}^d (1 + \phi_j)^{-1}$ is the most commonly used statistic. It is most useful if the underlying distributional assumptions appear to be met.
(2) Pillai-Bartlett trace (Pillai 1955),

$$\Lambda_{PB} = \text{trace}[(\mathbf{Q}_H - \mathbf{Q})\mathbf{Q}_H^{-1}] = \sum_{j=1}^d \left[\frac{\phi_j}{1 + \phi_j}\right].$$

This is more robust than Λ_W and is preferable with smaller sample sizes, unequal cell numbers, and unequal variances.
(3) The Lawley-Hotelling trace,

$$\begin{aligned}\Lambda_{LH} &= (n - p)\,\text{trace}[(\mathbf{Q}_H - \mathbf{Q})\mathbf{Q}^{-1}] \\ &= (n - p)\sum_{j=1}^d \phi_j.\end{aligned}$$

This test can generally be ignored as it is similar to Λ_W.

(4) Roy's (1953) greatest root test is based on the statistic $\Lambda_R = \phi_{\max}$, and it arises from the so-called union-intersection test. Here $\phi_{\max}$ is the largest eigenvalue of $(\mathbf{Q}_H - \mathbf{Q})\mathbf{Q}^{-1}$. We reject H if $\phi_{\max}$ is too large. This statistic is very sensitive to departures from the underlying distributional assumptions and should be used with caution. However $\phi_{\max}$ can be used to construct simultaneous confidence intervals, as we see later.

The above four statistics are translated into F-statistics in order to test the null hypothesis. In some cases, the F statistic is exact and in other cases it is approximate and good statistical packages will tell us whether the F is exact or approximate. Although the four statistics will generally differ, they produce identical F statistics in some cases. Because Roy's largest root is an upper bound on F, it is generally disregarded when it is significant and the others are not significant. All four test statistics are usually given in statistical computing packages.

9.4 Some Examples

We now apply the above theory to several examples. Univariate ANOVA methods generalize readily to multivariate methods since, in practice, we simply replace $\mathbf{y}'\mathbf{A}\mathbf{y} = \sum_i \sum_j a_{ij} y_i y_j$ by $\mathbf{Y}'\mathbf{A}\mathbf{Y} = \sum_i \sum_j a_{ij} \mathbf{y}_i \mathbf{y}_j'$, which means we simply replace $y_i y_j$ by $\mathbf{y}_i \mathbf{y}_j'$.

Example 9.1 (Randomized Block Design) We now consider the multivariate example given at the beginning of this chapter, namely

$$\boldsymbol{\theta}_{ij} = \boldsymbol{\mu} + \boldsymbol{\alpha}_i + \boldsymbol{\tau}_j,$$

where $\sum_i \boldsymbol{\alpha}_i = \mathbf{0}$ and $\sum_j \boldsymbol{\tau}_j = \mathbf{0}$. From (6.27) the least squares estimates are $\hat{\boldsymbol{\mu}} = \bar{\mathbf{y}}_{..}$, $\hat{\boldsymbol{\alpha}}_i = \bar{\mathbf{y}}_{i\cdot} - \bar{\mathbf{y}}_{..}$, and $\hat{\boldsymbol{\beta}}_j = \bar{\mathbf{y}}_{\cdot j} - \bar{\mathbf{y}}_{..}$. To test H that the $\boldsymbol{\alpha}_i$ are all zero, the univariate treatment sum of squares from Table 6.2 in Sect. 6.6 is $J\sum_i(\bar{y}_{i\cdot} - \bar{y}_{..})^2$ which now becomes $\mathbf{Q}_H - \mathbf{Q} = J\sum_i(\bar{\mathbf{y}}_{i\cdot} - \bar{\mathbf{y}}_{..})(\bar{\mathbf{y}}_{i\cdot} - \bar{\mathbf{y}}_{..})'$. The residual sum of squares $\sum_i \sum_j (y_{ij} - \bar{y}_{i\cdot} - \bar{y}_{\cdot j} + \bar{y}_{..})^2$ becomes

$$\mathbf{Q} = \sum_i \sum_j (\mathbf{y}_{ij} - \bar{\mathbf{y}}_{i\cdot} - \bar{\mathbf{y}}_{\cdot j} + \bar{\mathbf{y}}_{..})(\mathbf{y}_{ij} - \bar{\mathbf{y}}_{i\cdot} - \bar{\mathbf{y}}_{\cdot j} + \bar{\mathbf{y}}_{..})'.$$

Example 9.2 (Comparing Multivariate Means) We wish to compare the means $\boldsymbol{\mu}_i$ of I multivariate normal populations with common variance-covariance $\boldsymbol{\Sigma}$. For $i = 1, 2, \ldots, I$, let $\mathbf{y}_{ij}$ $(j = 1, 2, \ldots, J_i)$ be a sample of J_i observations from $N_d[\boldsymbol{\mu}_i, \boldsymbol{\Sigma}]$. In the univariate case we can use the normal equations obtained by differentiating $\sum_i \sum_j (y_{ij} - \mu_i)^2$ to obtain the least squares estimate $\hat{\mu}_i = \bar{y}_{i\cdot}$ of μ_i and $Q = \sum_i \sum_j (y_{ij} - \bar{y}_{..})^2$. Under $H : \mu_i = \mu$ for all i, we differentiate $\sum_i \sum_j (y_{ij} - \mu)^2$ to get $\hat{\mu}_H = \bar{y}_{..}$ and $Q_H = \sum_i \sum_j (y_{ij} - \bar{y}_{..})^2$. For the multivariate case we have $H : \boldsymbol{\mu}_i = \boldsymbol{\mu}$ for all i with

$$\mathbf{Q} = \sum_i \sum_j (\mathbf{y}_{ij} - \overline{\mathbf{y}}_{i\cdot})(\mathbf{y}_{ij} - \overline{\mathbf{y}}_{i\cdot})', \quad \mathbf{Q}_H = \sum_i \sum_j (\mathbf{y}_{ij} - \overline{\mathbf{y}}_{\cdot\cdot})(\mathbf{y}_{ij} - \overline{\mathbf{y}}_{\cdot\cdot})'$$

and $\mathbf{Q}_H - \mathbf{Q} = \sum_i \sum_j (\overline{\mathbf{y}}_{i\cdot} - \overline{\mathbf{y}}_{\cdot\cdot})(\overline{\mathbf{y}}_{i\cdot} - \overline{\mathbf{y}}_{\cdot\cdot})'$.

Example 9.3 (Regression Model) We return to the regression model considered in Sect. 9.1, namely $\boldsymbol{\Theta} = \mathbf{XB}$, where $\mathbf{X}$ is $n \times p$ of rank r. If $r = p$ then $\mathbf{B} = (\mathbf{X}'\mathbf{X})^{-1}\mathbf{X}'\boldsymbol{\Theta}$, $\hat{\boldsymbol{\Theta}} = \mathbf{X}\hat{\mathbf{B}}$, and from (9.5) $\hat{\mathbf{B}} = (\mathbf{XX}')^{-1}\mathbf{X}'\mathbf{Y}$. To test $\mathbf{AB} = \mathbf{0}$, where $\mathbf{A}$ is a known $q \times p$ of rank q, we wish to minimize $(\mathbf{Y} - \mathbf{XB})'(\mathbf{Y} - \mathbf{XB})$ subject to $\mathbf{0} = \mathbf{AB} = \mathbf{A}(\mathbf{XX}')^{-1}\mathbf{X}'\boldsymbol{\Theta} = \mathbf{A}_1\boldsymbol{\Theta}$. We therefore wish to minimize $(\mathbf{Y} - \boldsymbol{\Theta})'(\mathbf{Y} - \boldsymbol{\Theta})$ subject to the columns of $\boldsymbol{\Theta}$ lying in $\omega = \Omega \cap \mathcal{N}[\mathbf{A}_1]$. The least squares estimator of $\boldsymbol{\Theta}$ is now $\hat{\boldsymbol{\Theta}}_H = \mathbf{P}_\omega\mathbf{Y}$. From (4.7) we have

$$\begin{aligned}\mathbf{P}_\Omega - \mathbf{P}_\omega &= \mathbf{P}_\Omega\mathbf{A}_1'(\mathbf{A}_1\mathbf{P}_\Omega\mathbf{A}_1')^{-1}\mathbf{A}_1\mathbf{P}_\Omega \\ &= \mathbf{X}(\mathbf{X}'\mathbf{X})^{-1}\mathbf{A}'[\mathbf{A}(\mathbf{X}'\mathbf{X})^{-1}\mathbf{A}']^{-1}\mathbf{A}(\mathbf{X}'\mathbf{X})^{-1}\mathbf{X}'. \qquad (9.18)\end{aligned}$$

so that

$$\mathbf{Y}'(\mathbf{P}_\Omega - \mathbf{P}_\omega)\mathbf{Y} = \hat{\mathbf{B}}'\mathbf{A}'[\mathbf{A}(\mathbf{X}'\mathbf{X})^{-1}\mathbf{A}']^{-1}\mathbf{A}\hat{\mathbf{B}}.$$

Now using (9.18),

$$\begin{aligned}\mathbf{X}'\mathbf{X}\hat{\mathbf{B}}_H &= \mathbf{X}'\hat{\boldsymbol{\Theta}}_H \\ &= \mathbf{X}'\mathbf{P}_\omega\mathbf{Y} \\ &= \mathbf{X}'\mathbf{P}_\Omega\mathbf{Y} + \mathbf{X}'(\mathbf{P}_\omega - \mathbf{P}_\Omega)\mathbf{Y} \\ &= \mathbf{X}'\mathbf{X}\hat{\mathbf{B}} - \mathbf{A}'[\mathbf{A}(\mathbf{X}'\mathbf{X})^{-1}\mathbf{A}']^{-1}\mathbf{A}\hat{\mathbf{B}}, \qquad (9.19)\end{aligned}$$

so that

$$\hat{\mathbf{B}}_H = \hat{\mathbf{B}} - (\mathbf{X}'\mathbf{X})^{-1}\mathbf{A}'[\mathbf{A}(\mathbf{X}'\mathbf{X})^{-1}\mathbf{A}']^{-1}\mathbf{A}\hat{\mathbf{B}}.$$

If $\mathbf{X}$ is not of full rank, then the constraints $\mathbf{A}_i'\boldsymbol{\beta}^{(j)}$ must be estimable (see end of Sect. 3.4), that is the rows $\mathbf{a}_i'$ of $\mathbf{A}$ must be linear combinations of the rows of $\mathbf{X}$, or $\mathbf{A} = \mathbf{MX}$. Referring to Example 4.5 in Sect. 4.3 we have that

$$\mathbf{P}_\Omega - \mathbf{P}_\omega = \mathbf{X}(\mathbf{X}'\mathbf{X})^{-}\mathbf{A}'[\mathbf{A}(\mathbf{X}'\mathbf{X})^{-}\mathbf{A}']^{-1}\mathbf{A}(\mathbf{X}'\mathbf{X})^{-}\mathbf{X}',$$

along with $\mathbf{P}_\Omega = \mathbf{X}(\mathbf{X}'\mathbf{X})^{-}\mathbf{X}$. If $\mathbf{HB} = \mathbf{0}$ are identifiability constraints we can use $(\mathbf{X}'\mathbf{X})^{-} = (\mathbf{G}'\mathbf{G})^{-1}$, where $\mathbf{G}'\mathbf{G} = \mathbf{X}'\mathbf{X} + \mathbf{H}'\mathbf{H}$.

Example 9.4 (Regression Coefficients) We consider the following example with $\mathbf{X}$ having full rank, as we make use of it later. Suppose in the previous example we set

$\mathbf{A} = \mathbf{I}_p$ so that we are then testing $H : \mathbf{B} = \mathbf{0}$ (i.e., $\boldsymbol{\Theta} = \mathbf{0}$ and $\mathbf{P}_\omega = \mathbf{0}$). When H is true we have

$$\begin{aligned}
\mathbf{Q}_H - \mathbf{Q} &= \mathbf{Y}'(\mathbf{P}_\Omega - \mathbf{P}_\omega)\mathbf{Y} \\
&= \mathbf{Y}'\mathbf{P}_\Omega\mathbf{Y} \\
&= \mathbf{Y}'\mathbf{P}_\Omega^2\mathbf{Y} \\
&= \mathbf{Y}'\mathbf{X}(\mathbf{X}'\mathbf{X})^{-1}(\mathbf{X}'\mathbf{X})(\mathbf{X}'\mathbf{X})^{-1}\mathbf{X}'\mathbf{Y} \\
&= \hat{\mathbf{B}}'\mathbf{X}'\mathbf{X}\hat{\mathbf{B}} \sim W_d[p, \boldsymbol{\Sigma}].
\end{aligned}$$

If we replace $\mathbf{Y}$ by $\mathbf{Y} - \mathbf{X}\mathbf{B}$ in the above algebra, we get $(\hat{\mathbf{B}} - \mathbf{B})'\mathbf{X}'\mathbf{X}(\hat{\mathbf{B}} - \mathbf{B})$ which is now $W_d[p, \boldsymbol{\Sigma}]$ in general (irrespective of whether H is true or not). We shall use this result to construct simultaneous intervals in Sect. 9.6.

Example 9.5 (Orthogonal Hypotheses) Suppose we have hypotheses $H_i : \boldsymbol{\theta} \in \omega_i$, $(i = 1, 2, \ldots, k)$ that are orthogonal with respect to $G : \boldsymbol{\theta} \in \Omega$, so that we have $\omega_i^\perp \cap \Omega \perp \omega_j^\perp \cap \Omega$ for all $i, j, i \neq j$. We now ask which of the four test statistics supports the separate method of Chap. 6. If $\mathbf{Q}_{12\ldots k} - \mathbf{Q}$ is the hypothesis matrix for testing $\boldsymbol{\theta} \in \omega_1 \cap \omega_2 \cdots \cap \omega_k$, then from the end of Sect. 6.2,

$$\begin{aligned}
\Lambda_{LH} &= (n-p)\,\mathrm{trace}[(\mathbf{Q}_{12\ldots k} - \mathbf{Q})\mathbf{Q}^{-1}] \\
&= \sum_{i=1}^{k} (n-p)\,\mathrm{trace}[(\mathbf{Q}_i - \mathbf{Q})\mathbf{Q}^{-1}] \\
&= \sum_{i=1}^{k} \Lambda_{LH}^{(i)},
\end{aligned}$$

so that we have the additive property of the individual test statistics. None of the other three test statistics have this property.

Example 9.6 (Generalized Linear Hypothesis) The theory in this chapter can be generalized in several ways and we consider one generalization. We have the usual model $\mathbf{Y} = \mathbf{X}\mathbf{B} + \mathbf{U}$, where $\mathbf{X}$ is $n \times p$ of rank p and the rows of $\mathbf{U}$ are i.i.d $N_d[\mathbf{0}, \boldsymbol{\Sigma}]$, but H now takes the form $\mathbf{A}\mathbf{B}\mathbf{D} = \mathbf{0}$, where $\mathbf{A}$ is $q \times p$ of rank q $(q \leq p)$, $\mathbf{B}$ is $p \times d$, and $\mathbf{D}$ is $d \times v$ of rank v $(v \leq d)$. As the hypothesis reduces to $\mathbf{A}\mathbf{B} = \mathbf{0}$ when $\mathbf{D} = \mathbf{I}_d$, a reasonable procedure for handling H is to try and carry out the same reduction with a suitable transformation. We can do this by setting $\mathbf{Y}_D = \mathbf{Y}\mathbf{D}$ so that

$$\begin{aligned}
\mathbf{Y}_D &= \mathbf{X}\mathbf{B}\mathbf{D} + \mathbf{U}\mathbf{D} \\
&- \mathbf{X}\boldsymbol{\Phi} + \mathbf{U}_D,
\end{aligned}$$

say, where the rows of

$$\mathbf{U}_D = \begin{pmatrix} \mathbf{u}_1' \\ \mathbf{u}_2' \\ \vdots \\ \mathbf{u}_n' \end{pmatrix} \mathbf{D} = \begin{pmatrix} (\mathbf{D}'\mathbf{u}_1)' \\ (\mathbf{D}'\mathbf{u}_2)' \\ \vdots \\ (\mathbf{D}'\mathbf{u}_n)' \end{pmatrix}$$

are i.i.d. $N_v[\mathbf{0}, \mathbf{D}'\boldsymbol{\Sigma}\mathbf{D}]$. Since H is now $\mathbf{A}\boldsymbol{\Phi} = \mathbf{0}$, we can apply the general theory of this chapter with (cf. (9.18))

$$\begin{aligned} \mathbf{Q}_H - \mathbf{Q} &= \mathbf{D}'\mathbf{Y}'\mathbf{X}(\mathbf{X}'\mathbf{X})^{-1}\mathbf{A}'[\mathbf{A}(\mathbf{X}'\mathbf{X})^{-1}\mathbf{A}']^{-1}\mathbf{A}(\mathbf{X}'\mathbf{X})^{-1}\mathbf{X}'\mathbf{Y}\mathbf{D} \qquad (9.20)\\ &= (\mathbf{A}\hat{\mathbf{B}}\mathbf{D})'[\mathbf{A}(\mathbf{X}'\mathbf{X})^{-1}\mathbf{A}']^{-1}\mathbf{A}\hat{\mathbf{B}}\mathbf{D}, \end{aligned}$$

and

$$\mathbf{Q} = \mathbf{Y}_D'(\mathbf{I}_n - \mathbf{P}_\Omega)\mathbf{Y}_D = \mathbf{D}'\mathbf{Y}'(\mathbf{I}_n - \mathbf{P}_\Omega)\mathbf{Y}\mathbf{D}.$$

The only change is that $\mathbf{Y}$ is replaced by $\mathbf{Y}_D$ and d by v. Then $\mathbf{Q} \sim W_v[n-p, \mathbf{D}'\boldsymbol{\Sigma}\mathbf{D}]$ and, when H is true, $\mathbf{Q}_H - \mathbf{Q} \sim W_v[q, \mathbf{D}'\boldsymbol{\Sigma}\mathbf{D}]$. If $\mathbf{X}$ has less than full rank, say $r < p$, then the above theory still holds, with $(\mathbf{X}'\mathbf{X})^{-1}$ replaced by $(\mathbf{X}'\mathbf{X})^{-}$, and p by r.

It transpires that by an appropriate choice of $\mathbf{A}$ the above theory can be used to carry out tests on one or more multivariate normal distributions such as testing for linear constraints on a mean or comparing profiles of several normal distributions. An example of the former is given in Example 9.7 in the next section. Another generalization of the above model is to use $\mathbf{Y} = \mathbf{X}\boldsymbol{\Delta}\mathbf{K}' + \mathbf{U}$ along with $H : \mathbf{A}\boldsymbol{\Delta}\mathbf{D} = \mathbf{0}$. This model can be used for analyzing growth curves.

9.5 Hotelling's Test Statistic

If $x \sim N_d[\boldsymbol{\mu}, \boldsymbol{\Sigma}]$, $\mathbf{W} \sim W_d[m, \boldsymbol{\Sigma}]$, $\boldsymbol{\Sigma}$ is positive definite, and $\mathbf{x}$ is statistically independent of $\mathbf{W}$, then

$$T^2 = m(\mathbf{x} - \boldsymbol{\mu})'\mathbf{W}^{-1}(\mathbf{x} - \boldsymbol{\mu}) \quad (m \geq d)$$

is said to have a Hotelling's $T^2_{d,m}$ distribution. In particular

$$\frac{m - d + 1}{d}\frac{T^2}{m} \sim F_{d,m-d+1}.$$

When $d = 1$, T^2 reduces to t^2, where t has the t_m distribution.

In Sect. 9.3, if $q = 1$ so that $\mathbf{Q}_H - \mathbf{Q} \sim W_d[1, \boldsymbol{\Sigma}]$ when H is true, then we find that all four test statistics reduce to the same test. To see this we note first that there exists $\mathbf{u} \sim N_d[\mathbf{0}, \boldsymbol{\Sigma}]$ such that $\mathbf{Q}_H - \mathbf{Q} = \mathbf{u}\mathbf{u}'$ (by definition of the Wishart distribution), where $\mathbf{u}$ is statistically independent of $\mathbf{Q}$. Then, by A.4(i)

$$\text{rank}[(\mathbf{Q}_H - \mathbf{Q})\mathbf{Q}^{-1}] = \text{rank}[\mathbf{Q}_H - \mathbf{Q}] = 1$$

so that (9.17) has only one (non-zero) root that we can call $\phi_{\max}$. We see that the four statistics are $\Lambda_W = (1+\phi_{\max})^{-1}$, $\Lambda_{PB} = \phi_{\max}/(1+\phi_{\max})$, $\lambda_{LH} = (n-p)\phi_{\max}$, and $\Lambda_R = \phi_{\max}$ that are all monotonic functions of $\phi_{\max}$. Also, using A.1

$$\begin{aligned}
\Lambda_{LH} &= (n-p)\,\text{trace}[(\mathbf{Q}_H - \mathbf{Q})\mathbf{Q}^{-1}] &\quad (9.21)\\
&= (n-p)\,\text{trace}[\mathbf{u}\mathbf{u}'\mathbf{Q}^{-1}]\\
&= (n-p)\,\text{trace}[\mathbf{u}'\mathbf{Q}^{-1}\mathbf{u}]\\
&= (n-p)\mathbf{u}'\mathbf{Q}^{-1}\mathbf{u}\\
&= \mathbf{u}'\mathbf{S}^{-1}\mathbf{u} &\quad (9.22)\\
&= T^2,
\end{aligned}$$

where $T^2 \sim T^2_{d,n-p}$ and $\mathbf{S} = \mathbf{Q}/(n-p)$.

Example 9.7 (Testing for constraints on a multivariate normal mean) Let $\mathbf{y}_1, \mathbf{y}_2, \ldots, \mathbf{y}_n$ be i.i.d. $N_d[\boldsymbol{\mu}, \boldsymbol{\Sigma}]$ and suppose we wish to test $H : \mathbf{D}'\boldsymbol{\mu} = \mathbf{0}$, where $\mathbf{D}'$ is a known $q \times d$ matrix of rank q. Putting $\mathbf{Y}' = (\mathbf{y}_1, \mathbf{y}_2, \ldots, \mathbf{y}_n)$ and $\mathbf{XB} = \mathbf{1}_n\boldsymbol{\mu}'$, we have the linear model $\mathbf{Y} = \mathbf{XB} + \mathbf{U}$, where the rows of $\mathbf{U}$ are i.i.d. $N_d[\mathbf{0}, \boldsymbol{\Sigma}]$. The hypothesis H now becomes $\mathbf{0}' = \boldsymbol{\mu}'\mathbf{D} = \mathbf{BD}$, which is a special case of $\mathbf{ABD} = \mathbf{0}$ with $\mathbf{A} = 1$ in Example 9.6, in the previous section. Now from (9.20)

$$\begin{aligned}
\mathbf{Q}_H - \mathbf{Q} &= \mathbf{D}'\mathbf{Y}'\mathbf{1}_n(\mathbf{1}_n'\mathbf{1}_n)^{-1}(\mathbf{1}_n'\mathbf{1}_n)(\mathbf{1}_n'\mathbf{1}_n)^{-1}\mathbf{1}_n'\mathbf{Y}\mathbf{D}\\
&= n\mathbf{D}'\overline{\mathbf{y}}\,\overline{\mathbf{y}}'\mathbf{D},
\end{aligned}$$

and

$$\begin{aligned}
\mathbf{Q} &= \mathbf{D}'\mathbf{Y}'\{\mathbf{I}_n - \mathbf{1}_n(\mathbf{1}_n'\mathbf{1}_n)^{-1}\mathbf{1}_n'\}\mathbf{Y}\mathbf{D}\\
&= \mathbf{D}'\sum_i(\mathbf{y}_i - \overline{\mathbf{y}}_\cdot)(\mathbf{y}_i - \overline{\mathbf{y}}_\cdot)'\mathbf{D}\\
&= \mathbf{D}'\mathbf{Q}_y\mathbf{D},
\end{aligned}$$

say. However as $\text{rank}[\mathbf{Q} - \mathbf{Q}_H] = 1$, $\mathbf{Q} \sim W_q[n-1, \Sigma]$, and $\text{Var}[\overline{\mathbf{y}}] = \Sigma/n$, we can test H using (cf. (9.21) and (9.22) with $\mathbf{u} = \mathbf{D}'\overline{\mathbf{y}}$)

$$\begin{aligned} T^2 &= (n-1)\,\text{trace}[(\mathbf{Q}_H - \mathbf{Q})\mathbf{Q}^{-1}] \\ &= n(\mathbf{D}'\overline{\mathbf{y}})'[\mathbf{D}'\mathbf{S}\mathbf{D}]^{-1}\mathbf{D}'\overline{\mathbf{y}}, \end{aligned}$$

where $\mathbf{S} = \mathbf{Q}_y/(n-1)$.

9.6 Simultaneous Confidence Intervals

Suppose we have $\mathbf{Y} = \mathbf{X}\mathbf{B} + \mathbf{U}$ as before where $\mathbf{X}$ is $n \times p$ of rank p and $\mathbf{B} = (\beta_{ij})$. From Example 9.4 in Sect. 9.4, the least squares estimate of $\mathbf{B}$ is $\hat{\mathbf{B}} = (\mathbf{X}'\mathbf{X})^{-1}\mathbf{Y}$, and we consider testing $\mathbf{B} = \mathbf{0}$ as a means of constructing simultaneous confidence intervals for the β_{ij}. We note that H is true if and only if $H_{ab} : \mathbf{a}'\mathbf{B}\mathbf{b} = 0$ is true for all $\mathbf{a}$ and $\mathbf{b}$, so that we can write $H = \cap_a \cap_b H_{ab}$. Setting $\mathbf{y} = \mathbf{Y}\mathbf{b}$, $\boldsymbol{\beta} = \mathbf{B}\mathbf{b}$, and $\hat{\boldsymbol{\beta}} = (\mathbf{X}'\mathbf{X})^{-1}\mathbf{X}'\mathbf{y}$, we can test $H_{ab} : \mathbf{a}'\boldsymbol{\beta} = 0$ using the F-ratio (with $q = 1$)

$$F_{(a,b)} = \frac{Q_H - Q}{Q/(n-p)},$$

where (cf. (9.18))

$$\begin{aligned} Q_H - Q &= \mathbf{y}'\mathbf{X}(\mathbf{X}'\mathbf{X})^{-1}\mathbf{a}[\mathbf{a}'(\mathbf{X}'\mathbf{X})^{-1}\mathbf{a}]^{-1}\mathbf{a}'(\mathbf{X}'\mathbf{X})^{-1}\mathbf{X}'\mathbf{y} \\ &= \{\mathbf{a}'(\mathbf{X}'\mathbf{X})^{-1}\mathbf{X}'\mathbf{Y}\mathbf{b}\}^2/\{\mathbf{a}'(\mathbf{X}'\mathbf{X})^{-1}\mathbf{a}\} \\ &= \frac{(\mathbf{a}'\mathbf{L}\mathbf{b})^2}{\mathbf{a}'\mathbf{M}\mathbf{a}}, \end{aligned}$$

where $\mathbf{L} = (\mathbf{X}'\mathbf{X})^{-1}\mathbf{X}'\mathbf{Y} = \hat{\mathbf{B}}$ and $\mathbf{M} = (\mathbf{X}'\mathbf{X})^{-1}$. We also have

$$\begin{aligned} Q &= \mathbf{y}'(\mathbf{I}_n - \mathbf{P}_\Omega)\mathbf{y} \\ &= \mathbf{b}'\mathbf{Y}'(\mathbf{I}_n - \mathbf{P}_\Omega)\mathbf{Y}\mathbf{b} \\ &= \mathbf{b}'\mathbf{Q}\mathbf{b}. \end{aligned}$$

Using the union-intersection principle, a test of H has acceptance region

$$\begin{aligned} \cap_a \cap_b \{\mathbf{Y} : F(a,b) \le k\} &= \{\mathbf{Y} : \sup_{\mathbf{a},\mathbf{b} \neq \mathbf{0}} F_{(a,b)} \le k\} \\ &= \{\mathbf{Y} : \sup_{\mathbf{a},\mathbf{b}} \frac{(\mathbf{a}'\mathbf{L}\mathbf{b})^2}{(\mathbf{a}'\mathbf{M}\mathbf{a})(\mathbf{b}'\mathbf{Q}\mathbf{b})} \le \frac{k}{n-p} = k_1\} \qquad (9.23) \\ &= \{\mathbf{Y} : \phi_{\max} \le k_1\}. \end{aligned}$$

where $\phi_{\max}$ is the maximum eigenvalue of $\mathbf{M}^{-1}\mathbf{L}\mathbf{Q}^{-1}\mathbf{L}'$ (by A.21(ii)), that is of (see A.6)

$$\begin{aligned}\mathbf{L}'\mathbf{M}^{-1}\mathbf{L}\mathbf{Q}^{-1} &= \mathbf{Y}'\mathbf{X}(\mathbf{X}'\mathbf{X})^{-1}\mathbf{X}'\mathbf{X}(\mathbf{X}'\mathbf{X})^{-1}\mathbf{X}'\mathbf{Q}^{-1} \\ &= \hat{\mathbf{B}}'\mathbf{X}'\mathbf{X}\hat{\mathbf{B}}\mathbf{Q}^{-1} \\ &= (\mathbf{Q}_H - \mathbf{Q})\mathbf{Q}^{-1} \quad \text{(by Example 9.4).}\end{aligned}$$

We have therefore arrived at Roy's maximum root test again. Following Example 9.4, we can replace $\mathbf{L}$ $(= \hat{\mathbf{B}})$ by $\hat{\mathbf{B}} - \mathbf{B}$ to obtain the following:

$$\begin{aligned}1 - \alpha &= \Pr[\phi_{\max} \leq \phi_\alpha] \\ &= \Pr[|\mathbf{a}'(\hat{\mathbf{B}} - \mathbf{B})\mathbf{b}| \leq \{\phi_\alpha \mathbf{a}'(\mathbf{X}'\mathbf{X})^{-1}\mathbf{a} \cdot \mathbf{b}'\mathbf{Q}\mathbf{b}\}^{1/2} \text{ for all } \mathbf{a}, \mathbf{b}\ (\neq \mathbf{0}).\end{aligned}$$

We therefore have a set of multiple confidence intervals for all linear combinations of $\mathbf{B}$ given by

$$\mathbf{a}'\hat{\mathbf{B}}\mathbf{b} \pm \{\phi_\alpha \mathbf{a}'(\mathbf{X}'\mathbf{X})^{-1}\mathbf{a} \cdot \mathbf{b}'\mathbf{Q}\mathbf{b}\}^{1/2},$$

and the set has an overall confidence of $100(1-\alpha)\%$. If we set $\mathbf{a}$ and $\mathbf{b}$ equal to vectors with 1 in the ith and jth positions, respectively, and zeroes elsewhere, we include confidence intervals for all the β_{ij}. These intervals will tend to be very wide. If we wish to include a set of confidence intervals from testing $\mathbf{AB} = \mathbf{0}$, it transpires that using (9.18) we simply replace $\mathbf{a}$ by $\mathbf{A}'\mathbf{a}$ in the above theory. This gives us the set of confidence intervals

$$\mathbf{a}'\mathbf{A}\hat{\mathbf{B}}\mathbf{b} \pm \{\phi_\alpha \mathbf{a}'\mathbf{A}(\mathbf{X}'\mathbf{X})^{-1}\mathbf{A}'\mathbf{a} \cdot \mathbf{b}'\mathbf{Q}\mathbf{b}\}^{1/2}.$$

The largest root test of $\mathbf{AB} = \mathbf{0}$ will be significant if at least one of the above intervals does not contain zero.

References

Eaton, M. L., & Perlman, M. D. (1973). The non-singularity of generalized sample covariance matrices. *Annals of Statistics, 1*, 710–717.

Okamoto, M. (1973). Distinctness of the eigenvalues of a quadratic form in a multivariate sample. *Annal of Statistics, 1*, 763–765.

Pillai, K. C. S. (1955). Some new test criteria in multivariate analysis. *Annals of Mathematical Statistics, 26*, 117–121.

Roy, S. N. (1953). On a heuristic method of test construction and its use in multivariate analysis. *Annals of Mathematical Statistics, 24*, 220–238.

Seber, G. A. F. (1984). *Multivariate observations*. New York: Wiley. Also reproduced in paperback by Wiley in 2004.

Seber, G. A. F. (2008). *A matrix handbook for statisticians*. New York: Wiley.

Wilks, S. S. (1932). Certain generalizations in the analysis of variance. *Biometrika, 24*, 471–494.

Chapter 10
Large Sample Theory: Constraint-Equation Hypotheses

10.1 Introduction

Apart from Chap. 8 on nonlinear models we have been considering linear models and hypotheses. We now wish to extend those ideas to non-linear hypotheses based on samples of n independent observations $x_1, x_2, \ldots, x_n$ (these may be vectors) from a general probability density function $f(x, \boldsymbol{\theta})$, where $\boldsymbol{\theta} = (\theta_1, \theta_2, \ldots, \theta_p)'$ and $\boldsymbol{\theta}$ is known to belong to W a subset of $\mathbb{R}^p$. We wish to test the null hypothesis H that $\boldsymbol{\theta}_T$, the true value of $\boldsymbol{\theta}$, belongs to W_H, a subset of W, given that n is large. We saw in previous chapters that there are two ways of specifying H; either in the form of "constraint" equations such as $\mathbf{a}(\boldsymbol{\theta}) = (a_1(\boldsymbol{\theta}), a_2(\boldsymbol{\theta}), \ldots, a_q(\boldsymbol{\theta}))' = \mathbf{0}$, or in the form of "freedom" equations $\boldsymbol{\theta} = \boldsymbol{\theta}(\boldsymbol{\alpha})$, where $\boldsymbol{\alpha} = (\alpha_1, \alpha_2, \ldots, \alpha_{p-q})'$, or perhaps by a combination of both constraint and freedom equations. Although to any freedom-equation specification there will correspond a constraint-equation specification and vice versa, this relationship is often difficult to derive in practice, and therefore the two forms shall be dealt with separately in this and the next chapter.

We saw in Sect. 8.5 that three large-sample methods of testing H are available for the nonlinear model: the likelihood ratio test, the Wald test, and the Score (Lagrange multiplier) test. The same tests apply in the general situation of sampling from a probability density function. The choice of which method to use will depend partly on the ease of computation of the test statistic and therefore to some extent on the method of specification of W_H. We shall show how a non-linear hypothesis and non-normal model can be approximated, for large n, by a linear normal model and linear hypothesis. The normality arises from fact that a maximum likelihood estimate is asymptotically normally distributed. We shall then use this approximation to define the three test statistics mentioned above and show that they are equivalent asymptotically. In this chapter we shall consider just the constraint-equation form $\mathbf{a}(\boldsymbol{\theta}) = \mathbf{0}$ only so that $W_H = \{\boldsymbol{\theta} : \mathbf{a}(\boldsymbol{\theta}) = \mathbf{0} \text{ and } \boldsymbol{\theta} \in W\}$. The freedom-equation hypothesis will be considered in the next chapter. We have a slight change in

G.A.F. Seber, *The Linear Model and Hypothesis*, Springer Series in Statistics,
DOI 10.1007/978-3-319-21930-1_10

notation because of subscript complications and replace $\hat{\boldsymbol{\theta}}_H$ by $\tilde{\boldsymbol{\theta}}$, the restricted (by H) maximum likelihood estimate of $\boldsymbol{\theta}$.

10.2 Notation and Assumptions

Let $L(\boldsymbol{\theta}) = \log \prod_{i=1}^n f(x_i, \boldsymbol{\theta})$ represent the log likelihood function, and let $\hat{\boldsymbol{\theta}}$ and $\tilde{\boldsymbol{\theta}}$ be the maximum likelihood estimates of $\boldsymbol{\theta}$ for $\boldsymbol{\theta}$ in W and W_H, respectively. Although the maximum likelihood estimates depend on n we shall drop the latter from the notation for simplicity. The (expected) information matrix is denoted by $\mathbf{B}_{\boldsymbol{\theta}}$, where $\mathbf{B}_{\boldsymbol{\theta}}$ is the $p \times p$ with i, jth element

$$-\frac{1}{n} E\left[\frac{\partial^2 L(\boldsymbol{\theta})}{\partial \theta_i \partial \theta_j}\right] = -E\left[\frac{\partial^2 \log f(x, \boldsymbol{\theta})}{\partial \theta_i \partial \theta_j}\right].$$

Let $\mathbf{D}L(\boldsymbol{\theta})$ be the column vector with ith element $\partial L(\boldsymbol{\theta})/\partial \theta_i$, and let $\mathbf{A}_{\boldsymbol{\theta}}$ be the $q \times p$ matrix with i, jth element $\partial a_i(\boldsymbol{\theta})/\partial \theta_j$. For any function $g(\boldsymbol{\theta})$, $\mathbf{D}^2 g(\boldsymbol{\theta})$ is the matrix with i, jth element $\partial^2 g(\boldsymbol{\theta})/\partial \theta_i \partial \theta_j$.

We now assume that W, W_H, $f(x, \boldsymbol{\theta})$ and $\mathbf{a}(\boldsymbol{\theta})$ satisfy certain regularity assumptions which we list below (Silvey 1959). These are not the weakest assumptions we could use, but are perhaps the simplest for the development given here.

(i) $\boldsymbol{\theta}_T$, the true value of $\boldsymbol{\theta}$, is an interior point of W.
(ii) For every $\boldsymbol{\theta} \in W$, $z(\boldsymbol{\theta}) = \int (\log f(x, \boldsymbol{\theta})) f(x, \boldsymbol{\theta}_T) dx$ exists.
(iii) W is a convex compact subset of $\mathbb{R}^p$.
(iv) For almost all x, $\log f(x, \boldsymbol{\theta})$ is continuous on W.
(v) For almost all x and for every $\boldsymbol{\theta} \in W$, $\partial \log f(x, \boldsymbol{\theta})/\partial \theta_i$ exists for $i = 1, 2, \ldots, p$ and $|\partial \log f(x, \boldsymbol{\theta})/\partial \theta_i| < g(x)$ for $i = 1, 2, \ldots, p$, where $\int g(x) f(x, \boldsymbol{\theta}_T) dx < \infty$.
(vi) The function $\mathbf{a}(\boldsymbol{\theta})$ is continuous on W.
(vii) There exists a point $\boldsymbol{\theta}_* \in W_H$ such that $z(\boldsymbol{\theta}_*) > z(\boldsymbol{\theta})$ when $\boldsymbol{\theta} \in W_H$ and $\boldsymbol{\theta} \neq \boldsymbol{\theta}_*$.
(viii) $\boldsymbol{\theta}_*$ is an interior point of W_H.
(ix) The functions $a_i(\boldsymbol{\theta})$ possess first- and second-order partial derivatives that are continuous (and therefore bounded) on W.
(x) The order of operations of integration with respect to x and differentiation with respect to $\boldsymbol{\theta}$ are reversible; thus

$$0 = (\partial/\partial \theta_i)(1) = (\partial/\partial \theta_i) \int f(x, \boldsymbol{\theta}) dx = \int \partial f/\partial \theta_i dx$$

and, using a similar argument,

$$0 = \int \partial^2 f/\partial \theta_i \partial \theta_j dx.$$

(xi) For almost all x, $\log f(x, \boldsymbol{\theta})$ possesses continuous second-order partial derivatives in a neighborhood of $\boldsymbol{\theta}_T$. Also if $\boldsymbol{\theta}$ belongs to this neighborhood, then

$$|\partial^2 \log f(x, \boldsymbol{\theta})/\partial\theta_i\partial\theta_j| < G_1(x) \text{ for } i,j = 1,2,\ldots,p,$$

where

$$\int G_1(x)f(x, \boldsymbol{\theta}_T)dx < \infty.$$

(xii) For almost all x, $\log f(x, \boldsymbol{\theta})$ possesses third-order partial derivatives in a neighborhood of $\boldsymbol{\theta}_T$, and if $\boldsymbol{\theta}$ is in this neighborhood, then

$$|\partial^3 \log f(x, \boldsymbol{\theta})/\partial\theta_i\partial\theta_j\partial\theta_k| < G_2(x) \text{ for } i,j,k = 1,2,\ldots,p,$$

where

$$\int G_2(x)f(x, \boldsymbol{\theta}_T)dx < \infty.$$

(xiii) The matrix $\mathbf{A}_{\boldsymbol{\theta}}$ has rank q in the neighborhood of $\boldsymbol{\theta}_T$.

In the above assumptions, the statement "for almost all x" means "for all x except for a set of measure zero—the probability measure being defined by the (cumulative) distribution function of $f(x, \boldsymbol{\theta})$". Also these assumptions can be applied to discrete probability functions by writing the above integrals in the Stieltjes form.

The matrices $\mathbf{B}_T$, $\mathbf{B}_*$, $\hat{\mathbf{B}}$ and $\tilde{\mathbf{B}}$ denote that $\mathbf{B}_{\boldsymbol{\theta}}$ is evaluated at $\boldsymbol{\theta}_T$, $\boldsymbol{\theta}_*$, $\hat{\boldsymbol{\theta}}$, and $\tilde{\boldsymbol{\theta}}$ respectively, with the same notation for $\mathbf{A}_{\boldsymbol{\theta}}$. We have a similar assignment for $\mathbf{D}$; for example $\mathbf{D}L(\boldsymbol{\theta}_T)$ is $\mathbf{D}L(\boldsymbol{\theta})$ evaluated at $\boldsymbol{\theta}_T$.

As we shall be considering asymptotic theory we will need some definitions. Let $\{\mathbf{a}_n\}$ be a sequence of vectors. If $g(n)$ is a positive function of n, we say that $\mathbf{a}_n = \mathbf{o}[g(n)]$ if

$$\lim_{n\to\infty} \mathbf{a}_n/g(n) = \mathbf{0},$$

and $\mathbf{a}_n = \mathbf{O}[g(n)]$ if there exists a positive integer n_0 and positive constant M such that

$$\| \mathbf{a}_n \| < Mg(n) \quad \text{for } n > n_0.$$

Let $\{\mathbf{z}_n\}$ be a sequence of random vectors. We write

$$p\lim_{n\leftarrow\infty} \mathbf{z}_n = \mathbf{0} \quad \text{if, for every } \delta > 0, \quad \lim_{n\leftarrow\infty} \Pr[\| \mathbf{z}_n \| \leq \delta] = 1.$$

Also $\mathbf{z}_n = \mathbf{o}_p[g(n)]$ if

$$p \lim_{n\to\infty} \mathbf{z}_n/g(n) = \mathbf{0},$$

and $\mathbf{z}_n = \mathbf{O}_p[g(n)]$ if for each $\varepsilon > 0$ there exists a $c(\varepsilon)$ such that

$$\Pr[\| \mathbf{z}_n \| \leq c(\varepsilon)g(n)] \geq 1 - \varepsilon$$

for all values of n.

10.3 Positive-Definite Information Matrix

We make a further assumption, namely,

(xiv) The matrix $\mathbf{B}_{\theta}$ exists and is positive definite in a neighborhood of $\boldsymbol{\theta}_T$; also its elements are continuous functions of $\boldsymbol{\theta}$ there.

Assumptions (xiv), (ix), and (xiii), imply that $(\mathbf{AB}^{-1}\mathbf{A}')_{\theta}$ is positive definite (A.9(iv)) and its elements are continuous functions of $\boldsymbol{\theta}$ in the neighborhood of $\boldsymbol{\theta}_T$.

Let

$$\mathbf{d} = \frac{\partial \log f(x, \boldsymbol{\theta})}{\partial \boldsymbol{\theta}} \text{ and } \mathbf{d}_i = \frac{\partial \log f(x_i, \boldsymbol{\theta})}{\partial \boldsymbol{\theta}} \text{ for } i = 1, 2, \ldots, n.$$

A key part in the proof that follows in the next section depends on the asymptotic distribution of

$$\begin{aligned} \frac{1}{n}\mathbf{D}L(\boldsymbol{\theta}) &= \frac{1}{n}\sum_{i=1}^{n} \frac{\partial \log f(x_i, \boldsymbol{\theta})}{\partial \boldsymbol{\theta}} \\ &= \overline{\mathbf{d}}. \end{aligned}$$

By the multivariate central limit theorem $n^{1/2}\overline{\mathbf{d}}$ is asymptotically normally distributed with mean

$$\begin{aligned} \mathrm{E}[\mathbf{d}] &= \mathrm{E}\left[\frac{\partial \log f(x, \boldsymbol{\theta})}{\partial \boldsymbol{\theta}}\right] \\ &= \int \frac{1}{f}\frac{\partial f}{\partial \boldsymbol{\theta}} f dx \\ &= \frac{\partial}{\partial \boldsymbol{\theta}} \int f dx = \frac{\partial (1)}{\partial \boldsymbol{\theta}} = \mathbf{0}, \text{ by assumption (x)} \end{aligned}$$

and variance-covariance matrix

$$\text{Var}[\mathbf{d}] = \text{E}[\mathbf{dd}'],$$

by Theorem 1.5(vi) in Sect. 1.6.

If $\mathbf{B}_\theta = (b_{ij})$ then

$$\begin{aligned}
b_{ij} &= -\frac{1}{n}E\left[\frac{\partial^2 L(\boldsymbol{\theta})}{\partial\theta_i\partial\theta_j}\right] \\
&= -\int \frac{\partial^2 \log f}{\partial\theta_i\partial\theta_j} f dx \\
&= -\int \frac{\partial}{\partial\theta_i}\left(\frac{1}{f}\frac{\partial f}{\partial\theta_j}\right) f dx \\
&= \int \frac{1}{f^2}\frac{\partial f}{\partial\theta_i}\frac{\partial f}{\partial\theta_j} f dx - \int \frac{1}{f}\frac{\partial^2 f}{\partial\theta_i\partial\theta_j} f dx \\
&= E\left[\frac{\partial \log f}{\partial\theta_i}\frac{\partial \log f}{\partial\theta_j}\right] + 0 \quad \text{(by Assumption (x))} \\
&= (\text{E}[\mathbf{dd}'])_{ij}.
\end{aligned}$$

Hence $\sqrt{n}\mathbf{D}n^{-1}L(\boldsymbol{\theta})$ is asymptotically normally distributed with mean $\mathbf{0}$ and variance matrix $\mathbf{B}_\theta$. This will give us the normality assumption for our asymptotic linear model.

10.3.1 Maximum Likelihood Estimation

We now derive some maximum likelihood equations. From assumptions (ii) to (v) it can be shown, using the Strong Law of Large Numbers, that for almost all sequences $\{x\} = x_1, x_1, \ldots$, the sequence $x_n = n^{-1}L(\boldsymbol{\theta})$ converges to $z(\boldsymbol{\theta})$ uniformly with respect to $\boldsymbol{\theta}$ in W. Assumption (iii) ensures that any continuous function on W attains it supremum at some point in W. In particular, the function $L(\boldsymbol{\theta})$, for almost all x, attains its supremum in W at $\hat{\boldsymbol{\theta}}$, the maximum likelihood estimate. But from Wald (1949), $z(\boldsymbol{\theta}_T) > z(\boldsymbol{\theta})$ when $\boldsymbol{\theta} \neq \boldsymbol{\theta}_T$ and $\boldsymbol{\theta} \in W$, and therefore it can be shown that $\hat{\boldsymbol{\theta}} = w_n$ (as it depends on n) converges to $\boldsymbol{\theta}_T$ for almost all sequences $\{w\}$ as $n \to \infty$. In other words we say that $\hat{\boldsymbol{\theta}} \to \boldsymbol{\theta}_T$ with probability one as $n \to \infty$, which implies the weaker statement $p\lim(\hat{\boldsymbol{\theta}} - \boldsymbol{\theta}_T) = \mathbf{0}$. Since $\boldsymbol{\theta}_T$ is an interior point of W (assumption (i)), it follows that for n sufficiently large, $\hat{\boldsymbol{\theta}}$ will also be an interior point of W and will, by the usual laws of calculus, emerge as a solution of

$$\mathbf{D}n^{-1}L(\hat{\boldsymbol{\theta}}) = \mathbf{0}.$$

Using a Taylor expansion, we have from assumption (xii) and the above equation

$$\mathbf{0} = \mathbf{D}n^{-1}L(\boldsymbol{\theta}^T) + [\mathbf{D}^2 n^{-1}L(\boldsymbol{\theta}_T)](\hat{\boldsymbol{\theta}} - \boldsymbol{\theta}_T) + \mathbf{o}_p(1).$$

But by assumption (xi), the Law of Large Numbers, and assumption (x),

$$p \lim \mathbf{D}^2 n^{-1}L(\boldsymbol{\theta}_T) = \mathbf{D}^2 z(\boldsymbol{\theta}_T) = -\mathbf{B}_T.$$

Thus we can write

$$\mathbf{D}^2 n^{-1}L(\boldsymbol{\theta}_T) = -\mathbf{B}_T + \mathbf{o}_p(1) \tag{10.1}$$

and hence from the previous three equations

$$\hat{\boldsymbol{\theta}} - \boldsymbol{\theta}_T = \mathbf{B}_T^{-1}\mathbf{D}n^{-1}L(\boldsymbol{\theta}_T) + \mathbf{o}_p(1).$$

Since $n^{1/2}\mathbf{D}n^{-1}L(\boldsymbol{\theta}_T)$ is asymptotically $N_p[\mathbf{0}, \mathbf{B}_T]$, it follows from Theorem 1.5(iii) that

$$n^{1/2}(\hat{\boldsymbol{\theta}} - \boldsymbol{\theta}_T) \text{ is asymptotically } N_p[\mathbf{0}, \mathbf{B}_T^{-1}], \tag{10.2}$$

and since $\mathbf{B}_T^{-1}$ does not depend on n we have

$$n^{1/2}(\hat{\boldsymbol{\theta}} - \boldsymbol{\theta}_T) = \mathbf{O}_p(1). \tag{10.3}$$

We now turn out attention to $\tilde{\boldsymbol{\theta}}$ and first of all make one further assumption:

(xv) If H is not true then $\boldsymbol{\theta}_T$ is "near" W_1. This means that since $\boldsymbol{\theta}_T$ and $\boldsymbol{\theta}_*$ maximize $z(\boldsymbol{\theta})$ for $\boldsymbol{\theta}$ belonging to W and W_H, respectively, $\boldsymbol{\theta}_T$ will be "near" $\boldsymbol{\theta}_*$. We define what we mean by nearness by

$$n^{1/2}(\boldsymbol{\theta}_T - \boldsymbol{\theta}_*) = \mathbf{O}(1). \tag{10.4}$$

Assumption (xv) assumes that in testing H we now consider classes of alternatives $\boldsymbol{\theta}_T$ that tend to W_H as $n \to \infty$. We choose this class of alternatives as for a fixed alternative, $\boldsymbol{\theta}_T$, the powers of the tests considered will tend to unity as $n \to \infty$. This method using a limiting sequence of alternatives is usually referred to a Pitman's limiting power or Pitman's local power analysis. However, according to McManus (1991), the idea was first introduced by Neyman and then developed further by Pitman. This assumption (xv) now implies that assumptions (xi), (xiii), and (xiv) are valid in a neighborhood of $\boldsymbol{\theta}_*$, and from assumptions (iii) and (vi) it follows that W_H is a convex compact subset of W. Therefore, by a similar argument that led to Eq. (10.3) we have (Silvey 1959, p. 394) using (10.4)

$$n^{1/2}(\tilde{\boldsymbol{\theta}} - \boldsymbol{\theta}_*) = \mathbf{O}_p(1) \quad \text{and} \quad n^{1/2}(\tilde{\boldsymbol{\theta}} - \boldsymbol{\theta}_T) = \mathbf{O}_p(1). \tag{10.5}$$

In addition, as $\boldsymbol{\theta}_*$ is an interior point of W_H, $\tilde{\boldsymbol{\theta}}$ will be an interior point also, for large enough n, and will emerge as a solution of (cf. Sect. 1.10)

$$\mathbf{D}n^{-1}L(\tilde{\boldsymbol{\theta}}) + \tilde{\mathbf{A}}'\tilde{\boldsymbol{\mu}} = \mathbf{0} \tag{10.6}$$

and

$$\mathbf{a}(\tilde{\boldsymbol{\theta}}) = \mathbf{0}, \tag{10.7}$$

where $\boldsymbol{\mu}$ is the Lagrange multiplier.

Finally, from Eqs. (10.3) to (10.5) we see that $\boldsymbol{\theta}_T$, $\boldsymbol{\theta}_*$, $\hat{\boldsymbol{\theta}}$ and $\tilde{\boldsymbol{\theta}}$ are all "near" each other. Since $\mathbf{A}_{\boldsymbol{\theta}}$ and $\mathbf{B}_{\boldsymbol{\theta}}$ are continuous functions of $\boldsymbol{\theta}$ in the neighborhood of $\boldsymbol{\theta}_T$, we have from Taylor expansions

$$\hat{\mathbf{B}} = \mathbf{B}_T + \mathbf{O}_p(n^{-1/2}) \tag{10.8}$$

$$\mathbf{A}_* = \mathbf{A}_T + \mathbf{O}(n^{-1/2}) \tag{10.9}$$

and

$$\tilde{\mathbf{A}} = \mathbf{A}_T + \mathbf{O}_p(n^{-1/2}). \tag{10.10}$$

10.3.2 The Linear Model Approximation

Using the asymptotic results above, we can now show that our original model and hypothesis can be approximated by the linear model

$$\mathbf{z} = \boldsymbol{\phi} + \boldsymbol{\varepsilon},$$

where ε is $N_p[\mathbf{0}, \mathbf{I}_p]$, $\Omega = \mathbb{R}^p$, and the linear hypothesis

$$H : \omega = \mathcal{N}[(\mathbf{AV})_T],$$

where nonsingular $\mathbf{V}_T$ is defined later. The argument is as follows.

From (10.3) and (10.5) we have

$$n^{1/2}(\tilde{\boldsymbol{\theta}} - \hat{\boldsymbol{\theta}}) = \mathbf{O}_p(1)$$

so that using a Taylor expansion,

$$\mathbf{D}n^{-1}L(\tilde{\boldsymbol{\theta}}) = \mathbf{D}n^{-1}L(\hat{\boldsymbol{\theta}}) + [\mathbf{D}^2n^{-1}L(\hat{\boldsymbol{\theta}})](\tilde{\boldsymbol{\theta}} - \hat{\boldsymbol{\theta}}) + \mathbf{O}_p(n^{-1}).$$

Now $\mathbf{D}n^{-1}L(\hat{\boldsymbol{\theta}}) = \mathbf{0}$, and applying (10.1) to a neighborhood of $\boldsymbol{\theta}_T$ containing $\hat{\boldsymbol{\theta}}$ gives us, by (10.8) and the previous equation,

$$\begin{aligned}\mathbf{D}n^{-1}L(\tilde{\boldsymbol{\theta}}) &= -\hat{\mathbf{B}}(\tilde{\boldsymbol{\theta}} - \hat{\boldsymbol{\theta}}) + \mathbf{O}_p(n^{-1}) \\ &= -\mathbf{B}_T(\tilde{\boldsymbol{\theta}} - \hat{\boldsymbol{\theta}}) + \mathbf{O}_p(n^{-1}).\end{aligned} \tag{10.11}$$

Therefore from (10.6),

$$\tilde{\mathbf{A}}'n^{1/2}\tilde{\boldsymbol{\mu}} = -n^{1/2}\mathbf{D}n^{-1}L(\tilde{\boldsymbol{\theta}}) = \mathbf{O}_p(1),$$

which means that we can write

$$\tilde{\mathbf{A}}'n^{1/2}\tilde{\boldsymbol{\mu}} = \mathbf{A}_T'n^{1/2}\tilde{\boldsymbol{\mu}} + \mathbf{o}_p(1). \tag{10.12}$$

Thus using (10.11), (10.6) becomes

$$\mathbf{B}_T n^{1/2}(\hat{\boldsymbol{\theta}} - \boldsymbol{\theta}_*) - \mathbf{B}_T n^{1/2}(\tilde{\boldsymbol{\theta}} - \boldsymbol{\theta}_*) + \mathbf{A}_T' n^{1/2}\tilde{\boldsymbol{\mu}} = \mathbf{o}_p(1). \tag{10.13}$$

In the same way,

$$\begin{aligned}\mathbf{0} &= n^{1/2}[\mathbf{a}(\tilde{\boldsymbol{\theta}}) - \mathbf{a}(\boldsymbol{\theta}_*)] \\ &= \mathbf{A}_* n^{1/2}(\tilde{\boldsymbol{\theta}} - \boldsymbol{\theta}_*) + \mathbf{O}_p(n^{-1/2}) \\ &= \mathbf{A}_T n^{1/2}(\tilde{\boldsymbol{\theta}} - \boldsymbol{\theta}_*) + \mathbf{O}_p(n^{-1/2}), \quad \text{by (10.9)}.\end{aligned}$$

Therefore (10.7) becomes

$$\mathbf{A}_T n^{1/2}(\tilde{\boldsymbol{\theta}} - \boldsymbol{\theta}_*) = \mathbf{o}_p(1). \tag{10.14}$$

Now from (10.2),

$$n^{1/2}(\hat{\boldsymbol{\theta}} - \boldsymbol{\theta}_*) = n^{1/2}(\boldsymbol{\theta}_T - \boldsymbol{\theta}_*) + \boldsymbol{\delta}, \tag{10.15}$$

where $\boldsymbol{\delta}$ is asymptotically $N_p[\mathbf{0}, \mathbf{B}_T^{-1}]$, which reminds us of the linear model given in Example 2.6 in Sect. 2.4. As $\mathbf{B}_T$ is positive definite, so is $\mathbf{B}_T^{-1}$, and there exists a non-singular matrix $\mathbf{V}_T$ such that $\mathbf{B}_T^{-1} = \mathbf{V}_T\mathbf{V}_T'$ (A.9(vi) and (iii)). Using $\mathbf{B}_T = (\mathbf{V}_T')^{-1}\mathbf{V}_T^{-1}$, we put

$$\mathbf{z} = n^{1/2}\mathbf{V}_T^{-1}(\hat{\boldsymbol{\theta}} - \boldsymbol{\theta}_*) \tag{10.16}$$

$$\boldsymbol{\phi} = n^{1/2}\mathbf{V}_T^{-1}(\boldsymbol{\theta}_T - \boldsymbol{\theta}_*) \tag{10.17}$$

and

$$\tilde{\phi} = n^{1/2}\mathbf{V}_T^{-1}(\tilde{\theta} - \theta_*) \tag{10.18}$$

in Eq. (10.15) to give us

$$\text{Var}[\mathbf{z}] = \mathbf{V}_T^{-1}\text{Var}[\delta](\mathbf{V}_T^{-1})' = \mathbf{V}_T^{-1}(\mathbf{V}_T\mathbf{V}_T')(\mathbf{V}_T^{-1})' = \mathbf{I}_p,$$

and the linear model

$$\mathbf{z} = \phi + \varepsilon,$$

where ε is $N_p[\mathbf{0}, \mathbf{I}_p]$. Premultiplying (10.13) by $\mathbf{V}_T'$ and using (10.14) leads to

$$\mathbf{z} - \tilde{\phi} + [\mathbf{AV}]_T' n^{1/2}\tilde{\mu} = \mathbf{o}_p(1)$$

and

$$[\mathbf{AV}]_T\tilde{\phi} = \mathbf{o}_p(1).$$

But these are asymptotically the least squares equations for testing the linear hypothesis $[\mathbf{AV}]_T\phi = \mathbf{0}$. Thus our original model is asymptotically equivalent to a linear model with

$$G : \Omega = \mathbb{R}^p \quad \text{and} \quad H : \omega = \mathcal{N}[(\mathbf{AV})_T]. \tag{10.19}$$

10.3.3 The Three Test Statistics

Consider the linear model $\mathbf{z} = \phi + \varepsilon$, where ε is $N_p[\mathbf{0}, \mathbf{I}_p]$, $G : \phi \in \Omega$, and $H : \phi \in \Omega \cap \mathcal{N}[\mathbf{C}] = \omega$ for some matrix $\mathbf{C}$. The least squares estimate of ϕ is $\hat{\phi} = \mathbf{P}_\Omega\mathbf{z}$. To find the restricted least squares estimate $\tilde{\phi}$ we minimize $\| \mathbf{z} - \phi \|^2$ subject to $\mathbf{C}\phi = \mathbf{0}$ and $(\mathbf{I}_p - \mathbf{P}_\Omega)\phi = \mathbf{0}$. Introducing Lagrange multipliers $-2\boldsymbol{\lambda}$ and $-2\boldsymbol{\lambda}_1$ and using Sect. 1.10, we have to solve the following equations (cf. Theorem 4.5 in Sect. 4.3)

$$\mathbf{z} - \tilde{\phi} + (\mathbf{I}_p - \mathbf{P}_\Omega)\tilde{\boldsymbol{\lambda}} + \mathbf{C}'\tilde{\boldsymbol{\lambda}}_1 = \mathbf{0}, \tag{10.20}$$

$$(\mathbf{I}_p - \mathbf{P}_\Omega)\tilde{\phi} = \mathbf{0} \quad \text{and} \quad \mathbf{C}\tilde{\phi} = \mathbf{0}. \tag{10.21}$$

Premultiplying (10.20) by $\mathbf{P}_\Omega$ and using (10.21) leads to

$$\begin{pmatrix} \mathbf{I}_p & -\mathbf{P}_\Omega\mathbf{C}' \\ -\mathbf{C}\mathbf{P}_\Omega & \mathbf{0} \end{pmatrix}\begin{pmatrix} \tilde{\phi} \\ \tilde{\boldsymbol{\lambda}}_1 \end{pmatrix} = \begin{pmatrix} \mathbf{P}_\Omega\mathbf{z} \\ \mathbf{0} \end{pmatrix}.$$

By choosing $\mathbf{C}$ correctly (Sect. 4.3), $(\mathbf{C}\mathbf{P}_\Omega\mathbf{C}')^{-1}$ will exist, and inverting the above matrix (cf. A.17) we have

$$\begin{pmatrix} \tilde{\phi} \\ \tilde{\boldsymbol{\lambda}}_1 \end{pmatrix} = \begin{pmatrix} \mathbf{I}_p - \mathbf{P}_\Omega\mathbf{C}'(\mathbf{C}\mathbf{P}_\Omega\mathbf{C}')^{-1}\mathbf{C}\mathbf{P}_\Omega & -\mathbf{P}_\Omega\mathbf{C}'(\mathbf{C}\mathbf{P}_\Omega\mathbf{C}')^{-1} \\ -(\mathbf{C}\mathbf{P}_\Omega\mathbf{C}')^{-1}\mathbf{C}\mathbf{P}_\Omega & -(\mathbf{C}\mathbf{P}_\Omega\mathbf{C}')^{-1} \end{pmatrix} \begin{pmatrix} \mathbf{P}_\Omega\mathbf{z} \\ \mathbf{0} \end{pmatrix},$$

$$\tilde{\phi} = [\mathbf{P}_\Omega - \mathbf{P}_\Omega\mathbf{C}'(\mathbf{C}\mathbf{P}_\Omega\mathbf{C}')^{-1}\mathbf{C}\mathbf{P}_\Omega]\mathbf{z} = \mathbf{P}_\omega\mathbf{z}$$

and

$$\tilde{\boldsymbol{\lambda}}_1 = -(\mathbf{C}\mathbf{P}_\Omega\mathbf{C}')^{-1}\mathbf{C}\mathbf{P}_\Omega\mathbf{z}.$$

Since

$$\mathrm{Var}[\tilde{\boldsymbol{\lambda}}_1] = (\mathbf{C}\mathbf{P}_\Omega\mathbf{C}')^{-1}\mathbf{C}\mathbf{P}_\Omega\mathbf{I}_r\mathbf{P}_\Omega\mathbf{C}'(\mathbf{C}\mathbf{P}_\Omega\mathbf{C}')^{-1} = (\mathbf{C}\mathbf{P}_\Omega\mathbf{C}')^{-1},$$

we have

$$\begin{aligned} (\hat{\phi} - \tilde{\phi})'(\hat{\phi} - \tilde{\phi}) &= \mathbf{z}'(\mathbf{P}_\Omega - \mathbf{P}_\omega)^2\mathbf{z} \\ &= \mathbf{z}'(\mathbf{P}_\Omega - \mathbf{P}_\omega)\mathbf{z} \text{ (by Theorems 4.2 and 4.3)} \\ &= \mathbf{z}'\mathbf{P}_\Omega\mathbf{C}'(\mathbf{C}\mathbf{P}_\Omega\mathbf{C}')^{-1}\mathbf{C}\mathbf{P}_\Omega\mathbf{z} && (10.22) \\ &= (\mathbf{C}\hat{\phi})'(\mathbf{C}\mathbf{P}_\Omega\mathbf{C}')^{-1}\mathbf{C}\hat{\phi} && (10.23) \\ &= \tilde{\boldsymbol{\lambda}}_1'(\mathbf{C}\mathbf{P}_\Omega\mathbf{C}')\tilde{\boldsymbol{\lambda}}_1 \\ &= \tilde{\boldsymbol{\lambda}}_1'(\mathrm{Var}[\tilde{\boldsymbol{\lambda}}_1])^{-1}\tilde{\boldsymbol{\lambda}}_1, && (10.24) \end{aligned}$$

a slight generalization of Theorem 4.5 in Sect. 4.3. (Note that the scale factor of -2 applied to $\boldsymbol{\lambda}_1$ at the beginning of this section cancels out of the above expression.) As $\sigma^2 = 1$, the likelihood function is (cf. (3.12) in Sect. 3.9)

$$\ell(\boldsymbol{\theta}, 1) = (2\pi)^{-n/2} \exp\left\{-\frac{1}{2} \parallel \mathbf{y} - \boldsymbol{\theta} \parallel^2\right\},$$

so that the likelihood ratio is given by

$$\begin{aligned} \Lambda[H|G] &= \frac{\max_{\boldsymbol{\theta}\in\omega} \ell(\boldsymbol{\theta}, 1)}{\max_{\boldsymbol{\theta}\in\Omega} \ell(\boldsymbol{\theta}, 1)} \\ &= \frac{\exp\left\{\frac{1}{2}(\mathbf{z}'(\mathbf{I}_p - \mathbf{P}_\Omega)\mathbf{z}\right\}}{\exp\left\{\frac{1}{2}(\mathbf{z}'(\mathbf{I}_p - \mathbf{P}_\omega)\mathbf{z})\right\}} \end{aligned}$$

and therefore

$$-2\log\Lambda[H|G] = \mathbf{z}'(\mathbf{P}_\Omega - \mathbf{P}_\omega)\mathbf{z}. \qquad (10.25)$$

For testing H we use the statistic $\mathbf{z}'(\mathbf{P}_\Omega - \mathbf{P}_\omega)\mathbf{z}$ which has a chi-square distribution when H is true, and we reject H if this statistic is too large. From the above we see that this statistic can be expressed in three forms (10.23) to (10.25), and each form defines a different test principle. Thus we accept H if $\mathbf{C}\hat{\boldsymbol{\phi}}$ is "near enough" to zero (Wald principle), or if the Lagrange multiplier $\tilde{\boldsymbol{\lambda}}$ is "near enough " to zero (Lagrange multiplier or Score principle), or if the likelihood ratio is "near enough" to unity.

If we put $\mathbf{C} = [\mathbf{AV}]$ and $\Omega = \mathbb{R}^p$, then $\mathbf{P}_\Omega = \mathbf{I}_p$ and $\mathbf{C}$ is now function of $\boldsymbol{\theta}$. Equations (10.20) and (10.21) now become

$$\mathbf{z} - \tilde{\boldsymbol{\phi}} + [\tilde{\mathbf{A}}\tilde{\mathbf{V}}]'\tilde{\boldsymbol{\lambda}}_1 = \mathbf{0}$$

and

$$\tilde{\mathbf{A}}\tilde{\mathbf{V}}\tilde{\boldsymbol{\phi}} = \mathbf{0},$$

which are asymptotically equivalent to the equations obtained at the end of Sect. 10.3.2 when $\tilde{\boldsymbol{\lambda}}_1 = n^{1/2}\tilde{\boldsymbol{\mu}}$ and $\tilde{\mathbf{A}}\tilde{\mathbf{V}}$ is approximated by $\mathbf{A}_T\mathbf{V}_T$. Now $\hat{\boldsymbol{\phi}} = \mathbf{z}$, and using a Taylor expansion we have

$$\begin{aligned}
[\mathbf{AV}]_T\hat{\boldsymbol{\phi}} &= \mathbf{A}_T\mathbf{V}_T\mathbf{z} \\
&= n^{1/2}\mathbf{A}_T(\hat{\boldsymbol{\theta}} - \boldsymbol{\theta}_*) \quad \text{by (10.16)} \\
&= n^{1/2}\mathbf{a}(\hat{\boldsymbol{\theta}}) + \mathbf{o}_p(1) \quad (\text{since } \mathbf{a}(\boldsymbol{\theta}_*) = \mathbf{0}).
\end{aligned}$$

Also, by virtue of the remarks made after assumption (xiv),

$$[\mathbf{A}'\mathbf{B}^{-1}\mathbf{A}]_T = [\mathbf{A}'\mathbf{B}^{-1}\mathbf{A}]_{\hat{\theta}} + \mathbf{o}_p(1),$$

where the inverse of the matrix on the right-hand side will exist for n sufficiently large. Using (10.23) with $\mathbf{C} = (\mathbf{AV})_T$, $\mathbf{P}_\Omega = \mathbf{I}_p$, and $\mathbf{CC}' = [\mathbf{AB}^{-1}\mathbf{A}]_T$, and combining the above two results gives us

$$(\mathbf{C}\hat{\boldsymbol{\phi}})'(\mathbf{CP}_\Omega\mathbf{C}')^{-1}\mathbf{C}\hat{\boldsymbol{\phi}} = n\mathbf{a}'(\hat{\boldsymbol{\theta}})[\mathbf{A}'\mathbf{B}^{-1}\mathbf{A}]_{\hat{\theta}}^{-1}\mathbf{a}(\hat{\boldsymbol{\theta}}) + \mathbf{o}_p(1),$$

the so-called Wald test statistic. From (10.12),

$$\begin{aligned}
\mathbf{A}_T'n^{1/2}\tilde{\boldsymbol{\mu}} &= \tilde{\mathbf{A}}'n^{1/2}\tilde{\boldsymbol{\mu}} + \mathbf{o}_p(1), \\
&= -n^{1/2}\mathbf{D}n^{-1}L(\tilde{\boldsymbol{\theta}}) + \mathbf{o}_p(1),
\end{aligned}$$

and from (10.6) with $\mathbf{C} = \tilde{\mathbf{A}}\tilde{\mathbf{V}}$ and $\tilde{\boldsymbol{\lambda}}_1 = n^{1/2}\tilde{\boldsymbol{\mu}}$,

$$\begin{aligned}
\tilde{\boldsymbol{\lambda}}_1'\mathbf{CP}_\Omega\mathbf{C}'\tilde{\boldsymbol{\lambda}}_1 &= (\tilde{\mathbf{A}}'\tilde{\boldsymbol{\lambda}}_1)'\tilde{\mathbf{V}}\tilde{\mathbf{V}}'(\tilde{\mathbf{A}}'\tilde{\boldsymbol{\lambda}}_1) \\
&= n[\mathbf{D}n^{-1}L(\tilde{\boldsymbol{\theta}})]'\tilde{\mathbf{B}}^{-1}[\mathbf{D}n^{-1}L(\tilde{\boldsymbol{\theta}})] + \mathbf{o}_p(1),
\end{aligned}$$

the Score test statistic. Using a Taylor expansion and (10.1), we have

$$\begin{aligned} L(\tilde{\theta}) - L(\hat{\theta}) &= \mathbf{D}L(\hat{\theta})'(\tilde{\theta} - \hat{\theta}) \\ &\quad + \frac{1}{2}(\tilde{\theta} - \hat{\theta})'[\mathbf{D}^2 L(\hat{\theta})](\tilde{\theta} - \hat{\theta}) + \mathbf{o}_p(1) \\ &= 0 - \frac{1}{2}n(\tilde{\theta} - \hat{\theta})'\hat{\mathbf{B}}(\tilde{\theta} - \hat{\theta}) + \mathbf{o}_p(1), \end{aligned}$$

and therefore, from (10.16) and (10.18) with $\mathbf{P}_\Omega = \mathbf{I}_p$,

$$\begin{aligned} -2L[(\tilde{\theta}) - L(\hat{\theta})] &= n(\tilde{\theta} - \hat{\theta})'\hat{\mathbf{B}}(\tilde{\theta} - \hat{\theta}) + \mathbf{o}_p(1) \\ &= n(\hat{\mathbf{V}}^{-1}\tilde{\theta} - \hat{\mathbf{V}}^{-1}\hat{\theta})'(\hat{\mathbf{V}}^{-1}\tilde{\theta} - \hat{\mathbf{V}}^{-1}\hat{\theta}) + \mathbf{o}_p(1) \\ &= (\mathbf{z} - \tilde{\phi})'(\mathbf{z} - \tilde{\phi}) + \mathbf{o}_p(1) \\ &= \mathbf{z}'(\mathbf{I}_p - \mathbf{P}_\omega)\mathbf{z} + \mathbf{o}_p(1) \\ &= \mathbf{z}'(\mathbf{P}_\Omega - \mathbf{P}_\omega)\mathbf{z} + \mathbf{o}_p(1) \\ &= -2\log \Lambda[H|G] + \mathbf{o}_p(1), \end{aligned}$$

the likelihood ratio test statistic (see (10.25)). Thus the three statistics

$$\begin{aligned} &n\mathbf{a}'(\hat{\theta})[\hat{\mathbf{A}}\hat{\mathbf{B}}^{-1}\hat{\mathbf{A}}']^{-1}\mathbf{a}(\hat{\theta}), \\ &n^{-1}[\mathbf{D}L(\tilde{\theta})]'\tilde{\mathbf{B}}^{-1}[\mathbf{D}L(\tilde{\theta})], \quad \text{and} \\ &-2[L(\tilde{\theta}) - L(\hat{\theta})] \end{aligned}$$

are asymptotically distributed as χ^2_q when H is true. When H is false, but θ_T is near W_1, then the above linear approximation is valid and the three statistics have an asymptotic non-central chi-square distribution with non-centrality parameter (cf. (10.22) with $\mathbf{C} = [\mathbf{AV}]_T$)

$$\begin{aligned} \delta &= \mathrm{E}[\mathbf{z}']\mathbf{C}'(\mathbf{CC}')^{-1}\mathbf{C}\mathrm{E}[\mathbf{z}] \\ &= \phi'\mathbf{C}'(\mathbf{CC}')^{-1}\mathbf{C}\phi \\ &= n(\theta_T - \theta_*)'\mathbf{A}_T'(\mathbf{AB}^{-1}\mathbf{A}')_T^{-1}\mathbf{A}_T(\theta_T - \theta_*) \\ &\approx n(\mathbf{a}(\theta_T) - \mathbf{a}(\theta_*))'(\mathbf{AB}^{-1}\mathbf{A}')_T^{-1}(\mathbf{a}(\theta_T) - \mathbf{a}(\theta_*)), \end{aligned}$$

which is 0 when H is true, i.e., when $\theta_T = \theta_*$. When θ_T is not near W_1, the linear approximation can not be used and we can say nothing about the power of the test except that it will tend to unity as n tends to infinity. This is obvious since, for example, $\sqrt{n}\mathbf{a}(\hat{\theta})$ will be far from $\mathbf{0}$ when $\mathbf{a}(\hat{\theta})$ is not near $\mathbf{0}$.

Example 10.1 Suppose we have the above model and we wish to test the hypothesis $H : \boldsymbol{\theta}_T - \boldsymbol{\theta}_0 = \mathbf{0}$. To do this we adopt the method described in Example 2.5 in Sect. 2.4 where we essentially shift the origin. Recalling (10.16)–(10.18), we define $\boldsymbol{\phi}_0 = n^{1/2}\mathbf{V}_T^{-1}(\boldsymbol{\theta}_0 - \boldsymbol{\theta}_*)$ and consider the asymptotic model $\mathbf{w} = \boldsymbol{\eta} + \boldsymbol{\varepsilon}$, where

$$\begin{aligned}
\mathbf{w} &= \mathbf{z} - \boldsymbol{\phi}_0 \\
&= n^{1/2}\mathbf{V}_T^{-1}[(\hat{\boldsymbol{\theta}} - \boldsymbol{\theta}_*) - (\boldsymbol{\theta}_0 - \boldsymbol{\theta}_*)] \\
&= n^{1/2}\mathbf{V}_T^{-1}(\hat{\boldsymbol{\theta}} - \boldsymbol{\theta}_0), \\
\boldsymbol{\eta} &= \boldsymbol{\phi} - \boldsymbol{\phi}_0 = n^{1/2}\mathbf{V}_T^{-1}(\boldsymbol{\theta}_T - \boldsymbol{\theta}_0),
\end{aligned}$$

and $\mathbf{A}_T = \mathbf{I}_p$. We now test $H : \mathbf{A}_T\mathbf{V}_T\boldsymbol{\eta} = n^{1/2}(\boldsymbol{\theta}_T - \boldsymbol{\theta}_0) = \mathbf{0}$, which is equivalent to testing $\boldsymbol{\theta}_T = \boldsymbol{\theta}_0$. We have

$$\begin{aligned}
-2\log \Lambda[H|G] &= \mathbf{w}'(\mathbf{P}_\Omega - \mathbf{P}_\omega)\mathbf{w} \\
&= \mathbf{w}'(\mathbf{I}_p - \mathbf{0})\mathbf{w} \\
&= \mathbf{w}'\mathbf{w} \\
&= n(\hat{\boldsymbol{\theta}} - \boldsymbol{\theta}_0)'\mathbf{B}_T(\hat{\boldsymbol{\theta}} - \boldsymbol{\theta}_0) \\
&\approx n(\hat{\boldsymbol{\theta}} - \boldsymbol{\theta}_0)\mathbf{B}_{\hat{\boldsymbol{\theta}}}(\hat{\boldsymbol{\theta}} - \boldsymbol{\theta}_0),
\end{aligned}$$

the Wald test. Rao's score test readily follows from the above theory, namely $n^{-1}[\mathbf{D}L(\boldsymbol{\theta}_0)]'\mathbf{B}_{\boldsymbol{\theta}_0}^{-1}\mathbf{D}L(\boldsymbol{\theta}_0)$. All three statistics are asymptotically distributed as χ_p^2 when H is true.

10.4 Positive-Semidefinite Information Matrix

The following is based on Seber (1963). If $\mathbf{B}_T$ is a $p \times p$ positive-semidefinite matrix of rank $p - r_0$, then $\boldsymbol{\theta}_T$ is not identifiable and we introduce r_0 independent constraints to make $\boldsymbol{\theta}_T$ identifiable, namely

$$\mathbf{h}(\boldsymbol{\theta}_T) = (h_1(\boldsymbol{\theta}_T), h_2(\boldsymbol{\theta}_T), \ldots, h_{r_0}(\boldsymbol{\theta}_T))' = \mathbf{0}.$$

Let $\mathbf{H}_T$ be the $r_0 \times p$ matrix of rank r_0 with (i,j)th element $[\partial h_i(\boldsymbol{\theta})/\partial\theta_j]_{\boldsymbol{\theta}_T}$. Since $\mathbf{B}_T$ is positive semidefinite, there exists a $(p - r_0) \times p$ matrix $\mathbf{R}_T$ of rank $p - r_0$ such that $\mathbf{B}_T = (\mathbf{R}'\mathbf{R})_T$ (A.9(iii)). We now add a further assumption, namely

(xvi) $(\mathbf{B} + \mathbf{H}'\mathbf{H})_T$ is positive definite, that is the $p \times p$ matrix $\mathbf{G}_T = (\mathbf{R}', \mathbf{H}')_T'$ is of rank p and $\mathbf{H}_T$ has rank r_0.

It follows from the above assumption that $(\mathbf{G}'\mathbf{G})_T = (\mathbf{B} + \mathbf{H}'\mathbf{H})_T$ is nonsingular, $\mathbf{G}_T$ is nonsingular, and from $[\mathbf{G}(\mathbf{G}'\mathbf{G})^{-1}\mathbf{G}']_T = \mathbf{I}_p$ we get

$$\begin{bmatrix} \mathbf{R}(\mathbf{G}'\mathbf{G})^{-1}\mathbf{R}', & \mathbf{R}(\mathbf{G}'\mathbf{G})^{-1}\mathbf{H}' \\ \mathbf{H}(\mathbf{G}'\mathbf{G})^{-1}\mathbf{R}', & \mathbf{H}(\mathbf{G}'\mathbf{G})^{-1}\mathbf{H}' \end{bmatrix}_T = \begin{pmatrix} \mathbf{I}_{p-r_0} & \mathbf{0} \\ \mathbf{0} & \mathbf{I}_{r_0} \end{pmatrix}. \tag{10.26}$$

Assuming certain underlying assumptions (Silvey 1959), the maximum likelihood estimate of $\boldsymbol{\theta}_T$ is the solution of

$$\mathbf{D}n^{-1}L(\hat{\boldsymbol{\theta}}) + \hat{\mathbf{H}}'\hat{\boldsymbol{\lambda}}_0 = \mathbf{0} \tag{10.27}$$

and

$$\mathbf{h}(\hat{\boldsymbol{\theta}}) = \mathbf{0}.$$

Since $\hat{\boldsymbol{\theta}}$ is near $\boldsymbol{\theta}_T$ we can use the usual Taylor expansions

$$n^{1/2}\mathbf{D}n^{-1}L(\hat{\boldsymbol{\theta}}) = n^{1/2}\mathbf{D}n^{-1}L(\boldsymbol{\theta}_T) - n^{1/2}\mathbf{B}_T(\hat{\boldsymbol{\theta}} - \boldsymbol{\theta}_T) + \mathbf{o}_p(1), \tag{10.28}$$

$$\hat{\mathbf{H}} = \mathbf{H}_T + \mathbf{O}_p(n^{-1/2}),$$

and, since $\mathbf{h}(\boldsymbol{\theta}_T) = \mathbf{0}$,

$$\begin{aligned} \mathbf{0} &= n^{1/2}\mathbf{h}(\hat{\boldsymbol{\theta}}) - n^{1/2}\mathbf{h}(\boldsymbol{\theta}_T) \\ &= n^{1/2}\mathbf{H}_T(\hat{\boldsymbol{\theta}} - \boldsymbol{\theta}_T) + \mathbf{o}_p(1). \end{aligned} \tag{10.29}$$

Multiplying (10.29) by $\mathbf{H}_T'$, subtracting the result (zero) from the right-hand side of (10.28), and noting that $n^{1/2}\mathbf{D}n^{-1}L(\boldsymbol{\theta}_T)$ is asymptotically $N_p[\mathbf{0}, \mathbf{B}_T]$, we get from (10.27)

$$n^{1/2}(\mathbf{G}'\mathbf{G})_T(\hat{\boldsymbol{\theta}} - \boldsymbol{\theta}_T) - n^{1/2}\mathbf{H}_T'\hat{\boldsymbol{\lambda}}_0 = \boldsymbol{\delta}_1 + \mathbf{o}_p(1),$$

where $\boldsymbol{\delta}_1 \sim N_p[\mathbf{0}, \mathbf{B}_T]$. Multiplying the above equation on the left by $\mathbf{H}_T(\mathbf{G}'\mathbf{G})^{-1}$ gives us

$$n^{1/2}\mathbf{H}_T(\hat{\boldsymbol{\theta}} - \boldsymbol{\theta}_T) - n^{1/2}\mathbf{H}_T(\mathbf{G}'\mathbf{G})^{-1}\mathbf{H}_T'\hat{\boldsymbol{\lambda}}_0 = \mathbf{H}_T(\mathbf{G}'\mathbf{G})^{-1}\boldsymbol{\delta}_1 + \mathbf{o}_p(1).$$

Using (10.29) and (10.26) in the above equation leads to

$$\begin{aligned} n^{1/2}\hat{\boldsymbol{\lambda}}_0 &= -\mathbf{H}_T(\mathbf{G}'\mathbf{G})^{-1}\boldsymbol{\delta}_1 + \mathbf{o}_p(1) \\ &= -\mathbf{C}\boldsymbol{\delta}_1 + \mathbf{o}_p(1), \text{ say.} \end{aligned}$$

Now

$$\begin{aligned}
\mathrm{var}[n^{1/2}\hat{\boldsymbol{\lambda}}_0] &= \mathbf{C}\mathrm{var}[\boldsymbol{\delta}_1]\mathbf{C}' + o(1) \\
&= \mathbf{C}\mathbf{B}_T\mathbf{C}' + o(1) \\
&= \{\mathbf{H}(\mathbf{G}'\mathbf{G})^{-1}\mathbf{R}'\mathbf{R}(\mathbf{G}'\mathbf{G})^{-1}\mathbf{H}'\}_T + o(1) \\
&= \mathbf{o}(1) \quad \text{by (10.26)},
\end{aligned}$$

so that $n^{1/2}\hat{\boldsymbol{\lambda}}_0 = \mathbf{o}_p(1)$ and

$$n^{1/2}\mathbf{D}n^{-1}L(\hat{\boldsymbol{\theta}}) = \mathbf{o}_p(1). \tag{10.30}$$

Since $\boldsymbol{\theta}_T$ is near $\boldsymbol{\theta}_*$ we have from (10.28) and (10.30)

$$\mathbf{0} = \boldsymbol{\delta}_1 - \mathbf{B}_T n^{1/2}(\hat{\boldsymbol{\theta}} - \boldsymbol{\theta}_*) + \mathbf{B}_T n^{1/2}(\boldsymbol{\theta}_T - \boldsymbol{\theta}_*) + \mathbf{o}_p(1), \tag{10.31}$$

$$n^{1/2}\mathbf{H}_T(\hat{\boldsymbol{\theta}} - \boldsymbol{\theta}_*) = \mathbf{o}_p(1) \tag{10.32}$$

and

$$n^{1/2}\mathbf{H}_T(\boldsymbol{\theta}_T - \boldsymbol{\theta}_*) = \mathbf{o}(1). \tag{10.33}$$

10.4.1 Hypothesis Testing

The hypothesis of interest H is $\mathbf{a}(\boldsymbol{\theta}) = \mathbf{0}$ as in Sect. 10.2, and $\mathbf{A}_{\boldsymbol{\theta}}$ is the $q \times p$ matrix of rank q of corresponding derivatives. To find the restricted maximum likelihood estimate $\tilde{\boldsymbol{\theta}}$ we solve

$$n^{1/2}\mathbf{D}n^{-1}L(\tilde{\boldsymbol{\theta}}) + n^{1/2}\tilde{\mathbf{H}}'\tilde{\boldsymbol{\lambda}}_0 + n^{1/2}\tilde{\mathbf{A}}'\tilde{\boldsymbol{\lambda}}_1 = \mathbf{0}, \tag{10.34}$$

$$\mathbf{h}(\tilde{\boldsymbol{\theta}}) = \mathbf{0},$$

and

$$\mathbf{a}(\tilde{\boldsymbol{\theta}}) = \mathbf{0}.$$

As $\tilde{\boldsymbol{\theta}}$, $\boldsymbol{\theta}_*$, and $\hat{\boldsymbol{\theta}}$ are all near each other, we can carry out the usual Taylor expansions to get from (10.30)

$$\begin{aligned}
n^{1/2}\mathbf{D}n^{-1}L(\tilde{\boldsymbol{\theta}}) &= n^{1/2}\mathbf{D}n^{-1}L(\hat{\boldsymbol{\theta}}) - \mathbf{B}_T n^{1/2}(\tilde{\boldsymbol{\theta}} - \hat{\boldsymbol{\theta}}) + \mathbf{o}_p(1) \\
&= \mathbf{0} - \mathbf{B}_T n^{1/2}(\tilde{\boldsymbol{\theta}} - \hat{\boldsymbol{\theta}}) + \mathbf{o}_p(1) \\
&= -\mathbf{B}_T n^{1/2}(\tilde{\boldsymbol{\theta}} - \boldsymbol{\theta}_*) + \mathbf{B}_T n^{1/2}(\hat{\boldsymbol{\theta}} - \boldsymbol{\theta}_*) + \mathbf{o}_p(1),
\end{aligned} \tag{10.35}$$

along with

$$n^{1/2}\mathbf{H}_T(\tilde{\theta} - \theta_*) = \mathbf{o}_p(1),$$

and

$$n^{1/2}\mathbf{A}_T(\tilde{\theta} - \theta_*) = \mathbf{o}_p(1).$$

The rows of $\mathbf{A}_T$ are assumed to be linearly independent of the rows of $\mathbf{H}_T$. Since $\tilde{\theta}$ is close to θ_T we have

$$\tilde{\mathbf{H}} = \mathbf{H}_T + \mathbf{O}_p(n^{-1/2})$$

and

$$\tilde{\mathbf{A}} = \mathbf{A}_T + \mathbf{O}_p(n^{-1/2}).$$

Using the above equations and substituting (10.35) in (10.34) gives us

$$-\mathbf{B}_T n^{1/2}(\tilde{\theta}-\theta_*)+\mathbf{B}_T n^{1/2}(\hat{\theta}-\theta_*)+n^{1/2}\mathbf{H}_T'\tilde{\lambda}_0+n^{1/2}\mathbf{A}_T'\tilde{\lambda}_1+\mathbf{o}_p(1) = \mathbf{0}, \quad (10.36)$$

and setting $\mathbf{y} = n^{1/2}(\hat{\theta} - \theta_*)$, $\beta = n^{1/2}(\theta_T - \theta_*)$, and $\tilde{\beta} = n^{1/2}(\tilde{\theta} - \theta_*)$ in equations (10.31)–(10.33) we get the asymptotic linear model

$$\mathbf{B}_T\mathbf{y} = \mathbf{B}_T\beta + \delta_1,$$

where $\delta_1 \sim N_p[\mathbf{0}, \mathbf{B}_T]$, $\mathbf{H}_T\mathbf{y} = \mathbf{0}$, and $\mathbf{H}_T\beta = \mathbf{0}$. From (10.36), ω is given by

$$\mathbf{A}_{2T}\beta = \begin{pmatrix} \mathbf{H}_T \\ \mathbf{A}_T \end{pmatrix}\beta = \mathbf{0},$$

where $\mathbf{A}_{2T}$ is $(q + r_0) \times p$ of rank $q + r_0$ $(q + r_0 < p)$. Recalling that $\mathbf{B}_T = (\mathbf{R}'\mathbf{R})_T$, where $\mathbf{R}_T$ is $(p - r_0) \times p$ of rank r_0, we get

$$\mathbf{R}_T'\mathbf{R}_T\mathbf{y} = \mathbf{R}_T'\mathbf{R}_T\beta + \delta_1.$$

Since $(\mathbf{R}\mathbf{R}')_T$ is $p - r_0 \times p - r_0$ of rank $p - r_0$ it is nonsingular, and multiplying the above equation by $[(\mathbf{R}\mathbf{R}')^{-1}\mathbf{R}]_T$ we get the linear model

$$\mathbf{z} = \phi + \varepsilon,$$

where $\mathbf{z} = \mathbf{R}_T\mathbf{y}$, $\mathbf{H}_T\mathbf{y} = \mathbf{0}$, $\varepsilon = [(\mathbf{R}\mathbf{R}')^{-1}\mathbf{R}]_T\delta_1$,

$$\text{Var}[\varepsilon] = [(\mathbf{R}\mathbf{R}')^{-1}\mathbf{R}(\mathbf{R}'\mathbf{R})\mathbf{R}'(\mathbf{R}\mathbf{R}')^{-1}]_T = \mathbf{I}_{p-r_0},$$

$\mathbf{H}_T\boldsymbol{\beta} = \mathbf{0}$, and $\boldsymbol{\phi} = \mathbf{R}_T\boldsymbol{\beta}$. Considering the previous two equations, it follows from A.11 and the assumption that $(\mathbf{G}'\mathbf{G})_T$ is positive definite that $\boldsymbol{\beta}$ is identifiable so that $\boldsymbol{\phi}$ is not constrained and $\Omega = \mathbb{R}^{p-r_0}$. This also follows from the fact that since $(\mathbf{G}'\mathbf{G})_T^{-1}$ is a generalized (weak) inverse of $(\mathbf{R}'\mathbf{R})_T$ (cf., A.14(iii)) and $\mathbf{P}_\Omega$ is unique, we have

$$\mathbf{P}_\Omega = [\mathbf{R}(\mathbf{R}'\mathbf{R})^{-}\mathbf{R}']_T = [\mathbf{R}(\mathbf{G}'\mathbf{G})^{-1}\mathbf{R}']_T = \mathbf{I}_{p-r_0} \quad \text{by (10.26).}$$

Since $\mathbf{R}_T'\boldsymbol{\phi} = \mathbf{B}_T\boldsymbol{\beta}$ and $\mathbf{H}_T\boldsymbol{\beta} = \mathbf{0}$ we have $\mathbf{R}_T'\boldsymbol{\phi} = (\mathbf{B}_T + \mathbf{H}_T'\mathbf{H}_T)\boldsymbol{\beta}$ so that $\boldsymbol{\beta} = (\mathbf{G}'\mathbf{G})_T^{-1}\mathbf{R}_T'\boldsymbol{\phi}$. Also $\mathbf{A}_T\boldsymbol{\beta} = \mathbf{0}$ implies that $\omega = \{\boldsymbol{\phi} \mid \mathbf{C}\boldsymbol{\phi} = \mathbf{0}\}$, where $\mathbf{C} = \mathbf{A}_T(\mathbf{G}'\mathbf{G})_T^{-1}\mathbf{R}_T'$. Replacing $\boldsymbol{\phi}$ by $\mathbf{z}$ we get $\mathbf{y} = (\mathbf{G}'\mathbf{G})_T^{-1}\mathbf{R}_T'\mathbf{z}$. We now have the asymptotic linear model and hypothesis

$$\mathbf{z} = \boldsymbol{\phi} + \boldsymbol{\varepsilon}, \quad \Omega = \mathbb{R}^{p-r_0}, \quad \omega = \mathcal{N}[\mathbf{A}_T(\mathbf{G}'\mathbf{G})_T^{-1}\mathbf{R}_T']. \tag{10.37}$$

Referring to Section to (10.25) in Sect. 10.3.3 with $\mathbf{P}_\Omega = \mathbf{I}_p$, and using the result $\mathcal{N}[\mathbf{C}] = \mathcal{C}[\mathbf{C}']^\perp$ (Theorem 1.1 in Sect. 1.2), along with generalized (weak) inverses (A.14), gives us

$$\begin{aligned}
-2\log \Lambda[H|G] &= \mathbf{z}'(\mathbf{I}_{p-r_0} - \mathbf{P}_\omega)\mathbf{z} && (10.38)\\
&= \mathbf{z}'\mathbf{C}'(\mathbf{C}\mathbf{C}')^{-}\mathbf{C}\mathbf{z}\\
&= \mathbf{y}'\mathbf{R}_T'\mathbf{C}'(\mathbf{C}\mathbf{C}')^{-}\mathbf{C}\mathbf{R}_T\mathbf{y}\\
&= \mathbf{y}'\mathbf{B}_T(\mathbf{G}'\mathbf{G})_T^{-1}\mathbf{A}_T'[\mathbf{A}(\mathbf{G}'\mathbf{G})^{-1}\mathbf{B}(\mathbf{G}'\mathbf{G})^{-1}\mathbf{A}']_T^{-}\mathbf{A}_T(\mathbf{G}'\mathbf{G})_T^{-1}\mathbf{B}_T\mathbf{y}\\
&= \mathbf{y}'\mathbf{A}_T'[\mathbf{A}(\mathbf{G}'\mathbf{G})^{-1}\mathbf{B}(\mathbf{G}'\mathbf{G})^{-1}\mathbf{A}']_T^{-}\mathbf{A}_T\mathbf{y}\ (\text{as } \mathbf{H}_T'\mathbf{H}_T\mathbf{y} = \mathbf{0})\\
& && (10.39)\\
&\approx \mathbf{a}(\hat{\boldsymbol{\theta}})'(\mathrm{Var}[\mathbf{a}(\hat{\boldsymbol{\theta}})])_{\hat{\theta}}^{-}\mathbf{a}(\hat{\boldsymbol{\theta}}), && (10.40)
\end{aligned}$$

as $\mathbf{A}_T\mathbf{y} = n^{1/2}\mathbf{A}_T(\hat{\boldsymbol{\theta}} - \boldsymbol{\theta}_*) \approx n^{1/2}\mathbf{a}(\hat{\boldsymbol{\theta}})$ and

$$\begin{aligned}
\mathrm{Var}[\mathbf{A}_T\mathbf{y}] &\approx \mathbf{A}_T\mathrm{Var}[(\mathbf{G}'\mathbf{G})_T^{-1}\mathbf{R}_T'\mathbf{z}]\mathbf{A}_T'\\
&= [\mathbf{A}(\mathbf{G}'\mathbf{G})^{-1}\mathbf{R}'\mathbf{R}(\mathbf{G}'\mathbf{G})^{-1}\mathbf{A}']_T\\
&= [\mathbf{A}(\mathbf{G}'\mathbf{G})^{-1}\mathbf{B}(\mathbf{G}'\mathbf{G})^{-1}\mathbf{A}']_T.
\end{aligned}$$

Therefore the Wald statistic, (10.40), is asymptotically equivalent to the likelihood ratio statistic, which has a χ_q^2 distribution when H is true. It is shown later that the above variance expression actually has an inverse.

To complete the picture we consider another form of the Wald statistic, namely

$$n\mathbf{a}_2(\hat{\boldsymbol{\theta}})'[\mathbf{A}_2(\mathbf{G}'\mathbf{G})^{-1}\mathbf{A}_2']_{\hat{\theta}}^{-1}\mathbf{a}_2(\hat{\boldsymbol{\theta}}) \approx \mathbf{y}'\{\mathbf{A}_2'[\mathbf{A}_2(\mathbf{G}'\mathbf{G})^{-1}\mathbf{A}_2']^{-1}\mathbf{A}_2\}_{\hat{\theta}}\mathbf{y} = W_1,$$

say, where $\mathbf{y} = n^{1/2}(\hat{\theta} - \theta_*)$ and $\mathbf{A}_2' = (\mathbf{A}', \mathbf{H}')$. Then, since $\mathbf{H}_T\mathbf{y} = \mathbf{0}$,

$$\begin{aligned} W_1 &= \mathbf{y}' \left\{ (\mathbf{A}', \mathbf{H}') \left[\begin{pmatrix} \mathbf{A} \\ \mathbf{H} \end{pmatrix} (\mathbf{G}'\mathbf{G})^{-1} (\mathbf{A}', \mathbf{H}') \right]^{-1} \begin{pmatrix} \mathbf{A} \\ \mathbf{H} \end{pmatrix} \right\}_{\hat{\theta}} \mathbf{y} \\ &= \mathbf{y}' \left\{ (\mathbf{A}', \mathbf{0}) \begin{bmatrix} \mathbf{A}(\mathbf{G}'\mathbf{G})^{-1}\mathbf{A}' & \mathbf{A}(\mathbf{G}'\mathbf{G})^{-1}\mathbf{H}' \\ \mathbf{H}(\mathbf{G}'\mathbf{G})^{-1}\mathbf{A}' & \mathbf{H}(\mathbf{G}'\mathbf{G})^{-1}\mathbf{H}' \end{bmatrix}^{-1} \begin{pmatrix} \mathbf{Ay} \\ \mathbf{0} \end{pmatrix} \right\}_{\hat{\theta}} \mathbf{y} \\ &= \mathbf{y}'[\mathbf{A}'\mathbf{F}^{-1}\mathbf{A}]_{\hat{\theta}}\mathbf{y}, \end{aligned}$$

where $\mathbf{F}^{-1}$ is the matrix in the $(1,1)$ position in the inverse of the above matrix. From A.17

$$\begin{aligned} \mathbf{F} &= \mathbf{A}(\mathbf{G}'\mathbf{G})^{-1}\mathbf{A}' - \mathbf{A}(\mathbf{G}'\mathbf{G})^{-1}\mathbf{H}'\mathbf{H}(\mathbf{G}'\mathbf{G})^{-1}\mathbf{A}' \\ &= \mathbf{A}(\mathbf{G}'\mathbf{G})^{-1}(\mathbf{G}'\mathbf{G})(\mathbf{G}'\mathbf{G})^{-1}\mathbf{A}' - \mathbf{A}(\mathbf{G}'\mathbf{G})^{-1}\mathbf{H}'\mathbf{H}(\mathbf{G}'\mathbf{G})^{-1}\mathbf{A}' \\ &= \mathbf{A}(\mathbf{G}'\mathbf{G})^{-1}\mathbf{B}(\mathbf{G}'\mathbf{G})^{-1}\mathbf{A}'. \end{aligned}$$

Hence $W_1 = \mathbf{y}'\{\mathbf{A}'[\mathbf{A}(\mathbf{G}'\mathbf{G})^{-1}\mathbf{B}(\mathbf{G}'\mathbf{G})^{-1}\mathbf{A}']^{-1}\mathbf{A}'\}_{\hat{\theta}}\mathbf{y}$, which leads to (10.39) once again.

10.4.2 Lagrange Multipler Test

To apply the Lagrange multiplier test statistic we add $\mathbf{H}_T'\mathbf{H}_T\tilde{\beta} = \mathbf{0}$ and $\mathbf{H}_T'\mathbf{H}_T\mathbf{y} = \mathbf{0}$ to (10.36) to replace $\mathbf{B}_T$ by $(\mathbf{G}'\mathbf{G})_T$ and then multiply the resulting equation by $\mathbf{R}_T(\mathbf{G}'\mathbf{G})_T^{-1}$ to give us the approximate equation

$$-\mathbf{R}_T\tilde{\beta} + \mathbf{R}_T\mathbf{y} + n^{1/2}\mathbf{R}_T(\mathbf{G}'\mathbf{G})_T^{-1}\mathbf{H}_T'\tilde{\lambda}_0 + n^{1/2}\mathbf{R}_T(\mathbf{G}'\mathbf{G})_T^{-1}\mathbf{A}_T'\tilde{\lambda}_1 = \mathbf{0}$$

or, by (10.26),

$$-\tilde{\phi} + \mathbf{z} + \mathbf{0} + \mathbf{C}'n^{1/2}\tilde{\lambda}_1 = \mathbf{0}. \tag{10.41}$$

Hence, from (10.41) and $\tilde{\phi} = \mathbf{P}_\omega\mathbf{z}$, it follows from (10.38) that

$$\begin{aligned} \mathbf{z}'(\mathbf{I}_{p-r_0} - \mathbf{P}_\omega)\mathbf{z} &= (\mathbf{z} - \tilde{\phi})('\mathbf{z} - \tilde{\phi}) \\ &= n\tilde{\lambda}_1'\mathbf{CC}'\tilde{\lambda}_1 \\ &= n\tilde{\lambda}_1'[\mathbf{A}(\mathbf{G}'\mathbf{G})^{-1}\mathbf{R}'\mathbf{R}(\mathbf{GG}')^{-1}\mathbf{A}']_T\tilde{\lambda}_1 \\ &= n\tilde{\lambda}_1'[\mathbf{A}(\mathbf{G}'\mathbf{G})^{-1}\mathbf{B}(\mathbf{GG}')^{-1}\mathbf{A}']_T\tilde{\lambda}_1 \qquad (10.42) \\ &\approx n\tilde{\lambda}_1'[\mathbf{A}(\mathbf{G}'\mathbf{G})^{-1}\mathbf{B}(\mathbf{GG}')^{-1}\mathbf{A}']_{\tilde{\theta}}\tilde{\lambda}_1. \qquad (10.43) \end{aligned}$$

This is the Lagrange multiplier test statistic, based on $\tilde{\boldsymbol{\lambda}}_1$, which can also be written in the form of a score statistic as follows.

From continuity considerations, $\mathbf{G}'\mathbf{G}$ will be positive definite in a neighborhood of $\boldsymbol{\theta}_T$ (cf. A.9(viii)), so that it follows from (10.26) that $[\mathbf{R}(\mathbf{G}'\mathbf{G})^{-1}\mathbf{H}']_{\tilde{\theta}} = \mathbf{0}$. Multiplying (10.34) by $[\mathbf{R}(\mathbf{G}'\mathbf{G})^{-1}]_{\tilde{\theta}}$ we get

$$[\mathbf{R}(\mathbf{G}'\mathbf{G})^{-1}]_{\tilde{\theta}} n^{1/2}\mathbf{D}n^{-1}L(\tilde{\boldsymbol{\theta}}) + n^{1/2}[\mathbf{R}(\mathbf{G}'\mathbf{G})^{-1}\mathbf{A}']_{\tilde{\theta}}\tilde{\boldsymbol{\lambda}}_1 = \mathbf{0},$$

and from (10.43)

$$\begin{aligned} n\tilde{\boldsymbol{\lambda}}_1'[\mathbf{A}(\mathbf{G}'\mathbf{G})^{-1}\mathbf{R}'\mathbf{R}(\mathbf{G}'\mathbf{G})^{-1}\mathbf{A}']_{\tilde{\theta}}\tilde{\boldsymbol{\lambda}}_1 \\ = n\mathbf{D}n^{-1}L(\tilde{\boldsymbol{\theta}})'[(\mathbf{G}'\mathbf{G})^{-1}\mathbf{B}(\mathbf{G}'\mathbf{G})^{-1}]_{\tilde{\theta}}\mathbf{D}n^{-1}L(\tilde{\boldsymbol{\theta}}). \end{aligned} \tag{10.44}$$

Now $\mathbf{B}_T = [\mathbf{G}'\mathbf{G} - \mathbf{H}'\mathbf{H}]_T$ in (10.44) and from (10.35)

$$\mathbf{D}n^{-1}L(\tilde{\boldsymbol{\theta}})'[(\mathbf{G}'\mathbf{G})^{-1}\mathbf{H}']_T = -(\tilde{\boldsymbol{\theta}} - \hat{\boldsymbol{\theta}})'[\mathbf{R}'\mathbf{R}(\mathbf{G}'\mathbf{G})^{-1}\mathbf{H}']_T = \mathbf{0},$$

since $[\mathbf{R}(\mathbf{G}'\mathbf{G})^{-1}\mathbf{H}']_T = \mathbf{0}$ by (10.26). Hence (10.44) becomes

$$n^{-1}\mathbf{D}L(\tilde{\boldsymbol{\theta}})'[(\mathbf{G}'\mathbf{G})^{-1}]_{\tilde{\theta}}\mathbf{D}L(\tilde{\boldsymbol{\theta}}), \tag{10.45}$$

which is the well known score statistic. The only difference from the formula for the case when $\mathbf{B}_T$ is non-singular is to replace $\mathbf{B}_T$ by $(\mathbf{B} + \mathbf{H}'\mathbf{H})_T$.

Another form of the Lagrange multiplier test has been derived by Silvey (1959) and its derivation is instructive. Using a Taylor expansion (cf. (10.34)),

$$n^{1/2}\mathbf{D}n^{-1}L(\tilde{\boldsymbol{\theta}}) = n^{1/2}\mathbf{D}n^{-1}L(\boldsymbol{\theta}_T) - \mathbf{B}_T(\tilde{\boldsymbol{\theta}} - \boldsymbol{\theta}_T) + \mathbf{o}_p(1). \tag{10.46}$$

We now define $\tilde{\boldsymbol{\lambda}}_2 = (\tilde{\boldsymbol{\lambda}}_0', \tilde{\boldsymbol{\lambda}}_1')'$ and $\mathbf{A}_{2T} = (\mathbf{H}_T', \mathbf{A}_T')'$, and we now assume that H is true so that $\boldsymbol{\theta}_* = \boldsymbol{\theta}_T$ and $\tilde{\boldsymbol{\theta}}$ has an asymptotic mean of $\boldsymbol{\theta}_T$. We also have $\mathbf{H}_T(\tilde{\boldsymbol{\theta}} - \boldsymbol{\theta}_T) = \mathbf{o}_p(1)$ and $\mathbf{A}_T(\tilde{\boldsymbol{\theta}} - \boldsymbol{\theta}_T) = \mathbf{o}_p(1)$ so that $\mathbf{A}_{2T}(\tilde{\boldsymbol{\theta}} - \boldsymbol{\theta}_T) = \mathbf{o}_p(1)$. Substituting these expressions into (10.34) gives us

$$n^{1/2}(\mathbf{B}_T + \mathbf{H}_T'\mathbf{H}_T)(\tilde{\boldsymbol{\theta}} - \boldsymbol{\theta}_T) - n^{1/2}\mathbf{A}_{2T}'\tilde{\boldsymbol{\lambda}}_2 = n^{1/2}\mathbf{D}n^{-1}L(\boldsymbol{\theta}_T) + \mathbf{o}_p(1),$$

which can be approximately expressed in the form

$$n^{1/2}\begin{pmatrix} (\mathbf{G}'\mathbf{G})_T & -\mathbf{A}_{2T}' \\ -\mathbf{A}_{2T} & \mathbf{0} \end{pmatrix}\begin{pmatrix} \tilde{\boldsymbol{\theta}} - \boldsymbol{\theta}_T \\ \tilde{\boldsymbol{\lambda}}_2 \end{pmatrix} = \begin{pmatrix} \boldsymbol{\delta}_1 \\ \mathbf{0} \end{pmatrix},$$

where $\boldsymbol{\delta}_1$ is $N_p[\mathbf{0}, \mathbf{B}_T]$, and $\tilde{\boldsymbol{\lambda}}_2$ has approximately a zero mean. Inverting the matrix, we can now write

$$n^{1/2}\begin{pmatrix} \tilde{\boldsymbol{\theta}} - \boldsymbol{\theta}_T \\ \tilde{\boldsymbol{\lambda}}_2 \end{pmatrix} = \begin{pmatrix} \mathbf{U} & \mathbf{V}' \\ \mathbf{V} & \mathbf{W} \end{pmatrix}_T \begin{pmatrix} \boldsymbol{\delta}_1 \\ \mathbf{0} \end{pmatrix},$$

where from A.18

$$\begin{aligned} \mathbf{V} &= -[\mathbf{A}_2(\mathbf{G}'\mathbf{G})^{-1}\mathbf{A}_2']^{-1}\mathbf{A}_2(\mathbf{G}'\mathbf{G})^{-1}, \\ \mathbf{W} &= -[\mathbf{A}_2(\mathbf{G}'\mathbf{G})^{-1}\mathbf{A}_2']^{-1}, \quad \text{and} \\ \mathbf{V}\mathbf{A}_2' &= \mathbf{V}[\mathbf{H}', \mathbf{A}'] = -\mathbf{I}_{r_0+q} \quad \text{with} \quad \mathbf{V}\mathbf{H}' = -[\mathbf{I}_{r_0}, \mathbf{0}]'. \end{aligned}$$

Hence $n^{1/2}\tilde{\boldsymbol{\lambda}}_2 = \mathbf{V}_T\boldsymbol{\delta}_1$ so that

$$\begin{aligned} \text{Var}[n^{1/2}\tilde{\boldsymbol{\lambda}}_2] &= [\mathbf{V}\mathbf{B}\mathbf{V}']_T \\ &= [\mathbf{V}(\mathbf{G}'\mathbf{G} - \mathbf{H}'\mathbf{H})\mathbf{V}']_T \\ &= [\mathbf{W}\mathbf{A}_2(\mathbf{G}'\mathbf{G})^{-1}(\mathbf{G}'\mathbf{G})(\mathbf{G}'\mathbf{G})^{-1}\mathbf{A}_2'\mathbf{W}']_T - \mathbf{V}_T\mathbf{H}_T'\mathbf{H}_T\mathbf{V}_T' \\ &= -\mathbf{W}_T - \begin{pmatrix} \mathbf{I}_{r_0} & \mathbf{0} \\ \mathbf{0} & \mathbf{0} \end{pmatrix} \\ &= \mathbf{S}_T, \quad \text{say.} \end{aligned}$$

We now depart from Silvey's proof and show that (i) $-\mathbf{W}_T^{-1}$ is a weak inverse of $\mathbf{S}_T$ and (ii) trace$[-\mathbf{W}_T^{-1}\mathbf{S}_T] = q$. We first consider

$$\begin{aligned} -\begin{pmatrix} \mathbf{I}_{r_0} & \mathbf{0} \\ \mathbf{0} & \mathbf{0} \end{pmatrix} \mathbf{W}_T^{-1} \begin{pmatrix} \mathbf{I}_{r_0} & \mathbf{0} \\ \mathbf{0} & \mathbf{0} \end{pmatrix} &= \left\{ \begin{pmatrix} \mathbf{I}_{r_0} & \mathbf{0} \\ \mathbf{0} & \mathbf{0} \end{pmatrix} \begin{pmatrix} \mathbf{H} \\ \mathbf{A} \end{pmatrix} (\mathbf{G}'\mathbf{G})^{-1} (\mathbf{H}', \mathbf{A}') \begin{pmatrix} \mathbf{I}_{r_0} & \mathbf{0} \\ \mathbf{0} & \mathbf{0} \end{pmatrix} \right\}_T \\ &= \begin{pmatrix} \mathbf{I}_{r_0} & \mathbf{0} \\ \mathbf{0} & \mathbf{0} \end{pmatrix} \begin{pmatrix} \mathbf{H}(\mathbf{G}'\mathbf{G})^{-1}\mathbf{H}' & \mathbf{H}(\mathbf{G}'\mathbf{G})^{-1}\mathbf{A}' \\ \mathbf{A}(\mathbf{G}'\mathbf{G})^{-1}\mathbf{H}' & \mathbf{A}(\mathbf{G}'\mathbf{G})^{-1}\mathbf{A}' \end{pmatrix}_T \begin{pmatrix} \mathbf{I}_{r_0} & \mathbf{0} \\ \mathbf{0} & \mathbf{0} \end{pmatrix} \\ &= \begin{pmatrix} \mathbf{H}(\mathbf{G}'\mathbf{G})^{-1}\mathbf{H}') & \mathbf{0} \\ \mathbf{0} & \mathbf{0} \end{pmatrix}_T \\ &= \begin{pmatrix} \mathbf{I}_{r_0} & \mathbf{0} \\ \mathbf{0} & \mathbf{0} \end{pmatrix}, \end{aligned}$$

by (10.26). Using this result we then find that $\mathbf{S}_T(-\mathbf{W}_T^{-1})\mathbf{S}_T = \mathbf{S}_T$ so that $-\mathbf{W}_T^{-1}$ is a weak inverse of $\mathbf{S}_T$. Now using (10.26) again,

$$-\mathbf{S}_T\mathbf{W}_T^{-1} = \mathbf{I}_{r_0+q} + \begin{pmatrix} \mathbf{I}_{r_0} & \mathbf{0} \\ \mathbf{0} & \mathbf{0} \end{pmatrix} \mathbf{A}_{2T}(\mathbf{G}'\mathbf{G})_T^{-1}\mathbf{A}_{2T}'$$

$$= \mathbf{I}_{r_0+q} - \begin{pmatrix} \mathbf{H}_T(\mathbf{G}'\mathbf{G})_T^{-1}\mathbf{H}_T' & * \\ \mathbf{0} & \mathbf{0} \end{pmatrix}$$

$$= \mathbf{I}_{r_0+q} - \begin{pmatrix} \mathbf{I}_{r_0} & * \\ \mathbf{0} & \mathbf{0} \end{pmatrix}$$

and trace$[-\mathbf{S}_T\mathbf{W}_T^{-1}] = q$. We can now apply A.16 and then approximate $\mathbf{W}_T$ by $\tilde{\mathbf{W}}$ to prove that

$$-n\tilde{\boldsymbol{\lambda}}_2'\tilde{\mathbf{W}}^{-1}\tilde{\boldsymbol{\lambda}}_2 = n\tilde{\boldsymbol{\lambda}}_2'[\mathbf{A}_2(\mathbf{G}'\mathbf{G})^{-1}\mathbf{A}_2']_{\tilde{\theta}}\tilde{\boldsymbol{\lambda}}_2$$

is approximately distributed as χ_q^2 when H is true. This expression looks very different from (10.43), being based on $\tilde{\boldsymbol{\lambda}}_2$ rather than $\widetilde{\boldsymbol{\lambda}_1}$, and Silvey shows that r_0 of the transformed normal variables are identically zero. We note from (10.34) that

$$\mathbf{D}n^{-1}L(\tilde{\theta}) = -\tilde{\mathbf{A}}_2'\tilde{\boldsymbol{\lambda}}_2$$

so that our Lagrange Multiplier statistic above can be expressed in the form of the Score statistic

$$n^{-1}\mathbf{D}L(\tilde{\theta})'[(\mathbf{G}'\mathbf{G})^{-1}]_{\tilde{\theta}}\mathbf{D}L(\tilde{\theta}), \tag{10.47}$$

which is (10.45) again.

One other approach is worth mentioning. Assuming H to be true so that $\boldsymbol{\theta}_* = \boldsymbol{\theta}_T$ once again, we substitute (10.42) into (10.34) to get

$$n^{1/2}\mathbf{D}n^{-1}L(\theta_T) - \mathbf{B}_T n^{1/2}(\tilde{\theta} - \theta_T) + n^{1/2}\mathbf{H}_T'\tilde{\boldsymbol{\lambda}}_0 + n^{1/2}\mathbf{A}_T'\tilde{\boldsymbol{\lambda}}_1 = \mathbf{o}_p(1)$$

or the approximate equation

$$\boldsymbol{\delta}_1 \approx \mathbf{B}_T n^{1/2}(\tilde{\theta} - \theta_T) - n^{1/2}\mathbf{H}_T'\tilde{\boldsymbol{\lambda}}_0 - n^{1/2}\mathbf{A}_T'\tilde{\boldsymbol{\lambda}}_1.$$

Since $\mathbf{H}_T(\tilde{\theta} - \theta_T) = \mathbf{o}_p(1)$, we can replace $\mathbf{B}_T$ by $\mathbf{B}_T + \mathbf{H}_T'\mathbf{H}_T = (\mathbf{G}'\mathbf{G})_T$ and express the above equation in matrix form

$$n^{1/2}\begin{pmatrix} (\mathbf{G}'\mathbf{G})_T & -\mathbf{H}_T' & -\mathbf{A}_T' \\ -\mathbf{H}_T & \mathbf{0} & \mathbf{0} \\ -\mathbf{A}_T & \mathbf{0} & \mathbf{0} \end{pmatrix}\begin{pmatrix} \tilde{\theta} - \theta_T \\ \tilde{\boldsymbol{\lambda}}_0 \\ \tilde{\boldsymbol{\lambda}}_1 \end{pmatrix} \approx \begin{pmatrix} \boldsymbol{\delta}_1 \\ \mathbf{0} \\ \mathbf{0} \end{pmatrix}.$$

The matrix on the left hand side can be inverted using A.19 to get

$$n^{1/2}\tilde{\boldsymbol{\lambda}}_1 = (\mathbf{AMA}')_T^{-1}\mathbf{A}_T\mathbf{M}_T\boldsymbol{\delta}_1,$$

where

$$\begin{aligned}\mathbf{M}_T &= (\mathbf{G}'\mathbf{G})_T^{-1}[\mathbf{I}_p - \mathbf{H}'(\mathbf{H}(\mathbf{G}'\mathbf{G})^{-1}\mathbf{H}')^{-1}\mathbf{H}(\mathbf{G}'\mathbf{G})^{-1}]_T \\ &= (\mathbf{G}'\mathbf{G})_T^{-1}[\mathbf{I}_p - \mathbf{H}'\mathbf{H}(\mathbf{G}'\mathbf{G})^{-1}]_T \quad \text{by (10.26)} \\ &= (\mathbf{G}'\mathbf{G})_T^{-1}[\mathbf{I}_p - (\mathbf{G}'\mathbf{G} - \mathbf{B})(\mathbf{G}'\mathbf{G})^{-1}]_T \\ &= (\mathbf{G}'\mathbf{G})_T^{-1}\mathbf{B}_T(\mathbf{G}'\mathbf{G})_T^{-1}.\end{aligned}$$

Using $\mathbf{B}_T = [(\mathbf{G}'\mathbf{G})^{-1} - \mathbf{H}'\mathbf{H}]_T$ once again we find that $[\mathbf{MBM}]_T = \mathbf{M}_T$ and

$$\begin{aligned}\mathrm{Var}[n^{1/2}\tilde{\boldsymbol{\lambda}}_1] &= \{[\mathbf{AMA}']^{-1}\mathbf{AMBMA}'[\mathbf{AMA}']^{-1}\}_T \\ &= [\mathbf{AMA}']_T^{-1}.\end{aligned}$$

Thus

$$n\tilde{\boldsymbol{\lambda}}_1'(\mathrm{Var}[n^{1/2}\tilde{\boldsymbol{\lambda}}_1])^{-1}\tilde{\boldsymbol{\lambda}}_1 = n\tilde{\boldsymbol{\lambda}}_1'[\mathbf{A}(\mathbf{G}'\mathbf{G})^{-1}\mathbf{B}(\mathbf{G}'\mathbf{G})^{-1}\mathbf{A}']_T\tilde{\boldsymbol{\lambda}}_1,$$

which is the same as (10.42).

We see then that a major advantage of using the asymptotic linear model is that we have proved that the likelihood ratio, Wald, and Lagrange multiplier test statistics are all asymptotically equivalent as they are exactly equivalent for the asymptotic linear model.

Example 10.2 We revisit Example 10.1 at the end of Sect. 10.3.3 where we tested $\boldsymbol{\theta}_T - \boldsymbol{\theta}_0 = \mathbf{0}$, except we now have r_0 linear identifiability constraints $\mathbf{H}_T\boldsymbol{\theta}_T = \mathbf{0}$ (and $\mathbf{H}_T\boldsymbol{\theta}_0 = \mathbf{0}$). If $\boldsymbol{\beta}_0 = n^{1/2}(\boldsymbol{\theta}_0 - \boldsymbol{\theta}_*)$, then in the theory following (10.36) we replace $\mathbf{y}$ by

$$\mathbf{w} = \mathbf{y} - \boldsymbol{\beta}_0 = n^{1/2}[(\hat{\boldsymbol{\theta}} - \boldsymbol{\theta}_*) - (\boldsymbol{\theta}_0 - \boldsymbol{\theta}_*)] = n^{1/2}(\hat{\boldsymbol{\theta}} - \boldsymbol{\theta}_0),$$

and replace $\boldsymbol{\beta}$ by $\boldsymbol{\eta} = \boldsymbol{\beta} - \boldsymbol{\beta}_0 = n^{1/2}(\boldsymbol{\theta}_T - \boldsymbol{\theta}_0)$. Then $\mathbf{w} = \boldsymbol{\eta} + \boldsymbol{\varepsilon}$, $\mathbf{H}_T\mathbf{w} = \mathbf{0}$ and, since $\mathbf{H}_T\boldsymbol{\theta}_0 = \mathbf{0}$, we have $\mathbf{H}_T\boldsymbol{\eta} = \mathbf{0}$. We now wish to test $H : \boldsymbol{\eta} = \mathbf{0}$, given the identifiability constraints, so that $\mathbf{A}_T$ is an appropriately chosen $p - r_0 \times p$ matrix, depending on the formulation of $\mathbf{H}_T$ (an example is given in Sect. 12.3). Proceeding with the algebra we end up with the linear model $\mathbf{z} = \boldsymbol{\phi} + \boldsymbol{\varepsilon}$ where $\mathbf{z} = \mathbf{R}_T\mathbf{w}$, $\mathbf{H}_T\mathbf{w} = \mathbf{0}$, and $\boldsymbol{\phi} = \mathbf{R}_T\boldsymbol{\eta}$. From (10.37) we have $\Omega = \mathbb{R}^{p-r_0}$ and $\omega = \mathcal{N}[(\mathbf{A}_T\mathbf{G}'\mathbf{G})_T^{-1}\mathbf{R}_T']$. With $\mathbf{y}$ replaced by $\mathbf{w}$, the likelihood ratio test is given by (10.39) and (10.40), namely

$$\begin{aligned}&n(\hat{\boldsymbol{\theta}} - \boldsymbol{\theta}_0)'\{\mathbf{A}'[(\mathbf{A}(\mathbf{G}'\mathbf{G})^{-1}\mathbf{B}(\mathbf{G}'\mathbf{G})^{-1}\mathbf{A}']^{-1}\mathbf{A}\}_T(\hat{\boldsymbol{\theta}} - \boldsymbol{\theta}_0) \\ &\approx n(\hat{\boldsymbol{\theta}} - \boldsymbol{\theta}_0)'\{\mathbf{A}'[(\mathbf{A}(\mathbf{G}'\mathbf{G})^{-1}\mathbf{B}(\mathbf{G}'\mathbf{G})^{-1}\mathbf{A}']^{-1}\mathbf{A}\}_{\hat{\boldsymbol{\theta}}}(\hat{\boldsymbol{\theta}} - \boldsymbol{\theta}_0), \qquad (10.48)\end{aligned}$$

the Wald statistic. The Score statistic follows in a similar fashion. From (10.45) this statistic is given by

$$n^{-1}\mathbf{D}L(\theta_0)'[(\mathbf{G}'\mathbf{G})^{-1}]_{\theta_0}\mathbf{D}L(\theta_0). \tag{10.49}$$

10.5 Orthogonal Hypotheses

Suppose that we are interested in testing two hypotheses $H_i : \{\boldsymbol{\theta} : \boldsymbol{\theta} \in W; \mathbf{a}_i(\boldsymbol{\theta}) = \mathbf{0}\}$ $(i = 1, 2)$, namely $\boldsymbol{\theta} \in W_i$, given $G : \boldsymbol{\theta} \in W$. We first assume a nonsingular (and therefore positive-definite) expected information matrix $\mathbf{B}_T$, where $\mathbf{B}_T^{-1} = \mathbf{V}_T\mathbf{V}_T'$ and $\mathbf{V}_T$ is nonsingular. Given that $\boldsymbol{\theta}_T$ is close to $W_1 \cap W_2$ we can use our linear model approximation $\mathbf{z} = \boldsymbol{\phi} + \boldsymbol{\varepsilon}$, with (cf. (10.19)) $G : \Omega = \mathbb{R}^p$ and $\omega_i = \{\boldsymbol{\theta} : \boldsymbol{\theta} \in \mathcal{N}[\mathbf{C}_i]\}$, where $\mathbf{C}_i = (\mathbf{A}^{(i)}\mathbf{V})_T$ and $\mathbf{A}^{(i)} = (\partial \mathbf{a}_i/\partial \boldsymbol{\theta}')$. Now $\omega_i^{\perp} \cap \Omega = \omega_i^{\perp} = \mathcal{C}[\mathbf{V}_T'\mathbf{A}_T^{(i)\prime}]$ (by Theorem 1.1 in Sect. 1.2) so that we have orthogonal hypotheses if and only if $\omega_1^{\perp} \perp \omega_2^{\perp}$, that is if and only if

$$\mathbf{C}_1\mathbf{C}_2' = (\mathbf{A}^{(1)}\mathbf{V}\mathbf{V}'\mathbf{A}^{(2)\prime})_T = (\mathbf{A}^{(1)}\mathbf{B}^{-1}\mathbf{A}^{(2)\prime})_T = \mathbf{0}$$

for all $\boldsymbol{\theta}_T \in \omega_1 \cap \omega_2$. This result was given by Aitchison (1962, 246) using a different method. The following example is taken from Example 5 in his paper.

Example 10.3 The three dimensions of cuboids produced by a certain process are described by a random vector $\mathbf{x} = (x_1, x_2, x_3)'$, where the x_i are independently and identically distributed as the scaled Gamma distribution

$$f(x) = \frac{x^{k-1}\exp(-x/\theta)}{\theta^k\Gamma(k)} \quad (x > 0)$$

with known k. Here $W = \{\boldsymbol{\theta} : \theta_i > 0\ (i = 1, 2, 3)\}$. Two hypotheses of interest are $W_1 = \{\boldsymbol{\theta} : \boldsymbol{\theta} \in W; k^3\theta_1\theta_2\theta_3 = a^3\}$, the hypothesis that the average volume is a^3, and $W_2 = \{\boldsymbol{\theta} : \boldsymbol{\theta} \in W; \theta_1 = \theta_2 = \theta_3\}$, the hypothesis that the three dimension are equal. Assuming a large sample n of vector observations $\mathbf{x}_i = (x_{i1}, x_{i2}, x_{i3})'$ $(i = 1, 2, \ldots, n)$ from the Gamma distribution, we can use the large sample approximation described above. The likelihood function takes the form

$$\prod_{i=1}^{n}\prod_{j=1}^{3}\left\{\frac{x_{ij}^{k-1}e^{-(x_{ij}/\theta_j)}}{\theta_j^k\Gamma(k)}\right\}$$

and the log likelihood (without constants) is

$$L(\boldsymbol{\theta}) = \sum_i\sum_j(-x_{ij}/\theta_j - k\log\theta_j)$$

so that

$$\frac{\partial L}{\partial \theta_j} = \frac{\sum_i x_{ij}}{\theta_j^2} - \frac{nk}{\theta_j}.$$

Since $E[x_{ij}] = k\theta_j$,

$$-\frac{1}{n}E\left[\frac{\partial^2 L}{\partial\theta_i\partial\theta_j}\right] = \delta_{ij}\frac{k}{\theta_j^2}$$

giving us $\mathbf{B}^{-1} = k^{-1}\,\mathrm{diag}(\theta_1^2, \theta_2^2, \theta_3^2)$. We find that $\mathbf{A}_1 = k^3(\theta_2\theta_3, \theta_3\theta_1, \theta_1\theta_2)$, and

$$\mathbf{A}_2 = \begin{pmatrix} 1 & -1 & 0 \\ 1 & 0 & -1 \end{pmatrix},$$

so that $(\mathbf{A}_1\mathbf{B}^{-1}\mathbf{A}_2')_T = k^3\theta_1\theta_2\theta_3(\theta_1 - \theta_2, \theta_1 - \theta_3)_T = \mathbf{0}$, when $\boldsymbol{\theta}_T \in \omega_1 \cap \omega_2 \subset \omega_2$. We therefore have orthogonality. This completes the example.

When $\mathbf{B}_T$ $(= \mathbf{R}_T'\mathbf{R}_T)$ is singular we can use the theory of the previous section where we have identifiability constraints with derivative matrix $\mathbf{H}$. In this case, from (10.37), we replace $\mathbf{C}_i$ by the matrix $\mathbf{A}^{(i)}(\mathbf{G}'\mathbf{G})^{-1}\mathbf{R}_T$ so that the condition for orthogonality is

$$\mathbf{C}_1\mathbf{C}_2' = [\mathbf{A}^{(1)}(\mathbf{G}'\mathbf{G})^{-1}\mathbf{B}(\mathbf{G}'\mathbf{G})^{-1}\mathbf{A}^{(2)\prime}]_T = \mathbf{0} \tag{10.50}$$

for all $\boldsymbol{\theta} \in \omega_1 \cap \omega_2$, where $\mathbf{G}'\mathbf{G} = \mathbf{B} + \mathbf{H}'\mathbf{H}$. If we also assume either $[\mathbf{H}(\mathbf{G}'\mathbf{G})^{-1}\mathbf{A}^{(1)\prime}]_T = \mathbf{0}$ or $[\mathbf{H}(\mathbf{G}'\mathbf{G})^{-1}\mathbf{A}^{(2)\prime}]_T = \mathbf{0}$ then, by adding $\mathbf{H}'\mathbf{H}$ to $\mathbf{B}$ in (10.50), this condition reduces to $[\mathbf{A}^{(1)}(\mathbf{G}'\mathbf{G})^{-1}\mathbf{A}^{(2)\prime}]_T = \mathbf{0}$. This gives us sufficient conditions for orthogonality. We now apply the theory to Example 6 of Aitchison (1962).

Example 10.4 A random sample with replacement of n individuals is taken from a genetic population whose individuals belong to one or other of three types—dominant, hybrid, and recessive. We therefore have n independent multinomial trials with three categories having probabilities $\theta_1/\theta., \theta_2/\theta., \theta_3/\theta.$, where $\theta. = \theta_1 + \theta_2 + \theta_3$, for dominant, hybrid, and recessive types, respectively. Here $W = \{\boldsymbol{\theta} : \theta_i > 0; i = 1, 2, 3\}$. The first hypothesis of interest is H_1 that $\boldsymbol{\theta} \in W_1$, where $W_1 = \{\boldsymbol{\theta} : \boldsymbol{\theta} \in W; (\theta_1/\theta.)^{1/2} + (\theta_3/\theta.)^{1/2} = 1\}$, the hypothesis that the population is genetically stable, and the second hypothesis is H_2 that $\boldsymbol{\theta} \in W_2$, where $W_2 = \{\boldsymbol{\theta} : \boldsymbol{\theta} \in W; \theta_1 = \theta_3\}$, the hypothesis of equal proportions of dominants and recessives. Our identifiability constraint is $h(\boldsymbol{\theta}) = \theta_1 + \theta_2 + \theta_3 - 1 = 0$ so that $H = (1, 1, 1) = \mathbf{1}_3'$. Now we can use this constraint to replace W_1 by $W_1' = \{\boldsymbol{\theta} : \boldsymbol{\theta} \in W; \theta_1^{1/2} + \theta_3^{1/2} = 1\}$. The multinomial distribution is discussed in Chap. 12 and from (12.3) we have

$$\begin{aligned}\mathbf{G}'\mathbf{G} &= \mathbf{B} + \mathbf{H}'\mathbf{H}\\ &= \left(\frac{\delta_{jk}}{\theta_j} - 1\right) + \mathbf{1}_3\mathbf{1}_3'\\ &= \left(\frac{\delta_{jk}}{\theta_j}\right)\\ &= \text{diag}(\boldsymbol{\theta}^{-1})\end{aligned}$$

so that $(\mathbf{G}'\mathbf{G})^{-1} = \text{diag}(\theta_1, \theta_2, \theta_3)$. Then considering W_1',

$$\mathbf{A}_1 = \frac{1}{2}(\theta_1^{-1/2}, 0, \theta_3^{-1/2}),$$

and for W_2 we have $\mathbf{A}_2 = (1, 0, -1)$. Then

$$\begin{aligned}\mathbf{A}_1(\mathbf{G}'\mathbf{G})^{-1}\mathbf{A}_2' &= \frac{1}{2}(\theta_1^{1/2}, 0, \theta_3^{1/2})(1, 0, -1)'\\ &= \frac{1}{2}(\theta_1^{1/2} - \theta_3^{1/2}) = 0\end{aligned}$$

when $\theta_1 = \theta_3$ for $\theta \in \omega_1 \cap \omega_2$. Also

$$\begin{aligned}\mathbf{H}(\mathbf{G}'\mathbf{G})^{-1}\mathbf{A}_2' &= \mathbf{1}_3' \,\text{diag}(\theta_1, \theta_2, \theta_3)(1, 0, -1)'\\ &= \theta_1 - \theta_3 = 0,\end{aligned}$$

when $\theta_1 = \theta_3$. Hence ω_1 and ω_2 are orthogonal. This completes the example.

In concluding this section we now show that orthogonality of the hypotheses leads to a partitioning of test statistics. From Theorem 6.1 in Sect. 6.2 we have

$$Q_{12\ldots.k} - Q = \sum_{i=1}^{k}(Q_i - Q),$$

where $Q_{12\ldots k} - Q$ is the test statistic for testing $H_{12\ldots k} : \boldsymbol{\theta} \in w_1 \cap w_2 \cap \cdots \cap w_k$ and $Q_i - Q$ is the test statistic for testing $H_i : \boldsymbol{\theta} \in \omega_i$. Now

$$\begin{aligned}Q_i - Q &= \mathbf{y}'(\mathbf{P}_\Omega - \mathbf{P}_i)\mathbf{y}\\ &= \mathbf{y}'\mathbf{P}_{w_i^p}\mathbf{y},\end{aligned}$$

where $\mathbf{P}_\Omega = \mathbf{I}_p$ and $\omega_i^\perp \cap \Omega = \omega_i^\perp$. As $\mathbf{P}_{\omega_i^p}\mathbf{P}_{\omega_j^p} = \mathbf{0}\ (i \neq j)$ because of orthogonality, we have that the test statistic for $H_{12\ldots k}$ can be partitioned into independent test statistics for the individual hypotheses $H_i : \boldsymbol{\theta} \in \omega_i$. When $H_{12\ldots k}$ is true, all the test statistics have chi-square distributions.

References

Aitchison, J. (1962). Large sample restricted parametric tests. *Journal of the Royal Statistical Society B, 24*, 234–250.

McManus, D. A. (1991). Who invented local power analysis? *Econometric Theory, 7*, 265–268.

Seber, G. A. F. (1963). The linear hypothesis and maximum likelihood theory, Ph.D. thesis, University of Manchester

Silvey, S. D. (1959). The Lagrangian multiplier test. *Annals of Mathematical Statistics, 30*, 389–407.

Wald, A. (1949). A note on the consistency of the maximum likelihood estimate. *Annals of Mathematical Statistics, 20*, 595–601.

Chapter 11
Large Sample Theory: Freedom-Equation Hypotheses

11.1 Introduction

In this chapter we assume once again that $\boldsymbol{\theta} \in W$. However our hypothesis H now takes the form of freedom equations, namely $\boldsymbol{\theta} = \boldsymbol{\theta}(\boldsymbol{\alpha})$, where $\boldsymbol{\alpha} = (\alpha_1, \alpha_2, \ldots, \alpha_{p-q})'$. We require the following additional notation. Let $\boldsymbol{\Theta}_\alpha$ be the $p \times p - q$ matrix with (i,j)th element $\partial\theta_i/\partial\alpha_j$, which we assume to have rank $p - q$. As before, $L(\boldsymbol{\theta}) = \log \prod_{i=1}^n f(x_i, \boldsymbol{\theta})$ is the log likelihood function. Let $\mathbf{D}_\theta L(\boldsymbol{\theta})$ and $\mathbf{D}_\alpha L(\boldsymbol{\theta}(\boldsymbol{\alpha}))$ be the column vectors whose ith elements are $\partial L(\boldsymbol{\theta})/\partial\theta_i$ and $\partial L(\boldsymbol{\theta})/\partial\alpha_i$ respectively. As before, $\mathbf{B}_\theta$ is the $p \times p$ information matrix with i,jth element

$$-n^{-1}E_\theta\left[\frac{\partial^2 L(\boldsymbol{\theta})}{\partial\theta_i\partial\theta_j}\right] = -E\left[\frac{\partial^2 \log f(x, \boldsymbol{\theta})}{\partial\theta_i\partial\theta_j}\right],$$

and we add $\mathbf{B}_\alpha$, the $p - q \times p - q$ information matrix with i,jth element $-E[\partial^2 \log f(x, \boldsymbol{\theta}(\boldsymbol{\alpha}))/\partial\alpha_i\partial\alpha_j]$. To simplify the notation we use $[\cdot]_\alpha$ to denote that the matrix in square brackets is evaluated at $\boldsymbol{\alpha}$, for example

$$\mathbf{B}_\alpha = [\boldsymbol{\Theta}'\mathbf{B}_\theta\boldsymbol{\Theta}]_\alpha = \boldsymbol{\Theta}_\alpha'\mathbf{B}_{\theta(\alpha)}\boldsymbol{\Theta}_\alpha.$$

We note that

$$\mathbf{D}_\alpha L(\boldsymbol{\theta}) = \boldsymbol{\Theta}_\alpha'\mathbf{D}_\theta L(\boldsymbol{\theta}(\boldsymbol{\alpha})).$$

We use similar assumptions to those given in the previous chapter so that (10.4) and (10.5) still hold. This means that once again $\boldsymbol{\theta}_T$, $\boldsymbol{\theta}_* = \boldsymbol{\theta}(\boldsymbol{\alpha}_*)$, $\hat{\boldsymbol{\theta}}$ and $\tilde{\boldsymbol{\theta}} = \boldsymbol{\theta}(\tilde{\boldsymbol{\alpha}})$ are assumed to be all "near" each other. We consider two cases depending on whether $\mathbf{B}_{\theta_T}$ is positive definite or positive semi-definite. The theory in the next two sections is based on Seber (1964) with a couple of typos corrected and a change in notation.

G.A.F. Seber, *The Linear Model and Hypothesis*, Springer Series in Statistics,
DOI 10.1007/978-3-319-21930-1_11

11.2 Positive-Definite Information Matrix

As $\mathbf{B}_{\theta_T}$ is positive definite there exists a $p \times p$ nonsingular matrix $\mathbf{V}_T$ such that we have $\mathbf{B}_{\theta_T} = \mathbf{V}_T'\mathbf{V}_T$. We now show that our original model and hypothesis are asymptotically equivalent to the linear model

$$\mathbf{z} = \boldsymbol{\psi} + \boldsymbol{\varepsilon},$$

where $\boldsymbol{\varepsilon}$ is $N_p[\mathbf{0}, \mathbf{I}_p]$, $G : \Omega = \mathcal{C}[\mathbf{V}_T] = \mathbb{R}^p$ and $H : \omega = \mathcal{C}[\mathbf{V}_T\boldsymbol{\Theta}_{\alpha*}]$. In preparation for proving this result we need to find the least squares estimates for the above linear model. If $\mathbf{X} = \mathbf{V}_T\boldsymbol{\Theta}_{\alpha*}$, then under G we have $\hat{\boldsymbol{\psi}} = \mathbf{z}$ and under H the least squares (normal) equations are $\mathbf{X}'(\mathbf{z} - \tilde{\boldsymbol{\psi}})$ (cf. (3.2)). We use these equations below.

If $\hat{\boldsymbol{\theta}}$ and $\tilde{\boldsymbol{\alpha}}$ are the maximum likelihood estimates under G and H respectively and $\tilde{\boldsymbol{\theta}} = \boldsymbol{\theta}(\tilde{\boldsymbol{\alpha}})$, then these estimates are the solutions of

$$\mathbf{D}_{\theta} n^{-1} L(\hat{\boldsymbol{\theta}}) = \mathbf{0} \tag{11.1}$$

and

$$\mathbf{D}_{\alpha} n^{-1} L(\tilde{\boldsymbol{\theta}}) = \mathbf{0}. \tag{11.2}$$

As $\hat{\boldsymbol{\theta}}$ is near $\boldsymbol{\theta}_T$, a Taylor expansion of (11.1) give us

$$\mathbf{0} = \mathbf{D}_{\theta} n^{-1} L(\boldsymbol{\theta}_T) - \mathbf{B}_{\theta_T}(\hat{\boldsymbol{\theta}} - \boldsymbol{\theta}_T) + \mathbf{o}_p(1), \tag{11.3}$$

where $n^{1/2}\mathbf{D}_{\theta} n^{-1} L(\boldsymbol{\theta}_T)$ is asymptotically $N_p[\mathbf{0}, \mathbf{B}_{\theta_T}]$. Since $\boldsymbol{\theta}_*$ is near $\boldsymbol{\theta}_T$, (11.3) becomes

$$\boldsymbol{\delta} = \mathbf{B}_{\theta_T} n^{1/2}(\hat{\boldsymbol{\theta}} - \boldsymbol{\theta}_*) - \mathbf{B}_{\theta_T} n^{1/2}(\boldsymbol{\theta}_T - \boldsymbol{\theta}_*) + \mathbf{o}_p(1), \tag{11.4}$$

where $\boldsymbol{\delta}$ is $N_p[\mathbf{0}, \mathbf{B}_{\theta_T}]$. As $\tilde{\boldsymbol{\theta}}$ is near $\hat{\boldsymbol{\theta}}$ and therefore both near $\boldsymbol{\theta}_*$, we have a similar equation for (11.2), namely

$$\begin{aligned}
\mathbf{0} &= n^{1/2}\boldsymbol{\Theta}_{\tilde{\alpha}}' \mathbf{D}_{\theta} n^{-1} L(\tilde{\boldsymbol{\theta}}) \\
&= n^{1/2}\boldsymbol{\Theta}_{\alpha*}' \mathbf{D}_{\theta} n^{-1} L(\tilde{\boldsymbol{\theta}}) + \mathbf{o}_p(1) \\
&= -n^{1/2}\boldsymbol{\Theta}_{\alpha*}' \mathbf{B}_{\theta_T}(\tilde{\boldsymbol{\theta}} - \hat{\boldsymbol{\theta}}) + \mathbf{o}_p(1) \\
&= -n^{1/2}\boldsymbol{\Theta}_{\alpha*}' \mathbf{V}_T'\mathbf{V}_T[(\tilde{\boldsymbol{\theta}} - \boldsymbol{\theta}_*) - (\hat{\boldsymbol{\theta}} - \boldsymbol{\theta}_*)] + \mathbf{o}_p(1).
\end{aligned} \tag{11.5}$$

Multiplying (11.4) on the left by $\mathbf{V}_T'^{-1}$ and putting $\mathbf{z} = n^{1/2}\mathbf{V}_T(\hat{\boldsymbol{\theta}} - \boldsymbol{\theta}_*)$ and $\boldsymbol{\psi} = n^{1/2}\mathbf{V}_T(\boldsymbol{\theta}_T - \boldsymbol{\theta}_*)$, we get $\boldsymbol{\varepsilon} = \mathbf{z} - \boldsymbol{\psi}$, where $\boldsymbol{\varepsilon}$ is $N_p[\mathbf{0}, \mathbf{I}_p]$. If $\tilde{\boldsymbol{\psi}} = n^{1/2}\mathbf{V}_T(\tilde{\boldsymbol{\theta}} - \boldsymbol{\theta}_*)$ it follows from (11.5) that $\mathbf{X}'(\mathbf{z} - \tilde{\boldsymbol{\psi}}) = \mathbf{o}_p(1)$, where $\mathbf{X} = \mathbf{V}_T\boldsymbol{\Theta}_{\alpha*}$. We have therefore established that $\mathbf{z} = \boldsymbol{\psi} + \boldsymbol{\varepsilon}$, $G : \Omega = \mathbb{R}^p$, and $H : \omega = \mathcal{C}[\mathbf{V}_T\boldsymbol{\Theta}_{\alpha*}]$.

Unfortunately the above formulation is not very helpful in providing a test statistic, as it would be based $\mathbf{z}'(\mathbf{I}_p - \mathbf{P}_\omega)\mathbf{z}$, where

$$\mathbf{P}_\omega = \mathbf{V}_T\boldsymbol{\Theta}_{\alpha*}[\boldsymbol{\Theta}'_{\alpha*}\mathbf{B}_T\boldsymbol{\Theta}_{\alpha*}]^{-1}\boldsymbol{\Theta}'_{\alpha*}\mathbf{V}'_T.$$

Instead we can use the Score test based on the constraint equation specification from Sect. 10.3.3, namely

$$n^{-1}[\mathbf{D}_\theta L(\tilde{\theta})]'\tilde{\mathbf{B}}^{-1}[\mathbf{D}_\theta L(\tilde{\theta})],$$

where $\tilde{\theta} = \theta(\tilde{\alpha})$. We simply assume that a constraint equation exists for formulating the hypothesis, but we don't need to actually find it.

11.3 Positive-Semidefinite Information Matrix

Suppose that the $p \times p$ matrix $\mathbf{B}_{\theta_T}$ is positive semidefinite of rank $p - r_0$. We now find that $\boldsymbol{\theta}_T$ is not identifiable and we introduce r_0 identifiability constraints

$$\mathbf{h}(\boldsymbol{\theta}_T) = (h_1(\boldsymbol{\theta}_T), h_2(\boldsymbol{\theta}_T), \ldots, h_{r_0}(\boldsymbol{\theta}_T))' = \mathbf{0}.$$

Let $\mathbf{H}_T$ be the $r_0 \times p$ matrix of rank r_0 with (i,j)th element $\partial h_i/\partial\theta_j$ evaluated at $\boldsymbol{\theta}_T$. We make a further assumption.

(xvi) $\mathbf{B}_{\theta_T} + (\mathbf{H}'\mathbf{H})_T$ is positive definite. (We see below once again that this assumption follows naturally from the linear theory.)

Since $\mathbf{B}_{\theta_T}$ is positive semidefinite, there exists a $p - r_0 \times p$ matrix $\mathbf{R}_T$ of rank $p - r_0$ such that $\mathbf{B}_{\theta_T} = [\mathbf{R}'\mathbf{R}]_T$ (A.9(iii)). We will now show that our asymptotic linear model and hypothesis take the form (cf. Seber 1964, with a slight change in notation)

$$\mathbf{z} = \boldsymbol{\phi} + \boldsymbol{\varepsilon},$$

where $\boldsymbol{\varepsilon}$ is $N_{p-r_0}[\mathbf{0}, \mathbf{I}_{p-r_0}]$,

$$G : \Omega = \{\boldsymbol{\phi} \mid \boldsymbol{\phi} = \mathbf{R}_T\boldsymbol{\beta};\ \mathbf{H}_T\boldsymbol{\beta} = \mathbf{0}\} = \mathbb{R}^{p-r_0}$$

and

$$H : \omega = \{\boldsymbol{\phi} \mid \boldsymbol{\phi} = \mathbf{R}_T\boldsymbol{\Theta}_{\alpha*}\boldsymbol{\gamma};\ \mathbf{H}_T\boldsymbol{\Theta}_{\alpha*}\boldsymbol{\gamma} = \mathbf{0}\}.$$

Assumption (xvi) implies that $[\mathbf{B}_\theta + \mathbf{H}'\mathbf{H}]_T$ is of full rank p so that the $p \times p$ matrix $\mathbf{G}_T = [\mathbf{R}', \mathbf{H}']'_T$ has rank p and $\mathbf{H}_T$ has rank r_0, which are necessary and sufficient conditions for the constraints $\mathbf{H}_T\boldsymbol{\beta} = \mathbf{0}$ to be identifiable (see A.11). Since $(\mathbf{G}'\mathbf{G})_T$ is positive definite it follows from A.9(iv) that the $p-q \times p-q$ matrix

$\boldsymbol{\Theta}'_{\alpha*}\mathbf{G}'_T\mathbf{G}_T\boldsymbol{\Theta}_{\alpha*} = \mathbf{B}_{\alpha*} + (\mathbf{H}_T\boldsymbol{\Theta}_{\alpha*})'(\mathbf{H}_T\boldsymbol{\Theta}_{\alpha*})$ is also positive definite. This means that the constraints $\mathbf{H}_T\boldsymbol{\Theta}_{\alpha*}\gamma = \mathbf{0}$ in H above are necessary and sufficient conditions for γ to be identifiable.

The least squares estimates $\hat{\theta}$ and $\tilde{\theta}$ $(= \theta(\tilde{\alpha}))$ are respectively given by the solutions of

$$\mathbf{D}_\theta n^{-1}L(\hat{\theta}) + \mathbf{H}'_{\hat{\theta}}\hat{\boldsymbol{\lambda}}_0 = \mathbf{0}, \quad \mathbf{h}(\hat{\theta}) = \mathbf{0} \tag{11.6}$$

and

$$\mathbf{D}_\alpha n^{-1}L(\theta(\tilde{\alpha})) + \boldsymbol{\Theta}'_{\tilde{\alpha}}\mathbf{H}'_{\tilde{\theta}}\tilde{\boldsymbol{\lambda}}_1 = \mathbf{0}, \quad \mathbf{h}(\theta(\tilde{\alpha})) = \mathbf{0}, \tag{11.7}$$

where $\hat{\boldsymbol{\lambda}}_0$ and $\tilde{\boldsymbol{\lambda}}_1$ are the appropriate Lagrange multipliers for the identifiability constraints. From (10.30) we have that $n^{1/2}\hat{\boldsymbol{\lambda}}_0 = \mathbf{o}_p(1)$ and, using the same argument with $\boldsymbol{\alpha}$ the unknown vector parameter, we have that $n^{1/2}\tilde{\boldsymbol{\lambda}}_1 = \mathbf{o}_p(1)$. As $\boldsymbol{\theta}_T$ is near $\boldsymbol{\theta}_*$, we can use (10.30) and the following equations and replace (11.6) by

$$n^{1/2}\mathbf{D}_\theta n^{-1}L(\hat{\theta}) = \mathbf{o}_p(1), \tag{11.8}$$

$$\mathbf{H}_T n^{1/2}(\hat{\theta} - \theta_*) = \mathbf{o}_p(1) \quad \text{and} \quad \mathbf{H}_T n^{1/2}(\theta_T - \theta_*) = \mathbf{o}_p(1). \tag{11.9}$$

Similarly (11.7) can be replaced by

$$n^{1/2}\mathbf{D}_\alpha n^{-1}L(\theta(\tilde{\alpha})) = \mathbf{o}_p(1) \tag{11.10}$$

and $\mathbf{H}_T n^{1/2}(\tilde{\theta} - \theta_*) = \mathbf{o}_p(1)$, which can be approximated by

$$\mathbf{H}_{\theta_*}\boldsymbol{\Theta}_{\alpha*}n^{1/2}(\tilde{\alpha} - \alpha_*) = \mathbf{o}_p(1). \tag{11.11}$$

Now using a Taylor expansion for (11.10) we have

$$\begin{aligned}
\mathbf{0} &= n^{1/2}\boldsymbol{\Theta}'_{\tilde{\theta}}\mathbf{D}_\theta n^{-1}L(\tilde{\theta}) + \mathbf{o}_p(1), \\
&= n^{1/2}\boldsymbol{\Theta}'_{\alpha*}\mathbf{D}_\theta n^{-1}L(\tilde{\theta}) + \mathbf{o}_p(1) \\
&= n^{1/2}\boldsymbol{\Theta}'_{\alpha*}\mathbf{D}_\theta n^{-1}L(\hat{\theta}) - \boldsymbol{\Theta}'_{\alpha*}\mathbf{B}_{\theta_T}n^{1/2}(\tilde{\theta} - \hat{\theta}) + \mathbf{o}_p(1) \\
&= \mathbf{0} - \boldsymbol{\Theta}'_{\alpha*}\mathbf{B}_{\theta_T}n^{1/2}(\tilde{\theta} - \hat{\theta}) + \mathbf{o}_p(1) \quad \text{by (11.8)} \\
&= \boldsymbol{\Theta}'_{\alpha*}\mathbf{B}_{\theta_T}n^{1/2}(\hat{\theta} - \theta_*) - \boldsymbol{\Theta}'_{\alpha*}\mathbf{B}_{\theta_T}n^{1/2}(\tilde{\theta} - \theta_*).
\end{aligned} \tag{11.12}$$

Now setting $\mathbf{z} = \mathbf{R}_T n^{1/2}(\hat{\theta} - \theta_*)$, $\phi = \mathbf{R}_T n^{1/2}(\theta_T - \theta_*)$, $\tilde{\phi} = \mathbf{R}_T n^{1/2}(\tilde{\theta} - \theta_*)$, and recalling that $\mathbf{B}_{\theta_T} = [\mathbf{R}'\mathbf{R}]_T$, we have from (11.12)

$$\mathbf{\Theta}'_{\alpha*}\mathbf{R}'_T\mathbf{z} - \mathbf{\Theta}'_{\alpha*}\mathbf{R}'_T\tilde{\phi} = \mathbf{o}_p(1)$$

or

$$\mathbf{X}'(\mathbf{z} - \tilde{\phi}) = \mathbf{o}_p(1), \tag{11.13}$$

where $\mathbf{X} = \mathbf{R}_T\mathbf{\Theta}_{\alpha*}$. Referring to (10.31), we have

$$\mathbf{B}_{\theta_T} n^{1/2}(\hat{\theta} - \theta_*) - \mathbf{B}_{\theta_T} n^{1/2}(\theta_T - \theta_*) = \boldsymbol{\delta}_1 + \mathbf{o}_p(1),$$

where $\boldsymbol{\delta}_1$ is $N_p[\mathbf{0}, \mathbf{B}_{\theta_T}]$, a singular multivariate normal distribution. Now $\mathbf{R}_T\mathbf{R}'_T$ is $p - r_0 \times p - r_0$ of rank $p - r_0$, so it is nonsingular. Premultiplying the above equation by $[(\mathbf{R}\mathbf{R}')^{-1}\mathbf{R}]_T$ we get

$$\mathbf{R}_T n^{1/2}(\hat{\theta} - \theta_*) - \mathbf{R}_T n^{1/2}(\theta_T - \theta_*) = \boldsymbol{\varepsilon} + \mathbf{o}_p(1), \tag{11.14}$$

where $\boldsymbol{\varepsilon}$ is $N_{p-r_0}[\mathbf{0}, \mathbf{I}_{p-r_0}]$ as

$$\mathrm{Var}[\boldsymbol{\varepsilon}] = [(\mathbf{R}\mathbf{R}')^{-1}\mathbf{R}(\mathbf{R}'\mathbf{R})\mathbf{R}'(\mathbf{R}\mathbf{R}')^{-1}]_T = \mathbf{I}_{p-r_0}.$$

Putting all this together, we have from (11.14) the approximating linear model $\mathbf{z} = \phi + \boldsymbol{\varepsilon}$, where $\mathbf{z} = \mathbf{R}_T n^{1/2}(\hat{\theta} - \theta_*)$ and $\phi = \mathbf{R}_T n^{1/2}(\theta_T - \theta_*)$. From Sect. 10.4 the G model given there is the same as here so that $\Omega = \mathbb{R}^{p-r_0}$. Since (11.13) are the normal equations for H, we have $\omega = \{\phi \mid \phi = \mathbf{R}_T\mathbf{\Theta}_{\alpha*}\gamma\}$ with identifiability constraints $\mathbf{H}_T\mathbf{\Theta}_{\alpha*}\gamma = \mathbf{0}$. The likelihood ratio test statistic is (cf. (10.25))

$$-2\log\Lambda[H|G] = \mathbf{z}'(\mathbf{I}_{p-r_0} - \mathbf{P}_\omega)\mathbf{z},$$

where

$$\begin{aligned}\mathbf{P}_\omega &= \mathbf{R}_T\mathbf{\Theta}_{\alpha*}[\mathbf{\Theta}'_{\alpha*}(\mathbf{R}'_T\mathbf{R}_T + \mathbf{H}'_T\mathbf{H}_T)^{-1}\mathbf{\Theta}_{\alpha*}]^{-1}\mathbf{\Theta}'_{\alpha*}\mathbf{R}'_T \\ &= \mathbf{R}_T\mathbf{\Theta}_{\alpha*}[\mathbf{\Theta}'_{\alpha*}(\mathbf{G}'_T\mathbf{G}_T)^{-1}\mathbf{\Theta}_{\alpha*}]^{-1}\mathbf{\Theta}'_{\alpha*}\mathbf{R}'_T.\end{aligned}$$

As with the full rank case in Sect. 11.2, a more convenient approach is to use the the Score test, namely

$$n^{-1}[\mathbf{D}_\theta L(\tilde{\theta})]'(\tilde{\mathbf{B}} + \tilde{\mathbf{H}}'\tilde{\mathbf{H}})^{-1}[\mathbf{D}_\theta L(\tilde{\theta})]. \tag{11.15}$$

Reference

Seber, G. A. F. (1964). The linear hypothesis and large sample theory. *Annals of Mathematical Statistics, 35*(2), 773–779.

Chapter 12
Multinomial Distribution

12.1 Definitions

In this chapter we consider asymptotic theory for the multinomial distribution, which is defined below. Although the distribution used is singular, the approximating linear theory can still be used.

Let $\mathbf{e}_i$ be the r-dimensional vector with 1 in the ith position and 0 elsewhere. Let $\mathbf{y}$ be an r-dimensional random vector that takes the value $\mathbf{e}_i$ with probability $p_i/p.$ $(i = 1, 2, \ldots, k)$, where $p. = \sum_{i=1}^{r} p_i$. A random sample of n observations $\mathbf{y}_j$ $(j = 1, 2, \ldots, n)$ is taken from this multivariate discrete distribution giving the joint probability function

$$f(\mathbf{y}_1, \mathbf{y}_2, \ldots, \mathbf{y}_n) = \prod_{i=1}^{r} \left(\frac{p_i}{p.} \right)^{x_i}, \tag{12.1}$$

where x_i is the number of times $\mathbf{y}_j$ takes the value $\mathbf{e}_i$. We note that the probability function of the x_i is

$$f(x_1, x_2, \ldots, x_r) = \frac{n!}{\prod_{i=1}^{r} x_i!} \prod_{i=1}^{r} \left(\frac{p_i}{p.} \right)^{x_i}, \tag{12.2}$$

which is a (singular) multinomial distribution because $\sum_i x_i = n$. The p_i in (12.1) and (12.2) are not identifiable as we can replace p_i by ap_i without changing (12.2), so we need to add an identifiability constraint, namely $p. = 1$. The nonsingular multinomial distribution then takes the form

$$f(x_1, x_2, \ldots, x_{r-1}) = \frac{n!}{\prod_{i=1}^{r} x_i!} \prod_{i=1}^{r} p_i^{x_i},$$

with $x_r = n - \sum_{i=1}^{r-1} x_i$ and $p_r = 1 - \sum_{i=1}^{r-1} p_i$.

G.A.F. Seber, *The Linear Model and Hypothesis*, Springer Series in Statistics,
DOI 10.1007/978-3-319-21930-1_12

12.2 Test of $\mathbf{p} = \mathbf{p}_0$

If $\mathbf{p} = (p_1, p_2, \ldots, p_r)'$, then the log likelihood function from (12.1) is

$$\begin{aligned} L(\mathbf{p}) &= \log f(\mathbf{y}_1, \mathbf{y}_2, \ldots, \mathbf{y}_r) \\ &= \sum_{i=1}^{r} x_j \log p_i - n \log p.. \end{aligned} \tag{12.3}$$

To find the maximum likelihood estimate of $\mathbf{p}$ we differentiate $L(\mathbf{p}) + \lambda(p. - 1)$, where λ is a Lagrange multiplier (which we expect to be zero for an identifiability constraint). We have, differentiating with respect to p_i,

$$\frac{x_i}{p_i} - \frac{n}{p\cdot} + \lambda = 0 \quad \text{together with} \quad p. = 1.$$

Multiplying by p_i and summing on i gives us $n = \sum_i x_i = n - \lambda$, and $\lambda = 0$ as expected. Then the maximum likelihood estimate of p_i is $\hat{p}_i = x_i/n$ $(i = 1, 2, \ldots, r)$, and

$$\frac{\partial L}{\partial p_i} = \frac{x_i}{p_i} - \frac{n}{p\cdot},$$

so that

$$\frac{\partial^2 L}{\partial p_i^2} = -\frac{x_i}{p_i^2} + \frac{n}{p_\cdot^2} \quad \text{and} \quad \frac{\partial^2 L}{\partial p_i \partial p_j} = \frac{n}{p_\cdot^2} \quad (i \neq j).$$

Taking expected values gives us $E[x_i] = np_i/p.$,

$$c_{ij} = -E\left[\frac{\partial^2 L}{\partial p_i \partial p_j}\right] = n\frac{\delta_{ij}}{p_i} - n,$$

and our expected information matrix is

$$\mathbf{B}_{\mathbf{p}} = n^{-1}(c_{ij}) = (\frac{\delta_{ij}}{p_i} - 1), \tag{12.4}$$

where $\delta_{ij} = 1$ when $i = j$ and 0 otherwise.

Let

$$\begin{aligned} \hat{\mathbf{p}} &= (\hat{p}_1, \hat{p}_2, \ldots, \hat{p}_r)' \\ &= n^{-1}(x_1, x_2, \ldots, x_r)' \end{aligned}$$

$$= n^{-1} \sum_{i=1}^{r} x_i \mathbf{e}_i$$

$$= \bar{\mathbf{y}},$$

which has mean $\mathbf{p}$ and is asymptotically normal by the multivariate Central Limit Theorem. Now the variance-covariance matrix of $\mathbf{y}$ is

$$\boldsymbol{\Sigma}_{(r)} = \begin{pmatrix} p_1q_1 & -p_1p_2 & \cdots & -p_1p_r \\ -p_2p_1 & p_2q_2 & \cdots & -p_2p_r \\ \cdot & \cdot & \cdots & \cdot \\ -p_rp_1 & -p_rp_2 & \cdots & p_rq_r \end{pmatrix}$$

$$= \text{diag}(\mathbf{p}) - \mathbf{pp}', \tag{12.5}$$

where $\text{diag}(\mathbf{p})$ is $\text{diag}(p_1, p_2, \ldots, p_r)$. The matrix $\boldsymbol{\Sigma}_{(r)}$ is singular as $\boldsymbol{\Sigma}_{(r)}\mathbf{1}_r = \mathbf{0}$, and $n^{1/2}(\hat{\mathbf{p}} - \mathbf{p})$ is asymptotically $N_p[\mathbf{0}, \boldsymbol{\Sigma}_{(r)}]$.

Let $\mathbf{F} = \text{diag}(p_1^{-1}, p_2^{-1}, \ldots, p_r^{-1}) = \text{diag}(\mathbf{p}^{-1})$, say, then

$$\mathbf{F}\boldsymbol{\Sigma}_{(r)} = \text{diag}(\mathbf{p}^{-1})(\text{diag}(\mathbf{p}) - \mathbf{pp}')$$
$$= \mathbf{I}_r - \mathbf{1}_r\mathbf{p}',$$

and

$$\boldsymbol{\Sigma}_{(r)}\mathbf{F}\boldsymbol{\Sigma}_{(r)} = (\text{diag}(\mathbf{p}) - \mathbf{pp}')(\mathbf{I}_r - \mathbf{1}_r\mathbf{p}')$$
$$= \text{diag}(\mathbf{p}) - \mathbf{pp}'$$
$$= \boldsymbol{\Sigma}_{(r)},$$

so that $\mathbf{F}$ is a generalized inverse $\boldsymbol{\Sigma}_{(r)}^{-}$ of $\boldsymbol{\Sigma}_{(r)}$. Furthermore, we see that $\boldsymbol{\Sigma}_{(r)}\mathbf{F}\boldsymbol{\Sigma}_{(r)}\mathbf{F}\boldsymbol{\Sigma}_{(r)} = \boldsymbol{\Sigma}_{(r)}\mathbf{F}\boldsymbol{\Sigma}_{(r)}$ and

$$\text{trace}[\mathbf{F}\boldsymbol{\Sigma}_{(r)}] = \text{trace}[\mathbf{I}_r - \mathbf{1}_r\mathbf{p}']$$
$$= r - \text{trace}[\mathbf{p}'\mathbf{1}_r]$$
$$= r - 1.$$

It therefore follows from A.16 that

$$n(\hat{\mathbf{p}} - \mathbf{p})'\boldsymbol{\Sigma}_{(r)}^{-}(\hat{\mathbf{p}} - \mathbf{p}) = \sum_{i=1}^{r} \frac{(x_i - np_i)^2}{np_i} \tag{12.6}$$

is asymptotically χ^2_{r-1}. We can now test the hypothesis H that $\mathbf{p} = \mathbf{p}_0$, where we have $\mathbf{p}_0 = (p_{i0})$, using the so-called Pearson's goodness-of-fit statistic

$$X_0^2 = \sum_{i=1}^{r} \frac{(x_i - np_{i0})^2}{np_{i0}}, \tag{12.7}$$

which is approximately distributed as χ^2_{r-1} when H is true.

12.3 Score and Wald Statistics

We shall first see where the general theory of Sect. 10.4.1 with $p = r$ and $r_0 = 1$ fits into the picture. To simplify notation we drop the subscript "T" and let $\mathbf{p}_T = \mathbf{p}$, the true value of $\mathbf{p}$. Referring to Example 10.2 at the end of Sect. 10.4.2 where we test $H : \mathbf{p} = \mathbf{p}_0$, we have the linear model $\mathbf{w} = \boldsymbol{\eta} + \boldsymbol{\varepsilon}$, where $\mathbf{H}_\mathbf{p}\boldsymbol{\eta} = \mathbf{0}$ are the identifiability constraints and $\boldsymbol{\eta} = n^{1/2}(\mathbf{p}-\mathbf{p}_0)$. From (12.3), $\mathbf{B}_\mathbf{p} = \text{diag}(\mathbf{p}^{-1}) - \mathbf{1}_r\mathbf{1}_r'$ and since $\mathbf{1}_r'(\mathbf{p} - \mathbf{p}_0) = 0$, $\mathbf{H}_\mathbf{p} = \mathbf{1}_r'$ and $r_0 = 1$. Hence

$$(\mathbf{G}'\mathbf{G})^{-1}_{\mathbf{p}_0} = (\mathbf{B} + \mathbf{H}'\mathbf{H})^{-1}_{\mathbf{p}_0} = \text{diag}(\mathbf{p}_0)$$

and

$$\mathbf{D}L(\tilde{\mathbf{p}})' = \mathbf{D}L(\mathbf{p}_0)' = (\frac{x_1}{p_{10}} - n, \frac{x_2}{p_{20}} - n, \ldots, \frac{x_r}{p_{r0}} - n).$$

The Score statistic is therefore given by (cf. 10.49)

$$n^{-1}\mathbf{D}L(\boldsymbol{\theta}_0)'[(\mathbf{G}'\mathbf{G})^{-1}]_{\boldsymbol{\theta}_0}\mathbf{D}L(\boldsymbol{\theta}_0) = \sum_i \frac{(x_i - np_{i0})^2}{np_{i0}}, \tag{12.8}$$

which means that Pearson's goodness-of-fit statistic is also the Score (Lagrange Multiplier) statistic.

In a similar manner we can derive the Wald statistic. Since

$$\begin{aligned}(\mathbf{G}'\mathbf{G})^{-1}\mathbf{B}(\mathbf{G}'\mathbf{G})^{-1} &= \text{diag}\,\mathbf{p}[\text{diag}(\mathbf{p}^{-1}) - \mathbf{1}_r\mathbf{1}_r']\,\text{diag}(\mathbf{p}) \\ &= \text{diag}\,\mathbf{p} - \mathbf{p}\mathbf{p}',\end{aligned} \tag{12.9}$$

we have from (10.49) that the Wald statistic is

$$W = n(\hat{\mathbf{p}} - \mathbf{p}_0)'\{\mathbf{A}'[\mathbf{A}(\mathbf{G}'\mathbf{G})^{-1}\mathbf{B}(\mathbf{G}'\mathbf{G})^{-1}\mathbf{A}']^{-1}\mathbf{A}\}_{\hat{\mathbf{p}}}(\hat{\mathbf{p}} - \mathbf{p}_0)$$

where we can set $\mathbf{A} = (\mathbf{I}_{r-1}, \mathbf{0})$, an $r-1 \times r$ matrix. Now using (12.9)

$$\begin{aligned}\mathbf{A}(\mathbf{G}'\mathbf{G})^{-1}\mathbf{B}(\mathbf{G}'\mathbf{G})^{-1}\mathbf{A}' &= (\mathbf{I}_{r-1}, \mathbf{0})(\text{diag}(\mathbf{p}) - \mathbf{p}\mathbf{p}')(\mathbf{I}_{r-1}, \mathbf{0})' \\ &= \text{diag}(\mathbf{p}_{r-1}) - \mathbf{p}_{r-1}\mathbf{p}_{r-1}' \\ &= \boldsymbol{\Sigma}_{(r-1)},\end{aligned}$$

say, where $\mathbf{p}_{r-1} = (p_1, p_2, \ldots, p_{r-1})'$. From Seber (2008, result 10.27)

$$\boldsymbol{\Sigma}_{(r-1)}^{-1} = \text{diag}(\mathbf{p}_{r-1}) - p_r^{-1}\mathbf{1}_{r-1}\mathbf{1}_{r-1}'$$

and

$$\mathbf{A}'\boldsymbol{\Sigma}_{(r-1)}^{-1}\mathbf{A} = \text{diag}(\mathbf{p}_{r-1}^{-1}, 0) + p_r^{-1}(\mathbf{1}_{r-1}', 0)'(\mathbf{1}_{r-1}, 0).$$

Also

$$\begin{aligned}(\hat{\mathbf{p}} - \mathbf{p}_0)'\begin{pmatrix}\mathbf{1}_{r-1} \\ 0\end{pmatrix} &= \sum_{i=1}^{r-1}(\hat{p}_i - p_{0i}) \\ &= p_{0r} - \hat{p}_r,\end{aligned}$$

using the fact that $\sum_i p_i = 1$. Hence

$$\begin{aligned}W &= n(\hat{\mathbf{p}} - \mathbf{p}_0)'[\mathbf{A}'\boldsymbol{\Sigma}_{(r-1)}^{-1}\mathbf{A}]_{\hat{\mathbf{p}}}(\hat{\mathbf{p}} - \mathbf{p}_0) \\ &= \sum_{i=1}^{r-1}\frac{n(\hat{p}_i - p_{0i})^2}{\hat{p}_i} + \frac{n(\hat{p}_r - p_{0r})^2}{\hat{p}_r} \\ &= \sum_{j=1}^{r}\frac{(x_j - np_{j0})^2}{n\hat{p}_j}.\end{aligned}$$

We know from the general theory that the Score and Wald statistics are asymptotically equivalent to the likelihood ratio test, which is given by the likelihood ratio

$$\Lambda[H|G] = \frac{\prod_{i=1}^{r} p_{0i}^{x_i}}{\prod_{i=1}^{r} \hat{p}_i^{x_i}},$$

and corresponding test statistic

$$-2\log\Lambda = 2n\sum_{i=1}^{r}\hat{p}_i\log\left(\frac{\hat{p}_i}{p_{0i}}\right).$$

We have seen above how the asymptotic linear model can be used to provide the format of each of the three statistics and thereby prove their asymptotic equivalence, being all approximately distributed as χ^2_{r-1} when $\mathbf{p} = \mathbf{p}_0$. However, given the three statistics, another method can be used to prove their asymptotic equivalence described in Seber (2013, 41–44).

12.4 Testing a Freedom Equation Hypothesis

Suppose we wish to test a more general hypothesis such as $H : \mathbf{p} = \mathbf{p}(\boldsymbol{\alpha})$ where $\boldsymbol{\alpha}$ is $(r-q) \times 1$, as discussed in Chap. 11 with $p = r$. The likelihood ratio is then

$$\Lambda[H|G] = \frac{\prod_{i=1}^r p_i(\tilde{\alpha})^{x_i}}{\prod_{i=1}^r \hat{p}_i^{x_i}},$$

with test statistic

$$\begin{aligned} -2\log\Lambda &= 2n\sum_{i=1}^r \hat{p}_i \log\frac{\hat{p}_i}{p_i(\tilde{\alpha})} \\ &= 2n\sum_{i=1}^r \hat{p}_i \log\left(1 + \frac{\hat{p}_i - p_i(\tilde{\alpha})}{p_i(\tilde{\alpha})}\right) \\ &\approx \sum_{i=1}^r \frac{(x_i - np_i(\tilde{\alpha}))^2}{np_i(\tilde{\alpha})} \end{aligned} \qquad (12.10)$$

using the approximation $\log(1+y_i) \approx y_i - y_i^2/2$ for $|y_i| < 1$, where y_i converges to 0 in probability. The above statistic is asymptotically distributed as χ^2_q when H is true.

The Score statistic is readily obtained from (12.7); we simply replace $\mathbf{p}_0$ by $\tilde{\mathbf{p}}$ to get (12.10) again.

Example 12.1 (Test for Independence) Suppose we have a multinomial experiment giving rise to a two-way table consisting of I rows and J columns with x_{ij} the frequency of the (i,j)th cell. This is a single multinomial distribution with $r = IJ$ cells so that the log likelihood function is (cf. (12.3))

$$L(\mathbf{p}) = \sum_{i=1}^I \sum_{j=1}^J x_{ij} \log p_{ij} - n \log p_{\cdot\cdot},$$

where $n = \sum_i \sum_j x_{ij}$ and $p_{\cdot\cdot} = \sum_i \sum_j p_{ij}$. We can express the p_{ij} as a single vector

$$\mathbf{p} = (p_{11}, p_{12}, \ldots, p_{1J}, p_{21}, p_{22}, \ldots, p_{2J}, \ldots, p_{I1}, p_{I2}, \ldots, p_{IJ})'.$$

The hypothesis H of row and column independence is $p_{ij} = \alpha_i\beta_j$, where we have $\sum_{i=1}^{I} \alpha = 1$ and $\sum_{j=1}^{J} \beta_j = 1$. Therefore H takes the form $\mathbf{p} = \mathbf{p}(\boldsymbol{\theta})$, where $\boldsymbol{\theta} = (\alpha_1, \alpha_2, \ldots, \alpha_I, \beta_1, \beta_2, \ldots \beta_J)'$. Also, under H,

$$p_{i\cdot} = \alpha_i, \quad p_{\cdot j} = \beta_j \quad \text{and} \quad \sum_i \sum_j p_{ij} = 1,$$

so that H can also be expressed in the form $p_{ij} = p_{i\cdot}p_{\cdot j}$. We can therefore use the theory given above and test H using a chi-square statistic of the form (12.10), namely

$$\sum_{i=1}^{I} \sum_{j=1}^{J} \frac{(x_{ij} - np_{ij}(\tilde{\alpha}_i\tilde{\beta}_j))^2}{np_{ij}(\tilde{\alpha}_i\tilde{\beta}_j)}. \tag{12.11}$$

We need to find the maximum likelihood estimates $\tilde{\alpha}_i$ and $\tilde{\beta}_j$, and the degrees of freedom. Now the likelihood function (apart from constants) is

$$\ell(\boldsymbol{\alpha}, \boldsymbol{\beta}) = \prod_{i=1}^{I} \prod_{j=1}^{J} (\alpha_i\beta_j)^{x_{ij}} = \prod_i \alpha_i^{R_i} \prod_j \beta_j^{C_j},$$

where $R_i = \sum_j x_{ij}$ (the ith row sum) and $C_j = \sum_i x_{ij}$ (the jth column sum). If $L(\boldsymbol{\alpha}, \boldsymbol{\beta}) = \log \ell(\boldsymbol{\alpha}, \boldsymbol{\beta})$ and λ_1 and λ_2 are Lagrange multipliers we need to differentiate

$$L(\boldsymbol{\alpha}, \boldsymbol{\beta}) + \lambda_1 \left(\sum_i \alpha_i - 1 \right) + \lambda_2 \left(\sum_j \beta_j - 1 \right)$$

with respect to α_i and β_j. The estimates are then solutions of

$$\frac{R_i}{\alpha_i} + \lambda_1 = 0, \quad \sum_{i=1}^{I} \alpha_i = 1 \quad \text{and}$$

$$\frac{C_j}{\beta_j} + \lambda_2 = 0 \quad \sum_{j=1}^{J} \beta_j = 1.$$

Since $\lambda_1 = \lambda_2 = -\sum_i R_i = -\sum_j C_j = -n$, our maximum likelihood estimates are

$$\tilde{\alpha}_i = \frac{R_i}{n} \quad \text{and} \quad \tilde{\beta}_j = \frac{C_j}{n}.$$

Hence $p_{ij}(\tilde{\alpha}_i, \tilde{\beta}_j) = R_i C_j / n$ so that the Score test is, from (12.11),

$$\sum_{i=1}^{I} \sum_{j=1}^{J} \frac{(x_{ij} - R_i C_j / n)^2}{R_i C_j / n}.$$

Under H this statistic has approximately the chi-square distribution with degrees of freedom $IJ - 1 - (I - 1 + J - 1) = (I - 1)(J - 1)$ corresponding to the difference in the number of free parameters specifying Ω and ω.

12.5 Conclusion

In the last three chapters we have seen how general hypotheses about sampling from general distributions can, for large samples, be approximated by linear hypotheses about linear normal models. Here the normality comes from maximum likelihood estimates that are generally asymptotically normal. For the approximating linear model the three test statistics, the likelihood ratio, the Wald Test, and the Score (Lagrange Multiplier) test statistics are identical thus showing that for the original model they are asymptotically equal. Clearly the method used is a general one so that it can be used for other models as well. For example, one referee suggested structural equation, generalized linear, and longitudinal models as well as incorporating Bayesian and pure likelihood methods. The reader might like to try and extend the theory to other models as an extended exercise!

References

Seber, G. A. F. (2008). *A matrix handbook for statisticians*. New York: Wiley.

Seber, G. A. F. (2013). *Statistical models for proportions and probabilities* (Springer briefs in statistics). Heidelberg: Springer.

Appendix: Matrix Theory

In this appendix, conformable matrices are matrices that are the correct sizes when multiplied together. All matrices in this appendix are real, though many of the results also hold for complex matrices (see Seber 2008). Because of lack of uniformity in the literature on some definitions I give the following definitions.

A symmetric $n \times n$ matrix $\mathbf{A}$ is said to be non-negative definite (n.n.d.) if $\mathbf{x}'\mathbf{A}\mathbf{x} \geq 0$ for all $\mathbf{x}$, while if $\mathbf{x}'\mathbf{A}\mathbf{x} > 0$ for all $\mathbf{x} \neq \mathbf{0}$ we say that $\mathbf{A}$ is positive definite (p.d.). The matrix $\mathbf{A}$ is said to be positive semidefinite if it is non-negative definite and there exists $\mathbf{x} \neq \mathbf{0}$ such that $\mathbf{x}'\mathbf{A}\mathbf{x} = 0$, that is $\mathbf{A}$ is singular. A matrix $\mathbf{A}$ is said to be negative definite if $-\mathbf{A}$ is positive definite.

A matrix $\mathbf{A}^-$ is called a weak inverse of $\mathbf{A}$ if $\mathbf{A}\mathbf{A}^-\mathbf{A} = \mathbf{A}$. (We use the term weak inverse as the term generalized inverse has different meanings in the literature.)

Trace

Theorem A.1 *If* $\mathbf{A}$ *is* $m \times n$ *and* $\mathbf{B}$ *is* $n \times m$, *then*

$$\text{trace}[\mathbf{AB}] = \text{trace}[\mathbf{BA}] = \text{trace}[\mathbf{B}'\mathbf{A}'] = \text{trace}[\mathbf{A}'\mathbf{B}'].$$

Proof

$$\text{trace}[\mathbf{AB}] = \sum_{i=1}^{m}\sum_{j=1}^{n} a_{ij}b_{ji} = \sum_{j=1}^{n}\sum_{i=1}^{m} b_{ji}a_{ij} = \sum_{i=1}^{m}\sum_{j=1}^{n} b'_{ij}a'_{ji} = \sum_{n=1}^{n}\sum_{i=1}^{m} a'_{ji}b'_{ij}.$$

If $m = n$ and either $\mathbf{A}$ or $\mathbf{B}$ is symmetric then $\text{trace}[\mathbf{AB}] = \sum_{i=1}^{n} a_{ij}b_{ij}$.

G.A.F. Seber, *The Linear Model and Hypothesis*, Springer Series in Statistics,
DOI 10.1007/978-3-319-21930-1

Rank

Theorem A.2 *If* **A** *and* **B** *are conformable matrices, then*

$$\text{rank}[\mathbf{AB}] \leq \text{minimum}\{\text{rank}\,\mathbf{A}, \text{rank}\,\mathbf{B}\}.$$

Proof The ith row of **AB** is $\sum_j a_{ij}\mathbf{b}'_j$, where $\mathbf{b}'_j$ is the jth row of **B**. The rows of **AB** are therefore linear combinations of the rows of **B** so that the number of linearly independent rows of **AB** is less than or equal to those of **B**; thus rank[**AB**] $\leq$ rank[**B**]. Similarly, the columns of **AB** are linear combinations of the columns of **A**, so that rank[**AB**] $\leq$ rank[**A**].

Theorem A.3 *Let* **A** *be an* $m \times n$ *matrix with rank* r *and nullity* s*, where the nullity is the dimension of the null space of* **A**, *then*

$$r + s = \textit{number of columns of } \mathbf{A}.$$

Proof Let $\boldsymbol{\alpha}_1, \boldsymbol{\alpha}_2, \ldots, \boldsymbol{\alpha}_s$ be a basis for $\mathcal{N}[\mathbf{A}]$. Enlarge this set of vectors to give a basis $\boldsymbol{\alpha}_1, \boldsymbol{\alpha}_2, \ldots, \boldsymbol{\alpha}_r, \boldsymbol{\beta}_1, \boldsymbol{\beta}_2, \ldots, \boldsymbol{\beta}_t$ for $\mathbb{R}^n$. Every vector in $\mathcal{C}[\mathbf{A}]$ can be expressed in the form

$$\begin{aligned}\mathbf{Ax} &= \mathbf{A}\left(\sum_{i=1}^{s} a_i\boldsymbol{\alpha}_i + \sum_{j=1}^{t} b_j\boldsymbol{\beta}_j\right) \\ &= \sum_{j=1}^{t} b_j\mathbf{A}\boldsymbol{\beta}_j \\ &= \sum_{j=1}^{t} b_j\boldsymbol{\gamma}_j, \quad \text{say.}\end{aligned}$$

Now suppose that $\sum_{j=1}^{t} \mathbf{c}_j\boldsymbol{\gamma}_j = \mathbf{0}$, then

$$\mathbf{A}\left(\sum_{j=1}^{t} c_j\boldsymbol{\beta}_j\right) = \sum_{j=1}^{j} c_j\boldsymbol{\gamma}_j = \mathbf{0}$$

and $\sum c_j\boldsymbol{\beta}_j \in \mathcal{N}[\mathbf{A}]$. This is only possible if the c_j's are all zero so that the $\boldsymbol{\gamma}_j$ are linearly independent. Since every vector **Ax** in $\mathcal{C}[\mathbf{A}]$ can be expressed in terms of the $\boldsymbol{\gamma}_j$'s, the $\boldsymbol{\gamma}_j$'s form a basis for $\mathcal{C}[\mathbf{A}]$; thus $t = s$. Since $s + t = n$, our proof is complete.

Theorem A.4 *Let* $\mathbf{A}$ *be any matrix.*

(i) The rank of $\mathbf{A}$ *is unchanged when* $\mathbf{A}$ *is pre- or post-multiplied by a non-singular matrix.*
(ii) $\text{rank}[\mathbf{A}'\mathbf{A}] = \text{rank}[\mathbf{A}]$. *Since* $\text{rank}[\mathbf{A}'] = \text{rank}[\mathbf{A}]$ *this implies that* $\text{rank}[\mathbf{A}\mathbf{A}'] = \text{rank}[\mathbf{A}]$.

Proof

(i) If $\mathbf{Q}$ is a conformable non-singular matrix, then by A.2

$$\text{rank}[\mathbf{A}] \leq \text{rank}[\mathbf{AQ}] \leq \text{rank}[\mathbf{AQQ}^{-1}] = \text{rank}[\mathbf{A}]$$

so that $\text{rank}[\mathbf{A}] = \text{rank}[\mathbf{AQ}]$ etc.
(ii) $\mathbf{Ax} = \mathbf{0}$ implies that $\mathbf{A}'\mathbf{Ax} = \mathbf{0}$. Conversely, if $\mathbf{A}'\mathbf{Ax} = \mathbf{0}$ then $\mathbf{x}'\mathbf{A}'\mathbf{Ax} = 0$, which implies $\mathbf{Ax} = \mathbf{0}$. Hence the null spaces of $\mathbf{A}$ and $\mathbf{A}'\mathbf{A}$ are the same. Since $\mathbf{A}$ and $\mathbf{A}'\mathbf{A}$ have the same number of columns, it follows from A.3 that $\mathbf{A}$ and $\mathbf{A}'\mathbf{A}$ have the same ranks. Similarly, replacing $\mathbf{A}$ by $\mathbf{A}'$ and using $\text{rank}[\mathbf{A}] = \text{rank}[\mathbf{A}']$ we have $\text{rank}[\mathbf{A}'] = \text{rank}[\mathbf{AA}']$, and the result follows.

Theorem A.5 *If* $\mathbf{A}$ *is* $n \times p$ *of rank* p *and* $\mathbf{B}$ *is* $p \times r$ *of rank* r, *than* $\mathbf{AB}$ *has rank* r.

Proof We note that $n \geq p \geq r$. From A.4(ii), $\mathbf{A}'\mathbf{A}$ and $\mathbf{B}'\mathbf{B}$ are nonsingular. Multiplying $\mathbf{ABx} = \mathbf{0}$ on the left by $(\mathbf{B}'\mathbf{B})^{-1}\mathbf{B}'(\mathbf{A}'\mathbf{A})^{-1}\mathbf{A}'$ gives us $\mathbf{x} = \mathbf{0}$ so that the columns of $n \times r$ matrix $\mathbf{AB}$ are linearly independent. Hence $\mathbf{AB}$ has rank r.

Eigenvalues

Theorem A.6 *For conformable matrices, the nonzero eigenvalues of* $\mathbf{AB}$ *are the same as those of* $\mathbf{BA}$.

Proof Let λ be a nonzero eigenvalue of $\mathbf{AB}$. Then there exists $\mathbf{u}$ $(\neq \mathbf{0})$ such that $\mathbf{ABu} = \lambda\mathbf{u}$, that is $\mathbf{BABu} = \lambda\mathbf{Bu}$. Hence $\mathbf{BAv} = \lambda\mathbf{v}$, where $\mathbf{v} = \mathbf{Bu} \neq \mathbf{0}$ (as $\mathbf{ABu} \neq \mathbf{0}$), and λ is an eigenvalue of $\mathbf{BA}$. The argument reverses by interchanging the roles of $\mathbf{A}$ and $\mathbf{B}$

Theorem A.7 (Spectral Decomposition Theorem) *Let* $\mathbf{A}$ *be any* $n \times n$ *symmetric matrix. Then there exists an orthogonal matrix* $\mathbf{T}$ *such that* $\mathbf{T}'\mathbf{AT} = \text{diag}(\lambda_1, \lambda_2, \ldots, \lambda_n)$, *where the* λ_i *are the eigenvalues of* $\mathbf{A}$. *[For further details relating to this theorem see Seber (2008: 16.44).]*

Proof Most matrix books give a proof of this important result.

Theorem A.8 *If* $\mathbf{A}$ *is an* $n \times n$ *positive-definite matrix and* $\mathbf{B}$ *is a symmetric* $n \times n$ *matrix, then there exists a non singular matrix* $\mathbf{V}$ *such that* $\mathbf{V}'\mathbf{AV} = \mathbf{I}_n$ *and* $\mathbf{V}'\mathbf{BV} = \text{diag}(\gamma_1, \gamma_2, \ldots, \gamma_n)$, *where the* γ_i *are the roots of* $|\gamma\mathbf{A} - \mathbf{B}| = 0$, *(i.e., are the eigenvalues of* $\mathbf{A}^{-1}\mathbf{B}$ *(or* $\mathbf{BA}^{-1}$ *or* $\mathbf{A}^{-1/2}\mathbf{BA}^{-1/2}$*)).*

Proof There exists an orthogonal $\mathbf{T}$ such that $\mathbf{T}'\mathbf{A}\mathbf{T} = \mathbf{\Lambda}$, the diagonal matrix of (positive) eigenvalues of $\mathbf{A}$. Let $\mathbf{\Lambda}^{1/2}$ be the square root of $\mathbf{\Lambda}$, that is has diagonal elements $\lambda_i^{1/2}$, and let $\mathbf{R} = \mathbf{T}\mathbf{\Lambda}^{-1/2}$. Then $\mathbf{R}'\mathbf{A}\mathbf{R} = \mathbf{\Lambda}^{-1/2}\mathbf{T}'\mathbf{A}\mathbf{T}\mathbf{\Lambda}^{-1/2} = \mathbf{I}_n$. As $\mathbf{C} = \mathbf{R}'\mathbf{B}\mathbf{R}$ is symmetric, there exists an orthogonal matrix $\mathbf{S}$ such that $\mathbf{S}'\mathbf{C}\mathbf{S} = \text{diag}(\gamma_1, \gamma_2, \ldots, \gamma_n) = \mathbf{\Gamma}$, say, where the diagonal elements of $\mathbf{\Gamma}$ are the eigenvalues of $\mathbf{C}$. Setting $\mathbf{V} = \mathbf{R}\mathbf{S}$ we have $\mathbf{V}'\mathbf{A}\mathbf{V} = \mathbf{S}'\mathbf{R}'\mathbf{A}\mathbf{R}\mathbf{S} = \mathbf{I}_n$ and $\mathbf{V}'\mathbf{B}\mathbf{V} = \mathbf{S}'\mathbf{C}\mathbf{S} = \mathbf{\Gamma}$, where the γ_i are the roots of

$$0 = |\gamma\mathbf{I}_n - \mathbf{R}'\mathbf{B}\mathbf{R}| = |\gamma\mathbf{R}'\mathbf{A}\mathbf{R} - \mathbf{R}'\mathbf{B}\mathbf{R}| = |\mathbf{R}||\gamma\mathbf{A} - \mathbf{B}||\mathbf{R}'| = |\gamma\mathbf{A} - \mathbf{B}|,$$

that is of $|\gamma\mathbf{I}_n - \mathbf{A}^{-1}\mathbf{B}| = 0$. Using A.9(ii), we then apply A.6 to $\mathbf{A}^{-1/2}\mathbf{A}^{-1/2}\mathbf{B}$, which has the same eigenvalues as $\mathbf{A}^{-1/2}\mathbf{B}\mathbf{A}^{-1/2}$ to complete the proof.

Non-negative Definite Matrices

Theorem A.9 *Let* $\mathbf{A}$ *be an* $n \times n$ *matrix of rank* r $(r \leq n)$.

(i) $\mathbf{A}$ *is non-negative (positive) definite if and only if all its eigenvalues are non-negative (positive).*
(ii) *If* $\mathbf{A}$ *is non-negative (positive) definite, then exists a non-negative (positive) definite matrix* $\mathbf{A}^{1/2}$ *such that* $\mathbf{A} = (\mathbf{A}^{1/2})^2$.
(iii) $\mathbf{A}$ *is non-negative definite if and only if* $\mathbf{A} = \mathbf{R}\mathbf{R}'$ *where* $\mathbf{R}$ *is* $n \times n$ *of rank* r. *This result is also true if we replace* $\mathbf{R}$ *by an* $n \times r$ *matrix of rank* r. *If* $\mathbf{A}$ *is positive definite then* $r = n$ *and* $\mathbf{R}$ *is nonsingular.*
(iv) *If* $\mathbf{A}$ *is an* $n\times n$ *non-negative (positive) definite matrix and* $\mathbf{C}$ *is an* $n\times p$ *matrix of rank* p, *then* $\mathbf{C}'\mathbf{B}\mathbf{C}$ *is non-negative (positive) definite.*
(v) *If* $\mathbf{A}$ *is non-negative definite and* $\mathbf{C}'\mathbf{A}\mathbf{C} = \mathbf{0}$, *then* $\mathbf{A}\mathbf{C} = \mathbf{0}$; *in particular,* $\mathbf{C}'\mathbf{C} = \mathbf{0}$ *implies that* $\mathbf{C} = \mathbf{0}$.
(vi) *If* $\mathbf{A}$ *is positive definite then so is* $\mathbf{A}^{-1}$.
(vii) *If* $\mathbf{A}$ *is* $n\times p$ *of rank* p, *then* $\mathbf{A}'\mathbf{A}$ *is nonsingular and therefore positive definite.*
(viii) *If the elements of* $n \times n$ *matrix* $\mathbf{A}(\boldsymbol{\theta})$ *are continuous functions of* $\boldsymbol{\theta}$ *and* $\mathbf{A}(\boldsymbol{\theta}_0)$ *is positive definite, then it will be positive definite in a neighborhood of* $\boldsymbol{\theta}_0$.
(ix) *If* $\mathbf{A}$ *is non-negative definite (n.n.d.), then* trace[$\mathbf{A}$] *is the non-negative sum of the eigenvalues of* $\mathbf{A}$.

Proof

(i) Since $\mathbf{A}$ is symmetric there exists an orthogonal matrix $\mathbf{T}$ such that $\mathbf{T}'\mathbf{A}\mathbf{T} = \text{diag}(\lambda_1, \lambda_2, \ldots, \lambda_n) = \mathbf{\Lambda}$, where the λ_i are the eigenvalues of $\mathbf{A}$. Now $\mathbf{A}$ is n.n.d. if and only $\mathbf{x}'\mathbf{T}'\mathbf{A}\mathbf{T}\mathbf{x} = \lambda_1 x_1^2 + \lambda_2 x_2^2 + \cdots + \lambda_n x_n^2 \geq 0$, if and only if the λ_i are nonnegative as we can set $\mathbf{x} = \mathbf{e}_i$ for each i, where $\mathbf{e}_i$ has one for the ith element and zeros elsewhere.
(ii) From the previous proof,

$$\mathbf{A} = \mathbf{T}\mathbf{\Lambda}\mathbf{T}' = \mathbf{T}\mathbf{\Lambda}^{1/2}\mathbf{T}'\mathbf{T}\mathbf{\Lambda}^{1/2}\mathbf{T}' = (\mathbf{A}^{1/2})^2,$$

where $\mathbf{\Lambda}^{1/2} = \text{diag}(\lambda_1^{1/2}, \lambda_2^{1/2}, \ldots, \lambda_d^{1/2})$ is n.n.d. and $\mathbf{A}^{1/2} = \mathbf{T}\mathbf{\Lambda}^{1/2}\mathbf{T}'$ is n.n.d. by (iii).

(iii) Since $\mathbf{A}$ is positive semidefinite of rank r, we have from the proof of (i) that $\mathbf{T}'\mathbf{A}\mathbf{T} = \mathbf{\Lambda}$, where the eigenvalues λ_i are all positive for $i = 1, 2, \ldots, r$, say, and zero for the rest. Let $\mathbf{\Lambda}^{1/2} = \text{diag}(\lambda_1^{1/2}, \lambda_2^{1/2}, \ldots, \lambda_r^{1/2}, 0, \ldots, 0)$. Then $\mathbf{A} = \mathbf{T}\mathbf{\Lambda}^{1/2}\mathbf{\Lambda}^{1/2}\mathbf{T}' = \mathbf{R}\mathbf{R}'$, where $\mathbf{R} = \mathbf{T}\mathbf{\Lambda}^{1/2}$ has rank r. Conversely, if $\mathbf{A} = \mathbf{R}\mathbf{R}'$, then rank[$\mathbf{R}$] $= r =$ rank[$\mathbf{R}\mathbf{R}'$] $=$ rank[$\mathbf{A}$], and $\mathbf{x}'\mathbf{A}\mathbf{x} = \mathbf{x}'\mathbf{R}\mathbf{R}'\mathbf{x} = \mathbf{y}'\mathbf{y} \geq 0$, where $\mathbf{y} = \mathbf{R}'\mathbf{x}$. Hence $\mathbf{A}$ is positive semidefinite of rank r.

We can replace $\mathbf{R} = \mathbf{T}\mathbf{\Lambda}^{1/2}$ by the $n \times r$ matrix $\mathbf{T}_r\mathbf{\Lambda}_r^{1/2}$, where $\mathbf{\Lambda}_r^{1/2} = \text{diag}(\lambda_1^{1/2}, \ldots, \lambda_r^{1/2})$ and $\mathbf{T}_r$ consists of the first r columns of $\mathbf{T}$.

(iv) We note that $\mathbf{y}'\mathbf{C}'\mathbf{R}\mathbf{R}'\mathbf{C}\mathbf{y} = \mathbf{z}'\mathbf{z} \geq 0$, where $\mathbf{z} = \mathbf{R}'\mathbf{C}\mathbf{y}$. If $\mathbf{R}$ is nonsingular, $\mathbf{z} = \mathbf{0}$ if and only if $\mathbf{y} = \mathbf{0}$ as $\mathbf{C}$ has full column rank.

(v) We have from (iii) that $\mathbf{A} = \mathbf{R}\mathbf{R}'$ so that $\mathbf{0} = \mathbf{C}'\mathbf{R}\mathbf{R}'\mathbf{C} = \mathbf{B}'\mathbf{B}$ ($\mathbf{B} = \mathbf{R}'\mathbf{C}$), which implies that $\mathbf{b}_i'\mathbf{b}_i = 0$ and $\mathbf{b}_i = \mathbf{0}$ for every column $\mathbf{b}_i$ of $\mathbf{B}$. Hence $\mathbf{B} = \mathbf{0}$ and $\mathbf{A}\mathbf{C} = \mathbf{R}\mathbf{R}'\mathbf{C} = \mathbf{R}\mathbf{B} = \mathbf{0}$.

(vi) Using (iii),

$$\mathbf{A}^{-1} = (\mathbf{R}\mathbf{R}')^{-1} = \mathbf{R}'^{-1}\mathbf{R}^{-1} = \mathbf{R}^{-1'}\mathbf{R}^{-1} = \mathbf{S}'\mathbf{S},$$

say, where $\mathbf{S}$ is nonsingular. Hence $\mathbf{A}^{-1}$ is positive definite.

(vii) If $\mathbf{y} = \mathbf{A}\mathbf{x}$, then $\mathbf{x}'\mathbf{A}'\mathbf{A}\mathbf{x} = \mathbf{y}'\mathbf{y} \geq 0$ and $\mathbf{A}'\mathbf{A}$ is positive semi-definite. However by A.4(ii), the $p \times p$ matrix $\mathbf{A}'\mathbf{A}$ has rank p and is therefore non-singular and positive definite.

(viii) It is well-known that a matrix is positive definite if and only if all its leading minor determinants are positive (for a proof see Seber and Lee (2003, 461–462)). Now at $\boldsymbol{\theta}_0$ the ith leading minor determinant of $\mathbf{A}(\boldsymbol{\theta})$ is positive, so by continuity it will be positive in a neighborhood N_i of $\boldsymbol{\theta}_0$. Hence all the m leading minor determinants will be positive in the neighborhood $\mathcal{N} = \cap_{i=1}^m N_i$, and $\mathbf{A}(\boldsymbol{\theta})$ will be positive definite in $\mathcal{N}$.

(ix) This follows from the proof of (ii), with $\mathbf{T}$ orthogonal, that

$$\text{trace}[\mathbf{A}] = \text{trace}[\mathbf{T}\mathbf{\Lambda}\mathbf{T}'] = \text{trace}[\mathbf{T}'\mathbf{T}\mathbf{\Lambda}] = \text{trace}[\mathbf{\Lambda}] = \sum_i \lambda_i \geq 0,$$

by A.1 and (i).

Theorem A.10 *Let f be the matrix function*

$$f(\mathbf{\Sigma}) = \log|\mathbf{\Sigma}| + \text{trace}[\mathbf{\Sigma}^{-1}\mathbf{A}].$$

If the $d \times d$ matrix $\mathbf{A}$ is positive definite, then, subject to $\mathbf{\Sigma}$ being positive definite, $f(\mathbf{\Sigma})$ is minimized uniquely at $\mathbf{\Sigma} = \mathbf{A}$.

Proof Let $\lambda_1, \lambda_2, \ldots, \lambda_d$ be the eigenvalues of $\mathbf{\Sigma}^{-1}\mathbf{A}$, that is of $\mathbf{\Sigma}^{-1/2}\mathbf{A}\mathbf{\Sigma}^{-1/2}$ (by A.8). Since the latter matrix is positive definite (by A.9(iv)), the λ_i are positive.

Also, since the determinant of a symmetric matrix is the product of it eigenvalues

$$|(\boldsymbol{\Sigma}^{-1}\mathbf{A})| = |\boldsymbol{\Sigma}^{-1}||\mathbf{A}| = |\boldsymbol{\Sigma}^{-1/2}\mathbf{A}\boldsymbol{\Sigma}^{-1/2}| = \prod_i \lambda_i.$$

Hence

$$\begin{aligned} f(\boldsymbol{\Sigma}) - f(\mathbf{A}) &= \log|\boldsymbol{\Sigma}\mathbf{A}^{-1}| + \text{trace}[\boldsymbol{\Sigma}^{-1}\mathbf{A}] - \text{trace}\,\mathbf{I}_d \\ &= -\log|\boldsymbol{\Sigma}^{-1/2}\mathbf{A}\boldsymbol{\Sigma}^{-1/2}| + \text{trace}[\boldsymbol{\Sigma}^{-1/2}\mathbf{A}\boldsymbol{\Sigma}^{-1/2}] - d \\ &= -\log\prod_i \lambda_i + \sum_i \lambda_i - d \\ &= \sum_{i=1}^{d}(-\log\lambda_i + \lambda_i - 1) \geq 0, \end{aligned}$$

as $\log x \leq x - 1$ for $x > 0$. Equality occurs when each λ_i is unity, that is when $\boldsymbol{\Sigma} = \mathbf{A}$.

Identifiability Conditions

Theorem A.11 *Let* $\mathbf{X}$ *be an* $n \times p$ *matrix of rank* r, *and* $\mathbf{H}$ *a* $t \times p$ *matrix. Then the equations* $\boldsymbol{\theta} = \mathbf{X}\boldsymbol{\beta}$ *and* $\mathbf{H}\boldsymbol{\beta} = \mathbf{0}$ *have a unique solution for* $\boldsymbol{\beta}$ *for every* $\boldsymbol{\theta} \in \mathcal{C}[\mathbf{X}]$ *if and only if*

(i) $\mathcal{C}[\mathbf{X}'] \cap \mathcal{C}[\mathbf{H}'] = \mathbf{0}$, *and*

(ii) $\text{rank}[\mathbf{G}] = \text{rank}\begin{bmatrix}\mathbf{X}\\ \mathbf{H}\end{bmatrix} = p.$

Proof (Scheffé 1959: 17) We first of all find necessary and sufficient conditions for $\boldsymbol{\beta}$ to exist. Now $\boldsymbol{\beta}$ will exist if and only if

$$\boldsymbol{\phi} = \begin{pmatrix}\boldsymbol{\theta}\\ \mathbf{0}\end{pmatrix} = \begin{pmatrix}\mathbf{X}\\ \mathbf{H}\end{pmatrix}\boldsymbol{\beta} = \mathbf{G}\boldsymbol{\beta} \in \mathcal{C}[\mathbf{G}] \quad \text{for every } \boldsymbol{\theta} \in \mathcal{C}[\mathbf{X}].$$

This statement is equivalent to: every vector perpendicular to $\mathcal{C}[\mathbf{G}]$ is perpendicular to $\boldsymbol{\phi}$ for every $\boldsymbol{\theta} \in \mathcal{C}[\mathbf{X}]$. Let $\mathbf{a}' = (\mathbf{a}_X', \mathbf{a}_H')$ be any $n + t$ dimensional vector. Then

$$\begin{gathered} \mathbf{G}'\mathbf{a} = \mathbf{0} \Rightarrow \boldsymbol{\phi}'\mathbf{a} = 0 \quad \text{if and only if} \\ \mathbf{X}'\mathbf{a}_X + \mathbf{H}'\mathbf{a}_H = \mathbf{0} \Rightarrow \boldsymbol{\theta}'\mathbf{a}_X = 0 \quad \text{for every } \boldsymbol{\theta} \in \mathcal{C}[\mathbf{X}] \text{ if and only if} \\ \mathbf{X}'\mathbf{a}_X + \mathbf{H}'\mathbf{a}_H = \mathbf{0} \Rightarrow \mathbf{X}'\mathbf{a}_X = \mathbf{0} \text{ and hence } \mathbf{H}'\mathbf{a}_H = \mathbf{0}. \end{gathered}$$

Thus $\boldsymbol{\beta}$ will exist if and only if no linear combination of the rows of $\mathbf{X}$ is a linear combination of the rows of $\mathbf{H}$ except $\mathbf{0}$, or $\mathcal{C}[\mathbf{X}'] \cap \mathcal{C}[\mathbf{H}'] = \mathbf{0}$.

If $\boldsymbol{\beta}$ is to be unique, then the columns of $\mathbf{G}$ must be linearly independent so that $\text{rank}[\mathbf{G}] = p$.

We note that the theorem implies that $\text{rank}[\mathbf{H}]$ must be $p - r$ for identifiability, so we usually have $t = p - r$, with the rows of $\mathbf{H}$ linearly independent.

Idempotent Matrices

Theorem A.12 *Let $\mathbf{A}_1, \mathbf{A}_2, \ldots, \mathbf{A}_m$ be a sequence of $n \times n$ symmetric matrices such that $\sum_{i=1}^{m} \mathbf{A}_i = \mathbf{I}_n$. Then the following conditions are equivalent:*

(i) $\sum_{i=1}^{m} r_i = n$, *where* $r_i = \text{rank}[\mathbf{A}_i]$.
(ii) $\mathbf{A}_i\mathbf{A}_j = \mathbf{0}$ *for all* $i, j, i \neq j$.
(iii) $\mathbf{A}_i^2 = \mathbf{A}_i$ *for* $i = 1, 2, \ldots, m$.

Proof We first show that (i) implies (ii) and (iii). Since

$$\mathbf{y} = \mathbf{I}_n\mathbf{y} = \mathbf{A}_1\mathbf{y} + \mathbf{A}_2\mathbf{y} + \cdots + \mathbf{A}_m\mathbf{y}, \tag{A.1}$$

(i) implies that $\mathbb{R}^n = \mathcal{C}[\mathbf{A}_1] \oplus \cdots \oplus \mathcal{C}[\mathbf{A}_m]$. Let $\mathbf{y} \in \mathcal{C}[\mathbf{A}_j]$. Then the unique expression of $\mathbf{y}$ in the above form is

$$\mathbf{y} = \mathbf{0} + \cdots + \mathbf{y} + \cdots + \mathbf{0}. \tag{A.2}$$

Since Eqs. (A.1) and (A.2) must be equivalent as $\mathbf{y}$ has a unique decomposition into components in mutually exclusive subspaces, we have $\mathbf{A}_i\mathbf{y} = \mathbf{0}$ (all i, $i \neq j$) and $\mathbf{A}_j\mathbf{y} = \mathbf{y}$ when $\mathbf{y} \in \mathcal{C}[\mathbf{A}_j]$. In particular, for *any* $\mathbf{x}$, we have by putting $\mathbf{y} = \mathbf{A}_j\mathbf{x}$ that $\mathbf{A}_i\mathbf{A}_j\mathbf{x} = \mathbf{0}$ and $\mathbf{A}_j^2\mathbf{x} = \mathbf{A}_j\mathbf{x}$. Hence (ii) and (iii) are true.

That (ii) implies (iii) is trivial; we simply multiply $\sum_k \mathbf{A}_k = \mathbf{I}_n$ by $\mathbf{A}_i$.

If (iii) is true so that each $\mathbf{A}_i$ is idempotent, then $\text{rank}[\mathbf{A}_i] = \text{trace}[\mathbf{A}_i]$ (by A.13) and

$$\begin{aligned} n &= \text{trace}[\mathbf{I}_n] \\ &= \text{trace}[\sum_i \mathbf{A}_i] \\ &= \sum_i \text{trace}[\mathbf{A}_i] \\ &= \sum_i \text{rank}[\mathbf{A}_i] \\ &= \sum_i r_i, \end{aligned}$$

so that (iii) implies (i). This completes the proof.

Theorem A.13 *If $\mathbf{A}$ is an idempotent matrix (not necessarily symmetric) of rank r, then its eigenvalues are 0 or 1 and* $\text{trace}[\mathbf{A}] = \text{rank}[\mathbf{A}] = r$.

Proof As $\mathbf{A}^2 - \mathbf{A} = \mathbf{0}$, $\lambda^2 - \lambda = 0$ is the minimal polynomial. Hence its eigenvalues are 0 or 1 and $\mathbf{A}$ is diagonalizable. Therefore there exists a nonsingular matrix $\mathbf{R}$ such that

$$\mathbf{R}^{-1}\mathbf{A}\mathbf{R} = \begin{pmatrix} \mathbf{I}_r & \mathbf{0} \\ \mathbf{0} & \mathbf{0} \end{pmatrix},$$

since the rank is unchanged when pre- or post-multiplying by a nonsingular matrix (A.4(i)). Hence

$$\text{trace}[\mathbf{A}] = \text{trace}[\mathbf{A}\mathbf{R}\mathbf{R}^{-1}] = \text{trace}[\mathbf{R}^{-1}\mathbf{A}\mathbf{R}] = r.$$

When $\mathbf{A}$ is also symmetric we see from Theorem 1.4 in Sect. 1.5 that $\mathbf{R}$ is replaced by an orthogonal matrix.

Weak (Generalized) Inverse

Theorem A.14 *If* $\mathbf{A}$ *is any matrix with weak inverse* $\mathbf{A}^-$, *then* $\mathbf{A}\mathbf{A}^-$ *is idempotent and* $\text{trace}[\mathbf{A}\mathbf{A}^-] = \text{rank}[\mathbf{A}\mathbf{A}^-] = \text{rank}[\mathbf{A}]$.

Proof $(\mathbf{A}\mathbf{A}^-\mathbf{A})\mathbf{A}^- = \mathbf{A}\mathbf{A}^-$, so that $\mathbf{A}\mathbf{A}^-$ is idempotent. Now from A.4(ii),

$$\text{rank}[\mathbf{A}] = \text{rank}[\mathbf{A}\mathbf{A}^-\mathbf{A}] \leq \text{rank}[\mathbf{A}\mathbf{A}^-] \leq \text{rank}[\mathbf{A}].$$

Hence $\text{rank}[\mathbf{A}\mathbf{A}^-] = \text{rank}[\mathbf{A}] = \text{trace}[\mathbf{A}\mathbf{A}^-]$, by A.13.

Theorem A.15

(i) $(\mathbf{A}^-)'$ *is a weak inverse of* $\mathbf{A}'$, *which we can then describe symbolically as* $(\mathbf{A}^-)' = (\mathbf{A}')^-$. *(Technically* $\mathbf{A}^-$ *is s not unique as it represents a family of matrices.)*

(ii) If $\mathbf{X}$ *is* $n \times p$ *of rank* $r < p$ *and* $(\mathbf{X}'\mathbf{X})^-$ *is any weak inverse of* $\mathbf{X}'\mathbf{X}$, *then we have* $\mathbf{P} = \mathbf{X}(\mathbf{X}'\mathbf{X})^-\mathbf{X}'$ *is the unique projection matrix onto* $\mathcal{C}[\mathbf{X}]$, *so that it is symmetric and idempotent.*

(iii) If $\mathbf{G}$ *is defined in A.11, then* $(\mathbf{G}'\mathbf{G})^{-1}$ *is a weak inverse of* $\mathbf{X}'\mathbf{X}$.

Proof

(i) This is proved by taking the transpose of $\mathbf{A} = \mathbf{A}\mathbf{A}^-\mathbf{A}$.

(ii) Let $\Omega = \mathcal{C}[\mathbf{X}]$ and let $\boldsymbol{\theta} = \mathbf{X}\boldsymbol{\beta} \in \Omega$. Given the normal equations $\mathbf{X}'\mathbf{X}\boldsymbol{\beta} = \mathbf{X}'\mathbf{y}$, these have *a* solution $\hat{\boldsymbol{\beta}} = (\mathbf{X}'\mathbf{X})^-\mathbf{X}'\mathbf{y}$. If $\hat{\boldsymbol{\theta}} = \mathbf{X}\hat{\boldsymbol{\beta}}$, then

$$\begin{aligned}\hat{\boldsymbol{\theta}}'(\mathbf{y} - \hat{\boldsymbol{\theta}}) &= \hat{\boldsymbol{\beta}}'\mathbf{X}'(\mathbf{y} - \mathbf{X}\hat{\boldsymbol{\beta}}) \\ &= \hat{\boldsymbol{\beta}}'(\mathbf{X}'\mathbf{y} - \mathbf{X}'\mathbf{X}\hat{\boldsymbol{\beta}}) \\ &= 0.\end{aligned}$$

We therefore have an orthogonal decomposition of $\mathbf{y} = \hat{\theta} + \mathbf{y} - \hat{\theta}$ such that $\hat{\theta} \in \Omega$ and $(\mathbf{y} - \hat{\theta}) \perp \Omega$. Since $\hat{\theta} = \mathbf{X}\hat{\beta} = \mathbf{X}(\mathbf{X}'\mathbf{X})^{-}\mathbf{X}'\mathbf{y}$ and the orthogonal projection is unique, we must have $\mathbf{P}_{\Omega} = \mathbf{X}(\mathbf{X}'\mathbf{X})^{-}\mathbf{X}'$.

(iii) Using the normal equations $\mathbf{X}'\mathbf{X}\hat{\beta} = \mathbf{X}'\mathbf{y}$ and adding $\mathbf{H}\hat{\beta} = \mathbf{0}$ we have $(\mathbf{X}'\mathbf{X} + \mathbf{H}'\mathbf{H})\hat{\beta} = \mathbf{X}'\mathbf{y}$ so that $\mathbf{0} = \mathbf{H}\hat{\beta} = \mathbf{H}(\mathbf{G}'\mathbf{G})^{-1}\mathbf{X}'\mathbf{y}$ for all $\mathbf{y}$. Hence $\mathbf{H}(\mathbf{G}'\mathbf{G})^{-1}\mathbf{X}' = \mathbf{0}$ and

$$\mathbf{X}'\mathbf{X}(\mathbf{G}'\mathbf{G})^{-1}\mathbf{X}'\mathbf{X} = (\mathbf{X}'\mathbf{X} + \mathbf{H}'\mathbf{H})(\mathbf{G}'\mathbf{G})^{-1}\mathbf{X}'\mathbf{X} = \mathbf{X}'\mathbf{X}.$$

Theorem A.16 *Let* $\mathbf{y} \sim N_n[\mathbf{0}, \boldsymbol{\Sigma}]$, *where* $\boldsymbol{\Sigma}$ *is a nonnegative-definite matrix of rank* s. *If* $\boldsymbol{\Sigma}^{-}$ *is any weak inverse of* $\boldsymbol{\Sigma}$, *then* $\mathbf{y}'\boldsymbol{\Sigma}\ \mathbf{y} \sim \chi^2_s$.

Proof If $\mathbf{z} \sim N_n[\mathbf{I}_s, \mathbf{0}]$, then $\mathbf{y}$ has the same distribution as $\boldsymbol{\Sigma}^{1/2}\mathbf{z}$ (cf. A.9(ii)) since $\text{Var}[\mathbf{y}] = (\boldsymbol{\Sigma}^{1/2})^2 = \boldsymbol{\Sigma}$. Now $\mathbf{y}'\mathbf{A}\mathbf{y} = \mathbf{z}'\boldsymbol{\Sigma}^{1/2}\mathbf{A}\boldsymbol{\Sigma}^{1/2}\mathbf{z}$ is χ^2_r if $\boldsymbol{\Sigma}^{1/2}\mathbf{A}\boldsymbol{\Sigma}^{1/2}$ is idempotent (Theorem 1.10 in Sect. 1.9), where

$$r = \text{trace}[\boldsymbol{\Sigma}^{1/2}\mathbf{A}\boldsymbol{\Sigma}^{1/2}] = \text{trace}[\boldsymbol{\Sigma}\mathbf{A}]$$

(since $\text{trace}[\mathbf{CD}] = \text{trace}[\mathbf{DC}]$), that is if

$$\boldsymbol{\Sigma}^{1/2}\mathbf{A}\boldsymbol{\Sigma}^{1/2}\boldsymbol{\Sigma}^{1/2}\mathbf{A}\boldsymbol{\Sigma}^{1/2} = \boldsymbol{\Sigma}^{1/2}\mathbf{A}\boldsymbol{\Sigma}^{1/2}. \tag{A.3}$$

Multiplying the above equation on the left and right by $\boldsymbol{\Sigma}^{1/2}$ we get

$$\boldsymbol{\Sigma}\mathbf{A}\boldsymbol{\Sigma}\mathbf{A}\boldsymbol{\Sigma} = \boldsymbol{\Sigma}\mathbf{A}\boldsymbol{\Sigma}. \tag{A.4}$$

We now show that Eqs. (A.3) and (A.4) are equivalent conditions.

Let $\mathbf{B} = \mathbf{A}\boldsymbol{\Sigma}\mathbf{A} - \mathbf{A}$, then we need to show that $\boldsymbol{\Sigma}\mathbf{B}\boldsymbol{\Sigma} = \mathbf{0}$ implies that the matrix $\mathbf{D} = \boldsymbol{\Sigma}^{1/2}\mathbf{B}\boldsymbol{\Sigma}^{1/2} = \mathbf{0}$. Now $\mathbf{D}$ is symmetric and given $\boldsymbol{\Sigma}\mathbf{B}\boldsymbol{\Sigma} = \mathbf{0}$,

$$\begin{aligned} \text{trace}[\mathbf{D}^2] &= \text{trace}[\boldsymbol{\Sigma}^{1/2}\mathbf{B}\boldsymbol{\Sigma}^{1/2}\boldsymbol{\Sigma}^{1/2}\mathbf{B}\boldsymbol{\Sigma}^{1/2}] \\ &= \text{trace}[\boldsymbol{\Sigma}\mathbf{B}\boldsymbol{\Sigma}\mathbf{B}] \\ &= \mathbf{0}. \end{aligned}$$

However $\text{trace}[\mathbf{D}^2] = \sum_i \sum_j d_{ij}^2 = 0$ implies that $\mathbf{D} = \mathbf{0}$.

We now set $\mathbf{A} = \boldsymbol{\Sigma}^{-}$, then

$$\boldsymbol{\Sigma}\mathbf{A}\boldsymbol{\Sigma}\mathbf{A}\boldsymbol{\Sigma} = (\boldsymbol{\Sigma}\boldsymbol{\Sigma}^{-}\boldsymbol{\Sigma})\boldsymbol{\Sigma}^{-}\boldsymbol{\Sigma} = \boldsymbol{\Sigma}\boldsymbol{\Sigma}^{-}\boldsymbol{\Sigma} = \boldsymbol{\Sigma}\mathbf{A}\boldsymbol{\Sigma},$$

and the condition for idempotency is satisfied. We note that $r = \text{trace}[\boldsymbol{\Sigma}\mathbf{A}] = \text{trace}[\boldsymbol{\Sigma}\boldsymbol{\Sigma}^{-}]$ and, from A.14,

$$\text{rank}[\boldsymbol{\Sigma}\boldsymbol{\Sigma}^{-}] = \text{rank}[\boldsymbol{\Sigma}] = s = \text{trace}[\boldsymbol{\Sigma}\boldsymbol{\Sigma}^{-}].$$

Hence $r = s$ and $\mathbf{y}'\boldsymbol{\Sigma}^{-}\mathbf{y}$ is χ^2_s.

Inverse of a Partitioned Matrix

Theorem A.17 *If* **A** *and* **C** *are symmetric matrices and all inverses exist, then*

$$\begin{pmatrix} \mathbf{A} & \mathbf{B}' \\ \mathbf{B} & \mathbf{C} \end{pmatrix}^{-1} = \begin{pmatrix} \mathbf{F}^{-1} & -\mathbf{F}^{-1}\mathbf{G}' \\ -\mathbf{G}\mathbf{F}^{-1} & \mathbf{C}^{-1} + \mathbf{G}\mathbf{F}^{-1}\mathbf{G}' \end{pmatrix},$$

where $\mathbf{F} = \mathbf{A} - \mathbf{B}'\mathbf{C}^{-1}\mathbf{B}$ *and* $\mathbf{G} = \mathbf{C}^{-1}\mathbf{B}$.

Proof The result is proved by confirming that the matrix multiplied on the left by its inverse is the identity matrix.

Theorem A.18 *If* **A** *is positive definite and all inverses exist, then*

$$\begin{pmatrix} \mathbf{A} & \mathbf{B}' \\ \mathbf{B} & \mathbf{0} \end{pmatrix}^{-1} = \begin{pmatrix} \mathbf{A}^{-1} - \mathbf{A}^{-1}\mathbf{B}'(\mathbf{B}\mathbf{A}^{-1}\mathbf{B}')^{-1}\mathbf{B}\mathbf{A}^{-1} & \mathbf{A}^{-1}\mathbf{B}'(\mathbf{B}\mathbf{A}^{-1}\mathbf{B}')^{-1} \\ (\mathbf{B}\mathbf{A}^{-1}\mathbf{B}')^{-1}\mathbf{B}\mathbf{A}^{-1} & -(\mathbf{B}\mathbf{A}^{-1}\mathbf{B}')^{-1} \end{pmatrix}$$

Proof The matrix times its inverse is the identity matrix.

Theorem A.19 *Let* **A** *be an* $r \times r$ *positive definite matrix,* **B** *be an* $s_1 \times r$ *matrix of rank* s_1, **C** *be an* $s_2 \times r$ *matrix of rank* s_2, *and* $\mathcal{C}[\mathbf{B}'] \cap \mathcal{C}[\mathbf{C}'] = \mathbf{0}$ *so that* $\text{rank}[(\mathbf{B}', \mathbf{C}')] = s_1 + s_2$. *If*

$$\mathbf{Z} = \begin{pmatrix} \mathbf{A} & \mathbf{B}' & \mathbf{C}' \\ \mathbf{B} & \mathbf{0} & \mathbf{0} \\ \mathbf{C} & \mathbf{0} & \mathbf{0} \end{pmatrix},$$

then

$$\mathbf{Z}^{-1} = \begin{pmatrix} \mathbf{P} & \mathbf{Q}' & \mathbf{R}' \\ \mathbf{Q} & -\mathbf{Q}\mathbf{A}\mathbf{Q}' & -\mathbf{Q}\mathbf{A}\mathbf{R}' \\ \mathbf{R} & -\mathbf{R}\mathbf{A}\mathbf{Q}' & \mathbf{R}\mathbf{A}\mathbf{R}' \end{pmatrix},$$

where

$$\begin{aligned} \mathbf{P} &= \mathbf{M} - \mathbf{M}\mathbf{C}'(\mathbf{C}\mathbf{M}\mathbf{C}')^{-1}\mathbf{C}\mathbf{M}, \\ \mathbf{Q} &= (\mathbf{B}\mathbf{A}^{-1}\mathbf{B}')^{-1}\mathbf{B}\mathbf{A}^{-1}[\mathbf{I}_r - \mathbf{C}'(\mathbf{C}\mathbf{M}\mathbf{C}')^{-1}\mathbf{C}\mathbf{M}], \\ \mathbf{R} &= (\mathbf{C}\mathbf{M}\mathbf{C}')^{-1}\mathbf{C}\mathbf{M}, \quad \text{and} \\ \mathbf{M} &= \mathbf{M}' = \mathbf{A}^{-1}[\mathbf{I}_r - \mathbf{B}'(\mathbf{B}\mathbf{A}^{-1}\mathbf{B}')^{-1}\mathbf{B}\mathbf{A}^{-1}]. \end{aligned}$$

Proof Let

$$\mathbf{Z}^{-1} = \begin{pmatrix} \mathbf{P} & \mathbf{Q}' & \mathbf{R}' \\ \mathbf{Q} & \mathbf{T} & \mathbf{U}' \\ \mathbf{R} & \mathbf{U} & \mathbf{V} \end{pmatrix},$$

then from $\mathbf{Z}\mathbf{Z}^{-1} = \mathbf{I}_{r+s_1+s_2}$, we have a system of nine matrix equations, namely

$$\mathbf{AP} + \mathbf{B}'\mathbf{Q} + \mathbf{C}'\mathbf{R} = \mathbf{I}_r, \quad \mathbf{BP} = \mathbf{0}, \quad \mathbf{CP} = \mathbf{0}, \tag{A.5}$$

$$\mathbf{AQ}' + \mathbf{B}'\mathbf{T} + \mathbf{C}'\mathbf{U} = \mathbf{0}, \quad \mathbf{BQ}' = \mathbf{I}_{s_1}, \quad \mathbf{CQ}' = \mathbf{0}, \tag{A.6}$$

$$\mathbf{AR}' + \mathbf{B}'\mathbf{U}' + \mathbf{C}'\mathbf{V} = \mathbf{0}, \quad \mathbf{BR}' = \mathbf{0}, \quad \mathbf{CR}' = \mathbf{I}_{s_2}. \tag{A.7}$$

Now from Eq. (A.5)

$$\mathbf{P} = \mathbf{A}^{-1} - \mathbf{A}^{-1}\mathbf{B}'\mathbf{Q} - \mathbf{A}^{-1}\mathbf{C}'\mathbf{R} \tag{A.8}$$

and using $\mathbf{BP} = \mathbf{0}$ gives us

$$\mathbf{Q} = (\mathbf{BA}^{-1}\mathbf{B}')^{-1}\mathbf{BA}^{-1}(\mathbf{I}_r - \mathbf{C}'\mathbf{R}).$$

Substituting back into Eq. (A.8) and using $\mathbf{CP} = \mathbf{0}$ leads to

$$\mathbf{CM} - \mathbf{CMC}'\mathbf{R} = \mathbf{0}. \tag{A.9}$$

Since $\mathbf{A}^{-1}$ is positive definite, there exists a nonsingular $r \times r$ matrix $\mathbf{L}$ such that $\mathbf{A}^{-1} = \mathbf{L}'\mathbf{L}$ (A.9(iii)). Now $\mathbf{LB}'$ is $r \times s_1$ of rank s_1 so that

$$\begin{aligned} \mathbf{CMC}' &= \mathbf{CL}'[\mathbf{I}_r - \mathbf{LB}'(\mathbf{BL}'\mathbf{LB}')^{-1}\mathbf{BL}']\mathbf{LC}' \\ &= \mathbf{CL}'(\mathbf{I}_r - \mathbf{P}_{\mathcal{C}[\mathbf{LB}']})\mathbf{LC}' \\ &= \mathbf{CL}'\mathbf{P}_{\mathcal{N}[\mathbf{BL}']}(\mathbf{CL}')' \end{aligned}$$

by Theorem 1.1 in Sect. 1.2. Now Theorem 4.4 in Sect. 4.2 states that if $\mathbf{A}$ is $q \times n$ of rank q then $\text{rank}[\mathbf{P}_{\Omega}\mathbf{A}'] = q$ if and only if $\mathcal{C}[\mathbf{A}'] \cap \Omega^{\perp} = \mathbf{0}$. If $\Omega = \mathcal{N}[\mathbf{BL}']$ and $\mathbf{A} = \mathbf{CL}'$, an $s_2 \times r$ matrix of rank s_2, then

$$\mathcal{C}[\mathbf{A}'] \cap \Omega^{\perp} = \mathcal{C}[\mathbf{L}'\mathbf{C}] \cap \mathcal{C}[\mathbf{L}'\mathbf{B}] = \mathbf{0},$$

since $\text{rank}[\mathbf{B}', \mathbf{C}']$ is unchanged by premultiplying by $\mathbf{L}'$, a nonsingular matrix. Hence $\text{rank}[\mathbf{P}_{\mathcal{N}[\mathbf{BL}']}\mathbf{LC}'] = s_2$. As $\mathbf{P}_{\mathcal{N}[\mathbf{BL}']}$ is symmetric and idempotent, $\mathbf{CL}'\mathbf{P}_{\mathcal{N}[\mathbf{BL}']}\mathbf{P}_{\mathcal{N}[\mathbf{BL}']}\mathbf{LC}'$ is $s_2 \times s_2$ of rank s_2, and is therefore nonsingular, so that $\mathbf{CMC}'$ has an inverse. From Eq. (A.9)

$$\mathbf{R} = (\mathbf{CMC}')^{-1}\mathbf{CM},$$

and from Eq. (A.6)

$$\begin{aligned} \mathbf{0} &= \mathbf{QAQ}' + \mathbf{QB}'\mathbf{T} + \mathbf{QC}'\mathbf{U} \\ &= \mathbf{QAQ}' + \mathbf{I}_{s_1}\mathbf{T} \end{aligned}$$

so that $\mathbf{T} = -\mathbf{QAQ}'$. From premultiplying Eq. (A.7) by $\mathbf{Q}$, and then premultiplying (A.7) by $\mathbf{R}$, we obtain $\mathbf{U}' = -\mathbf{QAR}'$. Since $\mathbf{RC}' = \mathbf{I}_{s_2}$, $\mathbf{V} = \mathbf{RAR}'$.

Differentiation

Theorem A.20 *If $d/d\beta$ denotes the column vector with ith element $d/d\beta_i$, then:*

(i) $d(\mathbf{a}'\beta)/d\beta = \mathbf{a}$.
(ii) $d(\beta'\mathbf{A}\beta)/d\beta = 2\mathbf{A}\beta$.

Proof

(i) $d\sum_i a_i\beta_i/d\beta_i = a_i$.
(ii)

$$\begin{aligned} d(\beta'\mathbf{A}\beta)/d\beta_i &= d(\sum_i a_{ii}\beta_i^2 + \sum_i \sum_{j:j\neq i} a_{ij}\beta_i\beta_j/)d\beta_i \\ &= 2a_{ii}\beta_i + \sum_{j:j\neq i}(a_{ij} + a_{ji})\beta_j \\ &= 2\sum_j a_{ij}\beta_j. \end{aligned}$$

Inequalities

Theorem A.21

(i) If $\mathbf{D}$ *is positive definite, then for any* $\mathbf{a}$

$$\sup_{\mathbf{x}:\mathbf{x}\neq\mathbf{0}} \left\{ \frac{(\mathbf{a}'\mathbf{x})^2}{\mathbf{x}'\mathbf{D}\mathbf{x}} \right\} = \mathbf{a}'\mathbf{D}^{-1}\mathbf{a}.$$

(ii) If $\mathbf{M}$ *and* $\mathbf{N}$ *are positive definite, then*

$$\sup_{\mathbf{x},\mathbf{y},\mathbf{x}\neq\mathbf{0},\mathbf{y}\neq\mathbf{0}} \left\{ \frac{(\mathbf{x}'\mathbf{L}\mathbf{y})^2}{\mathbf{x}'\mathbf{M}\mathbf{x}\cdot\mathbf{y}'\mathbf{N}\mathbf{y}} \right\} = \theta_{\max},$$

where $\theta_{\max}$ *is the largest eigenvalue of* $\mathbf{M}^{-1}\mathbf{L}\mathbf{N}^{-1}\mathbf{L}'$, *and of* $\mathbf{N}^{-1}\mathbf{L}'\mathbf{M}^{-1}\mathbf{L}$.

Proof Proofs are given by Seber (1984: 527).

References

Scheffé, H. (1959). *The analysis of variance*. New York: Wiley.
Seber, G. A. F. (1984). *Multivariate observations*. New York: Wiley. Also reproduced in paperback by Wiley in 2004.
Seber, G. A. F. (2008). *A matrix handbook for statisticians*. New York: Wiley.
Seber, G. A. F., & Lee, A. J. (2003). *Linear regression analysis* (2nd ed.). New York: Wiley.

Index

A
Analysis of covariance, examples of, 24–25
Analysis of variance (ANOVA)
  examples of, 22–24, 67
  orthogonality in, 81
  table, 50
  table for randomized block design, 95
Autocorrelation, 39

B
Basis, vector space, 3

C
Cochran's theorem
  multivariate, 137
  univariate, 48
Column space, 2
Concentrated likelihood, 123
Concomitant variable, 104
Confidence intervals, 57
Constraint-equation hypothesis, 149
Contrasts, 55
Covariance of two vectors, 6
Critical region, 61

D
Dimension, 3

E
Eigenvalues, 191–192
Enlarging the model
  examples of, 103
  missing observations and, 111–116
  randomized block design and, 109
  regression extensions, 107
Estimable functions, 34, 54
Expectation of a random matrix, 6
Expected information matrix, 150

F
Factorial experiment, 55
Fitted values, 38
Freedom-equation hypothesis, 175–179, 186–188

G
Gauss-Markov theorem
  multivariate, 133
  univariate, 34
Generalized inverse, 33, 132
Generalized linear hypothesis, 143
Growth curves, 144

H
Hat matrix, 38
Hotelling's test statistic, 144–146
Hypothesis sum of squares, 50

I
Idempotent matrix, 195–196
  eigenvalues of, 5
  non-central chi-square and, 14
  orthogonal projection and, 28, 29
  symmetric, 5

G.A.F. Seber, *The Linear Model and Hypothesis*, Springer Series in Statistics,
DOI 10.1007/978-3-319-21930-1

Identifiability conditions, 18, 23, 32, 33, 161
multivariate, 132
Independent quadratic forms, 16
Indicator variables, 22
Inequalities, 200
Information matrix
definition of, 150
positive definite, 152, 176–177
positive semidefinite, 161–171, 177–179
Internally studentized residual, 39
Intersection of vector spaces, 3

L
Lagrange multiplier, 17, 32, 44, 110, 157
multiplier test, 127, 166–171
zero for identifiability conditions, 33
Least squares equations, 30
Least squares estimate, 27
Length of a vector, 1
Less than full rank model, 32–34
Likelihood ratio test, 47, 127, 185
Linear hypothesis, 21–26
definition of, 25
Linear independence, 3
Linear model approximation, 155
Linear regression
examples of, 21–22, 29, 30, 52
orthogonal hypotheses and, 78–81
Linear vector space, 2

M
Maximum likelihood estimation, 44–45
Minimum variance estimator, 34
Missing observations, 111–116
analysis of covariance for, 115
randomized block design and, 114
Moment generating function, 9
Multinomial distribution, 181
Multiple confidence intervals, 147
Multiple correlation coefficient, 42
Multivariate maximum likelihood estimates, 136
Multivariate models, 129–147
Multivariate normal distribution, 8
Multivariate randomized block design, 129, 141

N
Nested test method, 73
Non-central chi-square distribution, 9, 48, 160
Non-central F-distribution, 10, 49
Non-centrality parameter, 9
Non-linear regression
asymptotic theory for, 122
examples of, 117
large sample tests for, 127
Non-negative definite matrix, 5, 192
Normal equations, 30, 33
Nullity, 3
Null space, 2

O
Orthogonal complement, 2
Orthogonal decomposition, 74
Orthogonal hypotheses
definition of, 76
multivariate, 143
nonlinear, 171–174
p-factor layouts and, 89–93
randomized block design and, 93–96
regression models and, 78–81
two-factor layouts and, 81–89
Orthogonal matrix, 192
Orthogonal polynomials, 80
Orthogonal projection, 27, 48
Orthogonal transformation, 56
Orthonormal basis, 3, 27, 30, 48
Orthonormal contrasts, 55

P
Partitioned matrix, 198–200
p-factor layouts, 89–93
Pitman's limiting power, 154
Polynomial regression, 21
Positive-definite matrix, 189, 192, 193, 200
Positive semidefinite matrix, 189
Power of the F-test, 61–63
Projection matrix, 4

Q
Quadratically balanced design, 36
Quadratically balanced test, 67
Quadratic form
expectation of, 11
variance of, 12

R
Randomized block design, 93–96
Rank, 190–191
Residuals, 38

Residual sum of squares, 35
Robustness of F-test, 64–69
Row space, 2

S
Score test, 127, 184–186
Separate test method, 76
Serial correlation effect, 39
Simultaneous confidence intervals, 58
Sum of two vector spaces, 3
Systemic bias, 39

T
Test for independence, 186
Trace, 1, 189
Two-factor layouts, 81–89
Two-way ANOVA table, 88

U
Uniformly most powerful test, 61

V
Variance-covariance matrix, 6
Variance estimation, 35
Vector differentiation, 30, 200

W
Wald test, 51–55, 127, 165, 184–186
Weak inverse, 33, 197
Wishart distribution, 135

Printed in the United States
By Bookmasters